**In-Situ Transmission Electron
Microscopy Experiments**

In-Situ Transmission Electron Microscopy Experiments

Design and Practice

Renu Sharma

WILEY-VCH

Author

Dr. Renu Sharma
Arizona State University
Tempe
AZ 85287

Cover Image: Courtesy of Renu Sharma

■ All books published by **WILEY-VCH** are carefully produced. Nevertheless, authors, editors, and publisher do not warrant the information contained in these books, including this book, to be free of errors. Readers are advised to keep in mind that statements, data, illustrations, procedural details or other items may inadvertently be inaccurate.

Library of Congress Card No.: applied for

British Library Cataloguing-in-Publication Data
A catalogue record for this book is available from the British Library.

Bibliographic information published by the Deutsche Nationalbibliothek
The Deutsche Nationalbibliothek lists this publication in the Deutsche Nationalbibliografie; detailed bibliographic data are available on the Internet at <http://dnb.d-nb.de>.

© 2023 WILEY-VCH GmbH, Boschstraße 12, 69469 Weinheim, Germany

All rights reserved (including those of translation into other languages). No part of this book may be reproduced in any form – by photoprinting, microfilm, or any other means – nor transmitted or translated into a machine language without written permission from the publishers. Registered names, trademarks, etc. used in this book, even when not specifically marked as such, are not to be considered unprotected by law.

Print ISBN: 978-3-527-34798-8
ePDF ISBN: 978-3-527-83483-9
ePub ISBN: 978-3-527-83484-6
oBook ISBN: 978-3-527-83482-2

Typesetting   Straive, Chennai, India
Printing and Binding   CPI Group (UK) Ltd, Croydon, CR0 4YY

C107435_050523

Contents

Preface *xiii*
Acknowledgments *xvii*
List of Abbreviations *xix*
About the Author *xxiii*

1	**In-Situ TEM** *1*	
1.1	Introduction *1*	
1.2	General Scope of the Book *2*	
1.3	Why In-Situ TEM *3*	
1.4	TEM: Overview *4*	
1.4.1	Historical Perspective *4*	
1.4.2	Electron–Sample Interactions *4*	
1.4.3	Overview of Modern TEM *5*	
1.4.3.1	Electron Source or Electron Gun *5*	
1.4.3.2	Lenses *7*	
1.4.3.3	Lens Aberrations *7*	
1.4.3.4	Aberration Correctors *9*	
1.4.4	Data Acquisition Systems *9*	
1.4.4.1	Types of Detectors *9*	
1.5	TEM/STEM-Based Characterization Techniques *11*	
1.5.1	Diffraction *11*	
1.5.2	TEM Imaging Modes *12*	
1.5.3	STEM *14*	
1.5.4	Analytical TEM *14*	
1.5.4.1	Chemical Analysis *15*	
1.5.4.2	EFTEM *19*	
1.5.4.3	Spectrum Imaging (SI) *20*	
1.6	Other Techniques *20*	
1.6.1	Lorentz Microscopy *20*	
1.6.2	Holography *22*	
1.6.2.1	In-Line Holography *22*	
1.6.2.2	Off-Axis Holography *22*	
1.6.3	UEM and DTEM *23*	

1.7	Introduction to Different Stimuli Used for In-Situ TEM	*24*
1.7.1	Heating (Chapter 3)	*24*
1.7.2	Cooling (Cryo TEM – Chapter 4)	*24*
1.7.3	Interactions with Liquid/Electrochemistry (Chapter 6)	*24*
1.7.4	Interaction with Gas Environment/Catalysis (Chapter 7)	*25*
1.7.5	Other Stimuli Not Included in this Book	*25*
1.7.5.1	Mechanical Testing	*25*
1.7.5.2	Ion Radiation/Implantation	*25*
1.7.5.3	Biasing	*27*
1.7.5.4	Magnetization	*28*
1.8	Potential Limitations and Cautions	*29*
1.9	Take-Home Messages	*31*
	References	*31*

2 Experiment Design Philosophy *41*
2.1 General *41*
2.2 Choice of Technique and the Microscope *44*
2.2.1 Stimulus and Technique Selection *44*
2.2.2 Microscope Selection *45*
2.2.2.1 Operating Voltage *45*
2.2.2.2 TEM/STEM and Pole-Piece Gap *46*
2.2.2.3 Image Acquisition System and Detectors *46*
2.2.3 Development or Modification of New Tool *47*
2.3 TEM Holder Design and Selection *47*
2.4 Specimen Design and Preparation *48*
2.4.1 Direct Dispersion on a TEM Grid *48*
2.4.2 Sintering Pallets *49*
2.4.3 Ultramicrotomy *50*
2.4.4 Electropolishing *50*
2.4.5 Mechanical and Ion Milling *50*
2.4.6 Focused Ion Beam (FIB) *52*
2.4.7 Tripod Polishing *54*
2.4.8 Cryo Sample Preparation *54*
2.5 Guidelines for Experimental Setup *55*
2.5.1 Electron Beam Effects *55*
2.5.2 Choice of TEM Grid and Support Material *56*
2.5.2.1 Reactivity of Sample with Grid and/or Support Material *56*
2.5.2.2 Reactivity of TEM Grids Upon Heating *57*
2.5.2.3 Reactivity of TEM Grids in Gaseous Environment *58*
2.5.2.4 Reactivity of Liquids with the Windows *59*
2.5.2.5 Reactivity of Gases/Liquids with the TEM Holder Parts *59*
2.5.3 Purity of Gases *60*
2.5.4 Liquid Cell Experiments *62*
2.5.5 Experiments Using Other Stimuli *63*
2.6 Practical Example of Designing In-Situ TEM Experiment *63*

2.6.1	Growth of GaN Nanowires Using ETEM	63
2.6.2	Applications of Quantitative Data	64
2.6.2.1	Physical and Materials Science	66
2.6.2.2	Catalysis	67
2.7	Review	67
	References	68

3 In-Situ Heating 77

3.1	History	77
3.2	Currently Available Heating Holders	78
3.2.1	Direct Heating Holder	79
3.2.2	Indirect Heating Holders	79
3.2.2.1	Furnace Heating Holders	79
3.2.2.2	MEMS-Based Heating Holders	82
3.3	Experimental Considerations	84
3.3.1	General	84
3.3.2	Electron Beam	86
3.3.3	Sample Temperature at Nanoscale	88
3.3.4	Specimen Design and Selection	90
3.3.5	Thermal Drift	91
3.4	Select Applications	92
3.4.1	Dislocation Motion	93
3.4.2	Nucleation, Precipitation, and Crystallization	94
3.4.3	Sintering	98
3.4.4	Thermal Stability of Materials	100
3.4.4.1	Alloys	100
3.4.4.2	Core–Shell Structures	100
3.4.4.3	2-D Materials	102
3.4.5	Phase Transformation	102
3.4.6	Materials Synthesis	104
3.5	Limitations and Possibilities	105
3.6	Chapter Summary	106
	References	106

4 In-Situ Cryo-TEM 115

4.1	Historical Perspective	116
4.2	Specimen Holder Design and Function	116
4.3	Specimen Design and Preparation	119
4.4	Practical Aspects of Performing Cryogenic Cooling	121
4.5	Some Noteworthy Applications	122
4.5.1	Mitigating Radiation Damage	123
4.5.1.1	Structure of Polymers	124
4.5.1.2	Structure of MOF and Zeolites	125
4.5.1.3	Cryo-TEM for Energy Materials	126
4.5.1.4	Reactions in Liquids	128

4.5.1.5	Quantum and 2-D Materials	*129*
4.5.2	Phase Transformations Below RT	*132*
4.5.3	Correlative In-Situ Experiments at Low Temperature	*135*
4.5.3.1	Mechanical Testing	*135*
4.5.3.2	Magnetic Field	*136*
4.6	Benefits and Limitations	*137*
4.7	Chapter Summary	*138*
	References	*138*
5	**Designing Liquid and Gas Cell Holders**	***145***
5.1	Historical Perspective	*146*
5.2	Design Philosophy	*146*
5.3	Windows	*149*
5.3.1	Image Resolution: Thickness and Material Properties of the Windows	*149*
5.3.2	Strength and Flexibility	*150*
5.3.3	Tolerance for the Pressure Difference	*151*
5.3.4	Inert or Corrosion Resistant	*153*
5.4	Microfabricated Window Cell (Microchips)	*154*
5.4.1	Static Cells	*157*
5.4.2	Flow Cells	*159*
5.4.3	Incorporation of Other Stimuli	*161*
5.4.4	Monolithic Microchips	*162*
5.5	Examples of Modified Window Holders	*163*
5.5.1	Redesigning the Microchips for Commercial Holder	*164*
5.5.2	Modified Window Microchips and TEM Holder Combination	*166*
5.5.3	Non-window Cell Holder to Incorporate Other Stimuli	*166*
5.6	Take Home Message	*167*
	References	*168*
6	**In-Situ Solid–Liquid Interactions**	***173***
6.1	Historical Perspective	*173*
6.2	Holder Design and Selection	*175*
6.2.1	Closed Cells	*175*
6.2.1.1	Graphene Cells	*175*
6.2.1.2	Microfabricated Window Cell	*178*
6.2.2	Limitations of Closed Cells and Need for External Stimuli	*178*
6.2.3	Flow Reactors: Microfluidic Design	*178*
6.2.4	Electrochemical Cell: Biasing	*181*
6.2.5	Heating in Liquids	*182*
6.3	Specimen Design and Preparation	*184*
6.4	Data Acquisition	*185*
6.5	Practical Challenges	*185*
6.5.1	Sample Loading	*185*

6.5.2	Electron Beam Effects	187
6.5.3	Windows Bulging	188
6.5.4	Interaction of Sample with Windows	189
6.6	Select Examples of Applications	190
6.6.1	Nucleation and Growth of Nanoparticles	190
6.6.2	Corrosion/Oxidation	192
6.6.3	Galvanic Replacement Reactions	193
6.6.4	Growth of Core–Shell Nanoparticles	194
6.6.5	Soft Nanomaterials Analyzed by In-Situ Liquid TEM	195
6.6.6	Quantitative Electrochemical Measurements	197
6.6.7	Battery Research	198
6.6.7.1	Open Cell	199
6.6.7.2	Closed Liquid Cell	200
6.7	Limitations	201
6.8	Take-Home Messages	202
	References	203

7	**In-Situ Gas–Solid Interactions**	**215**
7.1	Historical Perspective	215
7.2	Current Strategies	218
7.2.1	Window Holders	218
7.2.1.1	Incorporation of Other Stimuli	221
7.2.1.2	Specimen Design and Preparation	221
7.2.1.3	Practical Challenges for Gas-Cell Holders	221
7.2.1.4	Review of Benefits and Limitations of Gas-Cell Holders	222
7.2.2	Environmental Microscopes (Open Cell)	223
7.2.2.1	ETEM Combined with Gas Injection Sample Holder	223
7.2.2.2	Differentially Pumped TEM	224
7.3	Gas Manifold Design and Construction	227
7.4	Practical Aspects of Performing Experiments in Gas Environment	228
7.4.1	Electron Beam Effects	229
7.4.2	Gas Pressure and Resolution	231
7.4.3	Sample Temperature and Cell Pressure	232
7.4.4	Anticontamination Device	233
7.5	Select Examples of Applications	234
7.5.1	Effect of Gas Environment on Catalyst Nanoparticles	234
7.5.2	Carbon Nanotube (CNT) Growth	236
7.5.3	Nanowire Growth	237
7.5.4	Electron-Beam-Induced Deposition	238
7.5.5	REDOX Reactions	239
7.5.6	Gas Adsorption Sites	241
7.6	Review of Benefits and Limitations	243
7.7	Take-Home Messages	244
	References	245

8	**Multimodal and Correlative Microscopy** *255*
8.1	Multimodal TEM *256*
8.1.1	Parallel Ion Electron Spectrometry (PIES) *257*
8.1.2	Hybrid Microscope *258*
8.1.3	Alternatives to Free Space Approach *260*
8.1.4	Introducing Light for Other Applications *263*
8.1.4.1	Through Sample Chamber Port *263*
8.1.4.2	Through Sample Holder *264*
8.1.5	Laser Alignment *269*
8.2	Correlative Approaches *269*
8.2.1	TEM and SEM *270*
8.2.2	Electron and X-ray Microscopies and Spectroscopies *272*
8.2.2.1	Portable Reactor for Various Platforms *274*
8.2.2.2	Independent Correlative Measurements *278*
8.3	Take Home Messages *280*
	References *280*

9	**Data Processing and Machine Learning** *285*
9.1	History of Image Simulation and Processing *285*
9.1.1	Image Simulations *286*
9.1.2	Image Processing *286*
9.2	Current Status *289*
9.2.1	Progress for Image Simulations *289*
9.2.2	Progress in Data (Image) Processing *290*
9.3	Data Management *291*
9.4	Data Processing and Machine Learning (ML) *292*
9.4.1	What Is Machine Learning? *293*
9.4.1.1	Unsupervised ML *293*
9.4.1.2	Supervised ML *294*
9.4.2	Motivation *296*
9.4.3	Current Status *298*
9.5	Select Applications *300*
9.5.1	Noise Reduction *300*
9.5.2	Structure Determination *301*
9.5.2.1	Diffraction Pattern Analysis *302*
9.5.2.2	Image Analysis *303*
9.5.2.3	Atomic Column Heights (3-D Structure) *305*
9.5.2.4	Other Applications *305*
9.6	Future Needs *307*
9.7	Limitations *309*
9.8	Take Home Messages *309*
	References *310*

10	**Future Vision** *317*	
10.1	Historical Aspect *318*	
10.2	Current Status *318*	
10.2.1	ETEM *318*	
10.2.2	UEM and DTEM *319*	
10.2.3	Stroboscopic TEM *319*	
10.2.4	PIES *319*	
10.3	Technical Challenges *319*	
10.3.1	List of Major Workshops *320*	
10.3.2	Open Challenges and Technical Roadmaps *323*	
10.3.2.1	Specific for Battery Research *323*	
10.3.2.2	Specific for Liquid-Cell TEM *324*	
10.3.2.3	Specific for Catalysis *324*	
10.3.2.4	Specific for Quantum Materials *325*	
10.4	Developing Relevant Strategies *326*	
10.4.1	Modifying Base TEM/STEM Unit *327*	
10.4.2	TEM Holders with Multiple Stimuli *332*	
10.4.3	Automation and Autonomous Operation *336*	
10.4.3.1	Automation *336*	
10.4.3.2	Autonomous Experiments *338*	
10.5	Take Home Messages *340*	
	References *340*	

Index *349*

Preface

The motivation to write this book originates from a book titled *Dynamic Experiments in Electron Microscope* by Butler and Hale published in 1981 that was very helpful at the initial stages of my journey in the field of in-situ TEM. Although this book included most of the stimuli we are currently using, remarkable advances have been made since its publication. Moreover, while working with collaborators with different background and/or training students/postdoc, I also realized the complexity of the multifaceted challenges we face before (planning), during and after (data analysis) in-situ experiments. We often learn from our mistakes, but they are not reported such that we can all learn from them collectively. Here I present a collective learning experience that I have gained through discussions and input from my peers who are acknowledged in a later section.

In-situ TEM has become an integral part of TEM to understand the reaction mechanisms during synthesis and functioning of materials at nanoscale. Experiments are performed in a TEM/STEM instrument using diverse stimuli in vacuum, liquid, or gaseous environments to mimic real-world conditions as much as possible. It is also used to make physical property measurements under specific stimulus and in controlled environment (liquid or gas). For example, measuring mechanical properties, such as strength and/or plasticity in H_2 environment as a function of temperature; or measuring electrical properties in an electrochemical cell (liquid environment) as a function of temperature; or measuring magnetic properties or behavior as a function of temperature, etc. The title of the book was chosen to make a distinction that this book covers the former, i.e. performing in-situ experiments, although the property measurements are briefly described in Chapter 1.

There are also two aspects of formal training that is required to perform successful in-situ TEM experiments and obtain unambiguous results. First and foremost, is the training required for understanding the functioning and data collection techniques on a TEM platform. Second requirement for in-situ or operando data collection is a deep understanding of physics and/or chemistry of material under observation. This need arises from the fact that during in-situ data collection, we are using TEM column as an experimental platform as well as nanoscale characterizing technique.

The book does **NOT** cover fundamental training required for successful operation of TEM-based instruments, but highlights the practical aspects of designing in-situ TEM experiments. The intended audience includes graduate students, postdoctoral fellow, and/or scientists, who are excited about entering the field of in-situ TEM and are trained in exploiting TEM-based techniques to their fullest extent. A brief synopsis and reference books about the fundamental aspect of TEM and related techniques are included. Also, references to a large volume of review articles and edited books, which include a wide span of fundamentals and applications of in-situ and/or operando TEM, are provided in the relevant chapters/sections. This book will make readers aware of the technique developments, possible pitfalls, and emphasizes on the "know how" for planning successful experiments. We include:

- **Foundational chapters:** Chapters 1 and 2 discuss the need for making in-situ TEM observations, review the TEM fundamentals, briefly describe TEM-based techniques (with appropriate references) and when to use them; introduction to various stimuli – their importance, limitation; best approaches to perform in-situ experiments; provide answers to some practical questions such as which microscope to use, what is suitable holder, should I use imaging (high resolution/low magnification), diffraction (SAED, CBED, or NED), STEM, or ATEM, overview of specimen preparation, experimental planning with respect to choice of specimen grids, holders, temperature, etc., and take-home messages.
- **In-situ technique Chapters 3 to 7:** Chapter 3 – heating and Chapter 4 – cooling, describe specific modifications to a TEM holder needed for temperature control, how to set up an in-situ experiment, applications, pros, and cons. Chapter 5 is dedicated to the microchip fabrication and window holder design as it is an integral and/or essential part of in-situ experiments in liquid (Chapter 6) or gas (Chapter 7) environment. Chapters 6 and 7 include the introduction, history, design consideration, applications, and limitations for experiments in liquids and gasses.
- **Multimodal and correlative microscopy:** Chapter 8 covers the multimodal experiments performed on modified TEM using various techniques and stimuli and correlative experiments that combine TEM with other characterization platforms, such as X-rays and/or photons under same or similar experimental conditions.
- **Data processing and machine learning:** Chapter 9 covers the need and methods to meet the big data challenge, includes the history of TEM data analysis such as image simulations and processing; introduction to machine learning and its application to TEM data processing; application of ML to automatic alignments and control of in-situ TEM experiments, a step toward running autonomous experiments.
- **Future vision:** Chapter 10 reviews the outcome of few major workshops aimed at defining challenges for in-situ TEM experiments; possible solutions, dream instruments, steps for increasing temporal resolution; integration of other techniques (combinatorial techniques); application of ML to run autonomous experiments.

This book does **NOT** include detailed insights for making property measurements but includes available technology and reference for them. It is hoped that the book will help in planning and performing successful in-situ experiments, develop new analytical methods, motivate to think outside the box for new technological advancements.

DISCLAIMER: *The book was written after retiring from NIST and the contents of this book are result of the research conducted in my personal time and the ideas are not supported by NIST in any form. – Renu Sharma*

November 4, 2022   *Renu Sharma*

Acknowledgments

It takes a village

– Hillary Rodham Clinton

First and foremost, I would like to acknowledge the contributions of Dr. Khalid Hattar (Sandia National Lab), starting from his encouragement, most of the interesting quotes (used throughout the book), ideas about the structure of the book and chapters, and multiple discussions during writing period of the book.

As one of my goals was to share the failures, lessons learnt, and remedies that are often not included in our publications, I needed to reach out to my peers to discuss their experiences. I am extremely thankful to the enthusiastic response I received from (alphabetic order) See Wee Chee, Peter Crozier, Katherine Jungjohann, Sergei Kalinin, Penghan Lu, Andrew Minor, Kristian Møhave, Joe Patterson, Søren B. Simonsen, Robert Sinclair, Eric Stach, Seiji Takeda, Michael Zachman, Jian-Ming (JM) Zuo. Their contributions are included in forms of a quote or note in different chapters. I understand that there are only so many hours in the day and some of you were not able to reach back to me; however, I know that you wanted to, and I am thankful to have you as my friends and peers.

I would like to acknowledge those who have reviewed various chapters and their comments; David Smith (Chapter 1); Robert Sinclair (Chapter 2); Vijayan Sriram (Chapter 3); Katherine Jungjohann (Chapter 4); Eric Satch (Chapter 5); Joe Patterson (Chapter 6); Peter Crozier (Chapter 7); See Wee Chee (Chapter 8), Joshua Taillon (Chapter 9); Raymond Unocic (Chapter 10). Moreover, I am also thankful for the stimulating discussions with Rafal Dunin-Borkowski, Joerg Jinschek, Marija Gajdardziska-Josifovska, June Lau, James LeBeau, Molly (Martha) McCartney Goetz Veser, Judith Yang, and Wei-Chang David Yang.

Last but not the least, I sincerely appreciate the ever-present moral support from my family, friends, collaborators, and peers.

November 4, 2022 *Renu Sharma*

List of Abbreviations

List of key terminology and abbreviations used in the book

1D	one dimensional
2D	two dimensional
3D	three dimensional
4D	four dimensional
ACN	artificial convolutional network
ACOM	automated crystal orientation mapping
AC-TEM	aberration corrected transmission electron microscope or microscopy
ADF	annular dark field
AI	artificial intelligence
ATEM	analytical transmission electron microscope or microscopy
BF	bright field
BOE	buffered oxide etch
CAD	computer-aided design
C_c	chromatic aberration constant
CBED	convergent beam electron diffraction
CCD	charge-coupled detector
CL	cathodoluminescence
CNN	convolutional neural network
CNT	carbon nanotube
C_s	spherical aberration constant
CVD	chemical vapor deposition
DE	direct electron
DF	dark field
DFT	density functional theory
DM	digital micrograph
DoE	Department of Energy
DP	diffraction pattern
DSC	differential scanning calorimetry
EBID	electron-beam-induced deposition
E-cell	environmental cell

EDP	electron diffraction pattern
EDS	energy-dispersive X-ray spectroscopy
EELS	electron energy loss spectroscopy
ELNES	energy-loss near-edge structure
ESTEM	environmental scanning transmission electron microscope or microscopy
ETEM	environmental transmission electron microscope or microscopy
EXAFS	extended X-ray absorption fine structure
FEA	finite element analysis
FEG	field-emission gun
FFT	fast Fourier transform
FIB	focused ion beam
GB	grain boundary
GC-MS	gas chromatograph–mass spectrometer
GIF	Gatan imaging filter
GPU	graphic processing unit
GSS	graphene sandwich superstructure
GVS	graphene veil superstructure
HAADF	high angle annular dark field (aka Z-contrast imaging)
HRTEM	high-resolution transmission electron microscope or microscopy
HVEM	high-voltage electron microscope
IBID	ion-beam-induced deposition
IBIL	ion-beam-induced luminescence
IPA	isopropanol alcohol
LPCVD	low-pressure chemical vapor deposition
LSPR	localized surface plasmon resonance
Maglev-TMP	magnetically levitated turbomolecular pump
MEMS	micro-electromechanical system
ML	machine learning
MOCVD	metal organic chemical vapor deposition
MWNT	multiwalled nanotubes
MWCNT	multiwalled carbon nanotube
NBED	nano beam electron diffraction
NDA	non-disclosure agreement
NED	nano electron diffraction
PED	precession electron diffraction
PL	photoluminescence
RGA	residual gas analyzer
RIE	reactive-ion etching
RT	room temperature
SAED or SAEDP	selected area electron diffraction or selected area electron diffraction pattern

SEM	scanning electron microscope or scanning electron microscope/microscopy
SIMS	secondary ion mass spectrometer or spectrometry
SLG	single-layer graphene
SNR	signal-to-noise resolution
STEM	scanning transmission electron microscope or microscopy
STM	scanning tunneling microscope or microscopy
STO	strontium titanate
STXM	scanning transmission X-ray microscope/microscopy
SWNT	single-walled nanotube
SWCNT	single-walled carbon nanotube
TEM	transmission electron microscope or microscopy
TGA	thermogravimetric analysis
TMP	turbomolecular pump
UHV	ultra-high vacuum
UV	ultra-violet
VLS	vapor–liquid–solid
VSS	vapor–solid–solid
XAS	X-ray absorption spectroscopy
XPS	X-ray photoelectron spectroscopy
XRD	X-ray diffraction
XTM	X-ray transmission electron microscope

About the Author

Renu Sharma is a retired project leader from the Nanoscale Imaging Group in the Physical Measurement Laboratory and is currently working as an emeritus associate with the Materials Science and Engineering group, both at the National Institute of Standards and Technology (NIST). She received a BS and BEd in Physics and Chemistry from Panjab University, India, and an MS and PhD degrees in Solid State Chemistry from the University of Stockholm, Sweden, where she had a Swedish Institute Fellowship. Renu joined the CNST/NIST in 2009, coming from Arizona State University (ASU), where she began as a faculty research associate in the Department of Chemistry and Biochemistry and the Center for Solid State Science and most recently served as a senior research scientist in the LeRoy Eyring Center for Solid State Science and as an affiliated faculty member in the School of Materials and Department of Chemical Engineering. She has been a pioneer in the development of environmental scanning transmission electron microscopy (E(S)TEM), combining atomic-scale dynamic imaging with chemical analysis to probe gas-solid reactions. She has applied this powerful technique to characterize the atomic-scale mechanisms underlying the synthesis and reactivity of nanoparticles (including catalysts), nanotubes, nanowires, inorganic solids, ceramics, semiconductors, and superconductor materials. She is a fellow of the Microscopy Society of America, has received a Bronze Medal of Service from the Department of Commerce for developing new measurement techniques, a Deutscher Akademischer Austauschdienst (DAAD) Faculty Research Fellowship, is a past President of the Arizona Imaging and Microanalysis Society, member of Editorial Board of *Nanomaterials*, and has given over 100 invited presentations, edited 1 book, published 3 book chapters, and over 200 research articles. At the NIST, she has established an advanced E(S)TEM measurement capability that combines Raman and cathodoluminescence spectroscopies with electron diffraction, electron spectroscopy, and high-resolution imaging for nanoscience research (ETEM Lab). She has advised several graduate students and postdoctoral researchers and also has Emeritus position at Arizona State University since 2010.

1

In-Situ TEM

1.1 Introduction

Transmission electron microscope (TEM) and related techniques (scanning transmission electron microscope [STEM], tomography, holography, Lorentz microscopy, etc.) are preferred methods to understand the atomic-scale structure and chemistry of materials, especially for nanomaterials. The size of grains, grain boundary structure, density of defects, dislocations, etc. control the materials properties making such information critical for their synthesis and applications. Beautiful atomic resolution images, electron diffraction patterns, and chemical maps provide unprecedented information about the structure and chemistry of the defects and grain boundaries in the material under investigation. Commonly employed procedure is to characterize the starting material (prenatal) and end products (postmortem) to deduce the pathway for chemical reactions or phase transformations occurring when the material is subjected certain stimuli, such as temperature, pressure, and/or mechanical stress. Main motivation behind this exercise is to be able to generate synthesis–structure–property relationships by identifying structure and chemistry of materials formed under different synthesis conditions and measuring their properties.

In-situ TEM observations provide a direct visualization of structural and chemical changes under synthesis or operational conditions of nanomaterials when they are subjected to relevant external stimuli. Note that (i) TEM requires thin samples such that most of electrons are transmitted through after interacting with the sample, (ii) we need high vacuum to avoid electron scattering by the gas molecules, (iii) electrons can be treated as particles and/or waves, (iv) image formation optics is quite similar to light microscopes. The restrictions imposed by first two points require us to find methods to make thin electron transparent samples and modify the TEM column or sample holder, to accommodate required experimental conditions. We will address these requirements in this book as described below.

> Seeing is believing but feeling is the truth.
> – Thomas Fuller

In-Situ Transmission Electron Microscopy Experiments: Design and Practice, First Edition. Renu Sharma.
© 2023 WILEY-VCH GmbH. Published 2023 by WILEY-VCH GmbH.

1.2 General Scope of the Book

During past couple of decades, materials word is continuously shrinking in size where the size of semiconductor chips or batteries has dropped down to nanometers. As a result, fabrication methods have adapted from synthesizing building blocks for future assembly to combining synthesis and assembly process into one step, putting stringent control on the fabrication process.

Therefore, TEM-based techniques have become a method of choice to understand and predict the desired synthesis or fabrication route for materials with desired properties. TEM community has recognized this need and responded accordingly. There has been an explosion in the breadth of combinatorial in-situ TEM techniques that are now readily available due to advanced microscope controls, the development of microfabricated TEM sample holders, and automated data handling. Moreover, monochromated electron source and aberration-corrected lenses have made it possible to use medium-voltage microscope (200–400 keV TEM) with a pole-piece gap of 5 mm to 7 mm, needed for inserting TEM holders equipped with heating, cooling, biasing, mechanical testing, liquid and gas containment, etc. for atomic-scale structural and chemical characterization. These advancements and rapidly changing demand have attracted many more scientists to participate in the field, with or without formal training for employing TEM platform for performing experiments. As a result, a symposium or session (sometimes more than two) on the in-situ TEM characterization is included in most of the major conferences in chemistry and materials field (American Chemical Society (ACS), American vacuum Society (AVS), Materials Research Society (MR S), American Institute of Chemical Engineers (AIChE)), beyond microscopy and microanalysis (M&M), and the number of publications has increased exponentially (Figure 1.1).

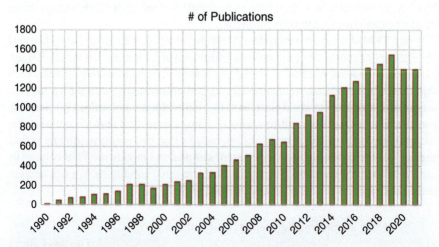

Figure 1.1 Number of publications reporting results obtained using in-situ or operando techniques in the last 40 years. Note the exponential growth since 2010. Source: Data from Web of Science.

1.3 Why In-Situ TEM

> Going from pretty pictures to touching and feeling
> – Murray Gibson

Let us first understand the meaning and difference between two commonly used terms: "in-situ" and "operando." First term "in-situ TEM" that we will commonly use throughout this book is also valid for in-situ STEM, in-situ analytical transmission electron microscope (ATEM) (EDS or EELS), etc. "In-situ" originates from Latin that means "in place" or "in position" and is used in many contexts. For us, it means TEM characterization of materials subjected to external stimuli at a specific "position or place" under synthesis or functioning conditions. Examples of external stimuli include, but are not limited to, temperature, gas or liquid environment, electrical biasing, magnetic or mechanical force, etc. to the material under observation using TEM-based platform. The term "operando" also originates from Latin and means "working," for example, we refer to the term in context of measuring the reactants, the product, and/or functionalities under working (functioning) conditions. In simple terms, while "in-situ" observations provide information about the changes under specific environmental conditions, the "operando" implies measuring the consequence of these conditions. In catalyst community, operando is strictly used to measure the kinetics as function of reaction variables, such as composition of reactants, products, nature (including loading, size) of catalyst/support, temperature, and pressure. Although in-situ TEM experiments have unveiled several catalytic mechanisms at atomic scale, most of the time, we do not operate under "working" reactor condition.

These are not strict definitions but are generally used by the materials, physical, and chemical scientists. In-situ TEM observations are more frequently used to follow reactivity of the material; however, operando measurements are relevant for catalysis, battery operations, and nanomaterial synthesis. Also, we generally use the term "in-situ" or "operando" when pursuing the changes with time, that may be termed as dynamic changes. In-situ characterization and measurements have following advantages:

- Same area or same nanoparticle is observed before, after, and during the reaction process such that all steps, including intermediate steps (if any), are identified.
- A careful design of the experiments leads to the observation and understanding of the relationship between morphological, structural, and chemical changes concurrently.
- Both the thermodynamic and the kinetic data of the reaction process may be obtained at nanometer or sub-nanometer scale.
- In-situ observations result in considerable time saving as the synthesis, property measurement, and characterization can be performed simultaneously.
- In-situ correlative light and electron microscopy (CLEM) is used to combine nanoscale and microscale characterization of the same sample subjected to same experimental conditions.

However, in-situ TEM experiments have their limitations and will be discussed later (Section 1.7).

1.4 TEM: Overview

This section is a brief reminder of some of the simple facts about the functioning of the TEM/STEM instruments. References to the books and articles are provided at the end of the chapter and should be consulted for detailed understanding of the electron optics, image formation, and chemical analysis.

1.4.1 Historical Perspective

After the advent of using coaxial magnetic coils to focus electron beam to a point, Ernst Ruska and Max Knoll, in 1931, built and demonstrated the first TEM, capable of magnifying objects to approximately 400 times, demonstrating the principles of electron microscopy [1]. The resolution limit of electron microscopes increased rapidly and, in 1939, and Siemens introduced first commercial instrument based on Ruska's design. Further development in the resolving power was slow, but a number of research groups worked on developing their own instruments leading to the formation of a number of commercial companies that took over the market from Siemens. Since then, there has been a steady development in improving the image as well as energy resolution. Moreover, several other applications that take advantage of the electron interaction with the sample such as diffraction and chemical analysis (energy-dispersive X-rays [EDS], electron energy-loss spectroscopy [EELS], cathodoluminescence [CL], etc.) have also been developed. To continue to take advantage of the TEM platform for in-situ observations, it is imperative for us to understand the principles of electron scattering and image formation in a TEM, basic design of TEM/STEM/ATEM instruments, and appreciate the developments over last 90 years that resulted in commercially available modern microscopes.

1.4.2 Electron–Sample Interactions

Electron–sample interactions can be divided into two main categories, elastic and inelastic (Figure 1.2). The collision of incident electrons with the nucleus is scattered by larger angles that results in the elastic scattering, whereas inelastic electrons scattering results from the collision of incident electron with the electrons cloud around the nucleus, including inner shell electrons, and is limited to small angles or small energy losses (below 50 eV). The elastic scattering contributes to the phase change in the transmitted electrons and is important for image formation as it contributes to the image contrast. In simple terms, elastic scattering allows us to acquire diffraction and images from the sample.

Apart from elastically and inelastically scattered electrons, the electron–sample interactions result in generating a host of signals, such as electron–hole pairs, back-scattered electrons, secondary electrons, Auger electrons, characteristic X-rays, and visible light (Figure 1.2). While secondary and back-scattered electrons are generally not used during TEM characterization, they are important for scanning electron microscope (SEM). We also ignore Auger electrons, but the light generated as a result of electron–hole recombination, i.e. CL, is often used to understand the semiconducting properties in materials. EELS and characteristic X-ray signal (EDS) are routinely used to obtain chemical composition of the sample.

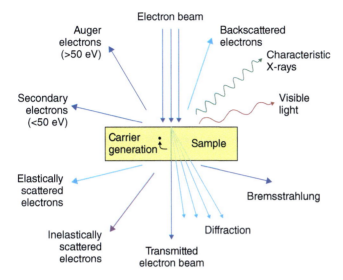

Figure 1.2 Distinct signals generated due to elastic and inelastic interactions of electrons with the sample. Most of the TEMs are equipped to collect electron diffraction, image, EELS and EDS data.

1.4.3 Overview of Modern TEM

Let us now revisit the ways TEMS/STEM instruments enable us to collect some or all the signals generated as a result of the electron/sample interaction shown in Figure 1.2. Image formation in TEM is based on the same optical principles as for a light microscope. The difference is that in a TEM, the electrons are used as a source of light and magnetic lenses are used for image formation and magnification. TEM can stand for transmission electron "microscope" or "microscopy," where one is an instrument (noun) and the other is a technique (adverb). For simplicity's sake, we will use the same acronym for both. We assume that the placement of the acronym in a sentence should be obvious enough to distinguish between "microscope" and "microscopy." Also, we use TEM for the basic system and specialized instruments or techniques, such as STEM or ATEM, can be considered as extension of the basic system. Although we will see that this simple explanation is not entirely correct as their design and functioning could have noticeable differences, i.e. a dedicated STEM has different optics than a TEM and an ATEM may have larger objective pole-piece gap. Here we will concentrate on TEM/STEM instruments as they are most effective for in-situ experiments. Figure 1.3a shows a simple ray diagram and location of various components of a basic TEM, and Figure 1.3b shows the detailed features of a modern double-corrected (C_s) TEM. A detailed description and functioning of each component are described in Sections 1.4.3.1–1.4.4.

1.4.3.1 Electron Source or Electron Gun

Electron gun can be considered as an "illumination source" for samples under observation and is an important component. High-energy electrons can be extracted from a thermoionic source such as a tungsten filament, or LaB$_6$ crystal, or field

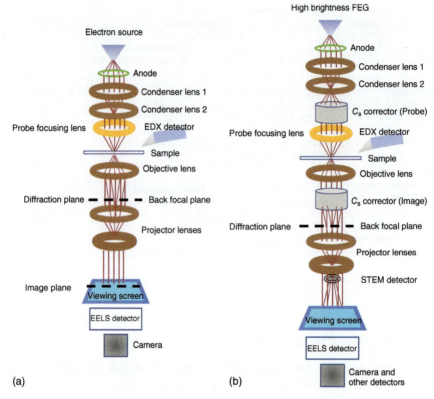

Figure 1.3 A schematic illustration of various components of (a) basic TEM and (b) double-corrected modern TEM/STEM with EDS, EELS, and STEM detectors.

emission from a pointed tungsten or tungsten tip coated by ZrO_2. Electrons are emitted from thermoionic sources by heating the source to a very high temperature. Electron density, brightness, coherence, and energy resolution of the illumination system are controlled by the electron source. Whereas thermoionic sources are easy to make and operate, they have low electron density, brightness, energy resolution and are incoherent. A field-emission source/gun (FEG) provides high brightness, energy resolution, and coherence, where the electrons are extracted from the source by applying an electric field either at room temperature (cold FEG) or at high temperature (Schottky). Simple but detailed information about the sources and their properties can be found in Chapter 5 of Ref. [2]. Moreover, small probes, on the order of sub-nanometer, can be formed using a FEG, which determines the spatial resolution for STEM imaging and atomic-scale chemical mapping [3].

> The type of electron gun is decided while constructing the microscope and cannot be changed afterward as the gun chamber, extraction and focusing systems are designed for specific source and are not easily interchangeable.

1.4.3.2 Lenses

Lenses are needed to focus the electrons on the sample and on the image screen after passing through the sample. In 1927, Busch successfully focused the electrons using electromagnets that motivated Ruska to design the first microscope in 1931 [1]. The same principle is used even today to construct lenses, where the magnetic field is generated by passing current through coils wound around a soft magnetic core. Since the magnetic field strength within the lens is function of current flowing through the electrostatic coils, both the focal length and the focal point can be controlled by varying the current. Following are the lenses used in a TEM:

I. **Condenser lens:** This lens is used to focus the beam of electrons, emitted from the gun to a point on the sample such as to form a point source. Modern TEMs have a set of condenser lenses that can be used to control the probe size and location of crossover along the optical axis.

II. **Objective lens:** Objective or image forming lens is located around the sample and focuses the transmitted electron beams on the image plane. The magnetic assembly generally consists of two parts, commonly known as pole pieces, placed above (top) and below (bottom) the sample, and is the most critical component that produces an axially symmetric magnetic field. The sample is located between these two pole pieces, collectively called objective lens, and defines the image resolution of the microscope. For optimum performance, the sample should be located at a specific position, called "eucentric height" between the two pole pieces. The gap between the pole pieces determines:

   o The achievable tilt angle of the holder tip, which is important, (i) to align the sample along one of the crystalline zone axes parallel to the incident beam, (ii) to collect highest X-ray signal for chemical analyses, and (iii) to obtain 3D images using tomography.
   o The space to insert objective aperture (OA) in the back focal plane of the sample or a CL detector.
   o The location of the sample within the gap for optimum resolution.

   Note that objective lens is probe forming lens in STEM mode for a TEM/STEM instrument, thus also determines the image resolution in STEM mode.

III. **Projector or magnifying lens:** Keep in mind that image resolution is determined by the objective lens and cannot be improved by employing other lenses; however, they can be used for image magnification that helps us see subtle changes in the image contrast. A set of electromagnetic lenses, known as projector lenses, provide this magnification. We can collect diffraction patterns or images by switching between image plane and back focal plane.

1.4.3.3 Lens Aberrations

Electromagnetic lenses, used to focus electrons, suffer from similar aberrations as the optical lenses, albeit they arise from different reasons. Spherical aberrations arise due to difference in the angle subtended by the electrons arriving at the lens from the electron source (for condenser lens) or from the sample (for objective lens). Electrons leaving a point object at large angle are scattered too strongly by the lens and brought

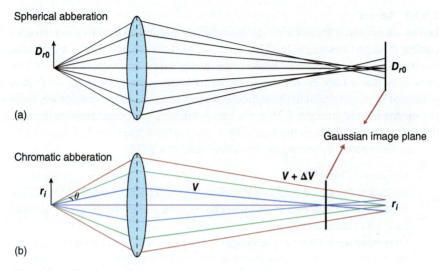

Figure 1.4 Simple ray diagrams showing the origin of spherical (a) and chromatic (b) aberrations.

to focus before the Gaussian plane giving rise to disc of confusion (Figure 1.4a). The diameter of the Gaussian image formed by point thus is given by

$$\delta = C_s \theta^3 \qquad (1.1)$$

where C_s is spherical aberration constant. The third-order dependence on θ implies that the beams scattered at high angles are most affected by aberrations. On modern instruments, the C_s ranges about 0.4 and 2.5 mm and can be corrected by using appropriate optics as will be explained under Section 1.4.3.4.

On the other hand, electrons with varying energy, ΔV, will also be focused on different points due to the small variation in the wavelength, which is the source of chromatic aberration (Figure 1.4b). It is important to note that modern high-voltage tanks, used to generate electrons, are very precise but not perfect. Following are the three important sources of fluctuation in the wavelength/voltage of the electrons:

- The energy spread of the electron leaving the filament.
- High voltage instabilities (typical $\Delta V_o/V_o$ is 2×10^{-6}/min; V_o is the operating voltage of the microscope).
- Varying energy losses in the specimen.

Currently, both aberrations can be corrected using a set of lenses that generate negative C_s or C_c such as the positive values of imaging lenses are canceled out. A monochromator is also employed to mitigate the energy spread of the electrons leaving the filament. As expected, achievable image and energy resolution of such microscopes is at sub-nanometer and below 1 eV, respectively.

Astigmatism Just as for optical lenses, a deviation from perfect circularity is commonly present for electromagnetic lenses and could be due to imperfections in the iron core, machining error, asymmetrical windings, dirty apertures, etc. As a result, a stretching in the image is observed as a point object is focused. This problem can be easily recognized by changing the focus as the stretching direction will change going from under focus to overfocus. Most microscopes are equipped with individual stigmator coils for all lenses to compensate for this distortion. Modern microscopes with aberration correctors also provide a precise correction of astigmatism for condenser and/or objective lenses.

> The lenses in a TEM have similar aberrations as optical lenses, but most of them can be corrected.

1.4.3.4 Aberration Correctors

Both the image and spectrum resolution are affected by the lens aberrations in a TEM [4–6]. Apart from C_s and C_c, described above, the energy of all electrons hitting the sample may not be the same, i.e. electron beam is not monochromatic. The high-voltage tank of most microscopes is designed to keep the energy fluctuation to a minimum value, and some sources are equipped with a monochromator. There are a number of ways, such as using electrostatic energy filter [7], or using a combination of electrostatic and magnetic quadruples [8], to filter out electrons with higher energy spread but often results in loss of intensity of the source. The energy resolution of a monochromated source may vary from 8 to 80 meV. We also find that the image resolution is also improved by using a monochromated source.

A TEM with C_s and C_c corrector for objective lens is also readily available [9, 10]. Both the cost and height of the TEM column increase with incorporation of correctors (Figure 1.3b).

1.4.4 Data Acquisition Systems

Electron–sample interactions result in generating various imaging and spectroscopy data that need to be collected for further analysis and documentation. Modern microscopes use digital cameras to record images, videos, diffraction patterns, EDS and EELS data. A microscope can be equipped with different digital detectors for recording TEM, STEM, annular dark-field (ADF), or bright-field (BF) images, or spectroscopy data such as EELS, EDS, or CL. We will look at them in detail as we work out the experimental plans.

1.4.4.1 Types of Detectors

Recording medium for image, DP, and spectroscopy data has different requirements. Whereas spectroscopy data have always been acquired using digital detectors, photographic film or image plates were used for recording images and DP due to their high dynamic range (pixel resolution) in the past. We should keep in mind that although recording medium to acquire DP and images may be the same, the acquisition process and time is different due to different intensity range present in the signals. The films have recently been replaced by electronic media such as phosphor/photomultiplier (PMT) or fiber-optic charge-coupled device (CCD). Here the phosphor is used as a scintillator to convert electrons into photons that are then

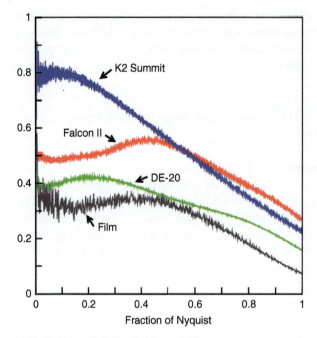

Figure 1.5 Measured DQE as a function of spatial frequency for the DE-20 (green), Falcon II (red), and K2 Summit (blue). The corresponding DQE of photographic film is shown in black. Source: McMullan et al. [13]/Elsevier/CC BY-3.0.

detected by a PMT or a CCD. PMT could be preferred for in-situ measurements due to their faster acquisition time, but their performance, as measured by their detective quantum efficiency (DQE), reduces at higher operating voltages (>120 keV) making CCD a better option. With continued rapid improvement in technology, a high DQE ≈ 0.7 and faster acquisition rate can be achieved for medium-voltage microscopes (200–300 keV) [11] by using direct electron detectors (DEDs) equipped with monolithic active pixel sensors. Furthermore, digital data acquisition enables data processing, including drift correction and adding a few frames to improve signal-to-noise ratio to obtain atomic resolution images at high time resolution [12]. The digital images are also compatible with machine learning approaches that improve our ability for unambiguous data analysis (Chapter 9). Figure 1.5 shows a comparison between three commercially available DED cameras and photographic film [13].

Apart from the DQE, time resolution, often termed as temporal resolution, is an important factor for in-situ measurements. Ideally, we need both high spatial and temporal resolution to decipher metastable steps of a chemical or physical process under investigation. Currently, ms time resolution is possible using DED cameras or another technology based on "pump and probe" idea (see Section 1.6.3). CMOS camera with scintillator [14] and pixel array detector [15] are a few other types of detectors available, and technology is improving at a fast pace, and cameras with both high spatial and temporal resolution are available for in-situ measurements.

1.5 TEM/STEM-Based Characterization Techniques

As we mentioned earlier, we assume that the readers of this book have thorough knowledge and adequate experience to operate TEM/STEM instruments. This section is a rapid review of TEM-based techniques that are available for in-situ measurements.

1.5.1 Diffraction

Electron diffraction is formed in the back focal plane of the objective lens and provides following structural information:

- **Crystallinity:** Diffuse rings in a DP indicate that the sample is amorphous (Figure 1.6a), spots on a ring pattern indicate polycrystalline nature (Figure 1.6b), and a 2D spot pattern is indicative of a single crystal (Figure 1.6d).
- **Structure of crystalline sample:** A single crystal sample can be aligned along one of its crystallographic zones to obtain structural information using selected area electron diffraction (SAD or SAED), convergent beam electron diffraction (CBED) (Figure 1.6e), nanobeam electron diffraction (NBED or NED), or nano area electron diffraction (NAED) (see below for details).
- **Defects:** A streaking or faint lines along the spot patterns (Figure 1.6c) indicate the presence of defects in the structure, which can then be further investigated using imaging.
- **Long-range order:** Presence of faint spots, or satellite spot, along with stronger spots, indicates the presence of long-range order along one of the crystallographic directions or perpendicular to viewing direction, respectively (Figure 1.6f).
- **Strain:** presence of faint lines along 2D diffraction spots is indicative of the presence of strain in a single crystalline sample. A number of methods to analyze and measure strain from electron diffraction patterns, especially from CBED or NBED, have been developed [17–19].

We often use a parallel electron beam and place an aperture, i.e. selected area aperture, to form diffraction pattern from a desired area (SAED or SAD). Such DP provides us information about the crystallinity and orientation of the sample. Some of the structural information, such as d-spacing and angle between planes in projection, is also obtained from these patterns. We can select the diffracted beams, using an objective aperture (OA) we want to use to form an image. For example, outer most diffraction spots arise from multiple scattering events, so we can improve image contrast by discarding them. We can also select one or two diffraction spots and discard central beam to form dark-field TEM image.

A CBED, which is in form of discs instead spots (Figure 1.6e), provides detailed structural information, such as crystal class and symmetry. Methods have been developed to use a CBED pattern to determine crystal structure by removing contribution of inelastically scattered electrons [20]. Nowadays, we can converge electron beam to nanometer size to obtain diffraction pattern from nanometer-size area, NBED or NED [21]. Interestingly, modern microscopes also allow us to obtain

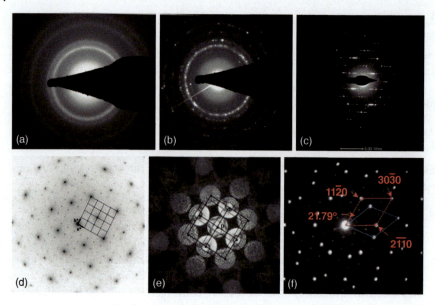

Figure 1.6 Selected area EDP with (a) continuous rings indicating amorphous nature of the sample, (b) rings with diffraction spots from polycrystalline sample, (c) streaks in ordered 2D pattern indicate presence of defects, (d) from single crystal Si oriented in [111] zone axis, (e) EBED from single crystal, (f) NED from a bicrystal α-Fe_2O_3, which can be indexed as two overlaying α-Fe_2O_3 platelets oriented along the ⟨0001⟩ zone axis, with a coincidence-site-lattice boundary with a twist angle of 21.79. Source: (d, e) Zhu et al. [16]/ from American Chemical Society.

electron diffraction by using parallel electron beam of nanometer size to select an area of interest and are known as NAEDs [21]. Nano-diffraction techniques allow as to meet the challenges of obtaining structural information from the building block of the nanoworld (Figure 1.6f). For example, chirality of individual single-walled nanotubes can be determined using NAED [22].

1.5.2 TEM Imaging Modes

The dual nature of electrons, as particles and waves, is used in understanding the image formation process. While the electrons scattered by the atoms or crystalline lattice propagate as transmitted electrons, the wave function makes the image formation similar to an optical microscope and fundamental mechanism also has some similarities. For example, the resolution can be defined by the Rayleigh criteria; separation between two-point objects should be more than 0.61λ, where λ is the wavelength of incident light. It implies that the image resolution can be improved by decreasing the wavelength, i.e. by increasing energy of the electrons. However, unlike light, electron scattering is both elastic and inelastic, and the Rayleigh's criteria do not hold as the resolution is impeded due to multiple scattering experienced by electrons while traveling through the sample and the magnetic lenses.

> The wavelength of the electron in such a microscope is only 1/20 of an angstrom. So it should be possible to see the individual atoms. What good would it be to see individual atoms distinctly?
>
> – Richard Feynman

In the high-resolution TEM, a parallel incident electron beam interacts strongly with the sample, forming multiple diffracted beams (as explained above) that are brought together by the objective lens such as they can interfere to create an image, and the structural information can be obtained from the exit electron wave function that has been attenuated by the multiple scattering events. We should keep in mind that the exit-wave carry the information about both the phase shifts and amplitude modification due to electron scattering by the sample. The phase information cannot be directly visualized as it is influenced by the lens aberrations and other imperfections; therefore, it is the amplitude of the electron wave that is recorded on the image plane.

However, we need the phase information of the exit waves to obtain structural information, which can be achieved by tuning the focus of the objective lens such that the phase of the wave is converted into amplitudes on the image plane. The combined effect of aberrations, objective aperture function, drift, and other instabilities on the image contrast can be mathematically treated as phase-contrast transfer function (CTF). Scherzer has shown that there is a specific value of defocus, depending on the properties of the microscope, where low spatial frequencies are transformed into image intensities with similar phase. This value, known as Scherzer defocus, also determines the resolution limit of the microscope. However, we should keep in mind that the information beyond Scherzer resolution can be obtained by using a highly coherent and monochromated electron beam generated by a FEG along with aberration correctors. Moreover, the phase difference can also be measured from interference patterns such as formed by using holography, where separate phase and amplitudes image are collected (Section 1.6.2).

As mentioned above, the location of the sample within the objective pole piece is critical for obtaining the best electron optical performance of the TEM. With appropriate choice of the microscope, imaging conditions, and modifications, TEM can be used to obtain morphological, structural, and chemical information. Some of them are described below as they are the ones that we utilize to follow dynamic changes during in-situ observations.

- **Low-magnification TEM:** is used to obtain morphological information such as size and shape of nanoparticles, particle size distribution, dislocations, measuring Burger's vectors.
- **High-resolution imaging:** atomic-scale information, including, but not limited to, from defects, grain boundaries, nanoparticles, etc.
- **Dark-field imaging:** objective aperture can be used to select the diffraction spots to form image that shows the corresponding region as bright, similar to STEM-ADF images.

- **Tomography:** 3-D images reconstructed from a series of images recorded at incremental positive and negative tilts (±70°). Full rotation of sample is possible by using needle-shaped sample geometry and a special TEM holder.

1.5.3 STEM

Image formation in a STEM is achieved by scanning a small electron probe across a thin sample. It is similar to SEM, except while secondary and/or back-scattered electrons are used for image formation in an SEM, forward scattered or transmitted electrons form a STEM image. On the other hand, reciprocity relationship can be found between TEM and STEM, where electron source and detector exchange their location while other components are the same. Also, while a parallel electron beam is used to form a TEM image, STEM image is formed by scanning focused electron probe, such that the STEM image resolution depends on the size of the electron probe. Integration with FEG source provides small probe size (0.1 nm or better in modern instruments) with enough intensity for imaging and chemical analysis. STEM has following advantages over TEM:

I. Electrons scattered at different angles contribute to formation of bright-field (BF), annular dark-field (ADF), or high-angle annular dark-field (HAADF) images and can be collected by using different detectors.
II. We can use the small probe to obtain atomic-scale chemical information using either EELS or EDS (chemical mapping).
III. Probe can be focused anywhere along the thickness (z-direction) of the sample, thereby providing a possibility to obtain 3-D information.
IV. HAADF images are sensitive to the atomic number (Z) of the atoms and are used to obtain chemical information directly from the intensity distribution in the atomic-resolution images.

Although there are dedicated STEM instruments still available, most of the modern microscopes can operate in both TEM and STEM modes. However, STEM is not generally used for in-situ dynamic characterization as the time resolution for acquiring images is longer than that for TEM imaging.

1.5.4 Analytical TEM

The strength of TEM/STEM is not only that we can obtain high-resolution images with 0.05–0.1 nm image resolution, with one-to-one correspondence between structure and the image, but also in the fact that we can obtain chemical information with spatial resolution of 0.1 nm to 10 nm. Spatial resolution of a TEM for obtaining chemical information depends on the probe formation characteristics of the TEM, i.e. the probe (spot) size and the current in the probe. As mentioned earlier, modern TEM has possibility to form a probe as small as 0.1 nm with enough beam current (0.2–0.3 pA) to generate enough EELS and/or X-ray signal to be collected in reasonable time (2 to 30 seconds time frame).

As explained above and shown in Figure 1.2, when a fast incident electron traverses through a thin sample, it may experience an inelastic scattering event in which energy and momentum from the incident electron are transferred to electrons in the solid that will result in an electronic excitation event. During this process, electrons in the solid undergo transitions to vacant states above the Fermi level, and the atoms of the solid are ionized. The energy lost by the incident electron, ΔE, is then equal to the difference in energy between the initial and final states of the electron(s) being excited in the solid. A collective interaction of both inner- and outer-shell electrons may be involved in this excitation process. It is this fundamental interaction that gives rise to the EELS and the EDS signal and is used for chemical analysis.

1.5.4.1 Chemical Analysis

EDS The vacancy created in one of the inner-shell levels of an atom during the ionization process by fast-moving electrons may subsequently be filled by an electron moving down from an upper shell, and the energy difference is emitted as X-rays. The X-rays are labeled according to the initial and final states of the electron making the transition to fill the vacancy. For example, a vacancy in the K or L shell gives rise to the production of K- or L- X-rays. Greek letters are used to identify the initial state of the electron making the transition. For example, L to K transitions give rise to K_α X-rays, and M to K transitions give rise to K_β X-rays. The K_β X-rays are always of higher energy than the K_α X-rays. X-rays produced through these processes are characteristic of the element involved and are always emitted isotopically. With the help of suitable detector, EDS is used for chemical analysis of the sample. An example of EDS data collected from CdTe lamella in STEM mode is shown in Figure 1.7.

In addition to these characteristics X-rays, non-characteristic X-rays are also generated by the fast electron through a process known as Bremsstrahlung. Classically, this process can be considered in terms of a fast electron being slowed down by the field of the atoms in the sample. Here, it is this change in electron velocity that gives rise to X-ray emission. Bremsstrahlung gives a continuous background in the X-ray spectrum and must be removed to quantify the characteristic X-ray emission. The energy of the X-ray produced is directly related to the element present in the sample and the orbital level involved in the ionization process.

It is important to note that EDS detectors are sensitive to heat (due to infrared radiation) and gaseous environment and therefore not suitable for in-situ measurements

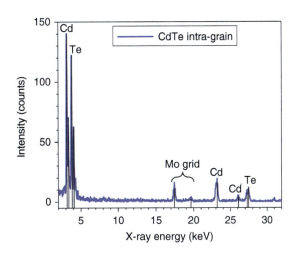

Figure 1.7 EDS from a CdTe solar cell lamella, collected in STEM mode, peaks from the TEM grid material (Mo in this case) are commonly present in the spectra. Courtesy: Wei-Chang (David) Yang, NIST.

under these conditions. However, the MEMS-based heating and gas cell holders have addressed this limitation [23, 24], although they have other problems as will be discussed in Chapters 3 and 7 in detail. Alternatively, chemical changes from the same area can be observed before and after reaction to directly correlate the effect of heat or environment.

Quantitative X-ray Analysis The EDS data are quite easy to collect and interpret and are often used for quantitative analysis. The first step in quantifying thin-film EDS data involves removal of the background to leave only the characteristic peaks of interest. The intensity of the remaining X-ray peaks can be expressed in terms of the atomic concentration, X-ray production factors, and instrumental parameters. For a thin film, the intensity of a characteristic X-ray from element A, I_A can be written as

$$I_A \propto I n_A \sigma_A \omega_A \varepsilon \tag{1.2}$$

where I is the total number of incident electrons, n_A is the number of atoms of element A per unit area, σ_A is the ionization cross section for producing a vacancy, ω is the florescence yield, and ε is the detector efficiency. For many applications, the quantity of interest is the relative concentration of one element with respect to another. The relative concentration of two elements then can be obtained using following equation:

$$C_{AB} = k_{AB} \frac{I_A}{I_B} \tag{1.3}$$

where we have combined all the terms involving X-ray generation and the detector efficiency into the quantity k_{AB}. This simple form is known as the Cliff–Lorimer equation for thin film X-ray analysis. It can be used to quantify X-ray spectra obtained from thin film provided that:

i. The electrons lose only a small fraction of their energy in the film and backscattering can be ignored.
ii. The specimen detector geometry is such that effects due to X-ray absorption and fluorescence can be ignored.
iii. Coherent scattering effects, which occur within crystals (such as channeling), have not been included in the derivation of the above two expressions. Whenever possible, crystals should be tilted away from strong diffracting conditions to avoid systematic errors in the analysis.

For a fixed accelerating voltage and detector geometry, the k-factor for each pair of elements is constant. The simplest way to determine the k-factor for a given microscope is to measure them directly using standards of known composition. Numerous tabulations of k-factors are now also available in the literature. It is also possible to determine k-factors using known values of the X-ray generation parameters for each element. In this case, it is also necessary to know the form of the efficiency function, ε, for the detector involved. It is generally assumed that the most accurate k-factors are obtained by measurements from standards. However, the standard-less approach can be useful when suitable standards are not available for the elements of interest.

EELS As explained above, during X-ray emission, an electron from one of the higher orbitals falls into the lower energy state to fill the vacancy. The energy lost by electrons during initial excitation to create the vacancy is also characteristic of the atom involved and allows elemental analysis to be performed. By adding a magnetic spectrometer beneath the viewing chamber, the energy distribution of electrons that have passed through a thin specimen in a TEM is measured (Figure 1.3) to obtain the EELS data. Another method, not very commonly available, is by using an in-column energy filter (Omega filter), located below the sample, to collect EELS data. The spectral energy resolution is largely determined by the energy span of the electron source: i.e. 1–2 eV for a thermionic source, 0.5–0.7 eV for FEG, and 0.01–0.1 eV for an FEG with a monochromator. Figure 1.8a shows various types of EELS peaks arising from interactions of electrons with atoms in the sample. Most of the high-energy electrons are transmitted through the sample with negligible loss of energy and constitute zero loss peak that has the highest intensity. Its full width at half maxima (FWHM) is the measure of energy resolution that is controlled by the source as well as by the spectrometer. Elastic scattering of the electrons by the nucleus gives rise to low-loss peaks such as plasmon peaks, and their number increases with the thickness of the sample. Low-loss region may also have peaks that can be attributed to vibrational and phonon modes, localized surface plasmon resonance, and bandgap of the material. Inelastic scattering results from the interaction with electron cloud around the sample, and the energy lost is generally higher than 50 eV (Figure 1.8a). Electron–electron interactions result in knocking out electrons from the outer or inner shell, and the energy as well as shape of these peaks is specific to the element as well as to the bonding environment (Figure 1.8b) (see Egerton's book [27] for more info). Since the electron beam can be focused into a probe of sub-nm dimensions

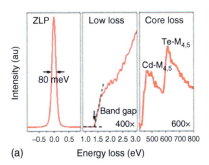
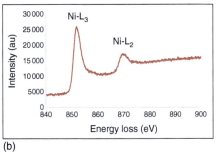

Figure 1.8 (a) Zero-loss, low-loss, and the core-loss regions of EEL spectra acquired in STEM mode from CdTe solar cell lamella using dual EELS detector. The intensities from low-loss and core-loss regions are magnified by 400× and 600×, respectively, for visualization purpose. FWHM of ZL peak is a measure of energy resolution of the source, low-loss region can be used measure the bandgap (shown here), plasmon, and LSPR energies (not shown here but check Figure 3.5 for plasmon energy applications). Courtesy: Wei-Chang (David) Yang, NIST. (b) Core-loss region showing $Ni-L_2$ and $Ni-L_1$ peaks, often termed as white lines. For most of the transition metals, white-line ratio and the onset of core-loss peaks can be used to determine their oxidation state. Source: Adapted from Refs. [25, 26].

at the specimen, this spectral information can be obtained with very good spatial resolution [6, 28]. We can use the EELS to obtain following information:

Thickness of the Sample The energy-loss spectrum can be employed to measure local specimen thickness. This relies on the fact that for incoherent scattering, the relative intensity of the zero-loss peak decreases monotonically with increasing thickness. The sample thickness t can be written as

$$t = \lambda L_n(I_{tot}/I_o) \tag{1.4}$$

where I_{tot} is the total integrated spectral intensity, I_o intensity of the zero-loss peak, and λ is a parameter called the inelastic mean-free path. This equation is quite simple to use and can be used to measure sample thickness between 5 and 500 nm.

Qualitative Elemental Analysis The inner-shell excitations show up as ionization edges, and their onset position corresponds to the ionization potential of the electron involved in the Coulomb interaction between the incident electrons and atomic electrons. For thin samples, most of the electrons do not lose energy, and zero-loss peak is much stronger than inner-shell excitation edges. Therefore, it is easy to identify edges from thin samples. EELS is very effective to identify light elements such as carbon, oxygen, and nitrogen.

Quantitative Elemental Analysis Quantitative analysis using EELS is not as straightforward as for EDS analysis and generally should be used only if the same information cannot be obtained from the EDS data. The intensity distributed on $I_A(E)$ in an inner-shell edge from element A in a thin film can be written as

$$I_A(E) = IN_A \frac{d\sigma_A}{dE} \tag{1.5}$$

where I is the number of incident electrons, and N_A is the number of atoms of type A per unit area. The quantity $d\sigma_A/dE$ is known as the energy-differential cross section (See Egerton's book for more info [27]), which is difficult to quantify, therefore making quantitative analysis challenging.

Bonding Information The onset energy, shape, and number of peaks in core-loss spectra are fingerprints of electron densities in the valence band. Therefore, depending upon the spectral resolution of the microscope, they contain the information about the bonding with the first, second, and other nearest-neighbor atoms. Careful examination of ionization edges reveals the band structure as oscillations within the first 20 or 30 eV of the edge onset, which are referred as near-edge structures. A comparison between theoretical model and experimental spectra can be used to obtain bonding information of the element.

Detailed interpretation of near-edge structure remains a topic of ongoing research, but some general applications are outlined below:

- Changes in the shape of ionization edge structure signify a change in the chemical environment around a particular element, i.e. bonding.

- For 3d and 4d transition metals, the intensity of the peaks at the threshold (the so-called white lines, Figures 1.8b and 4.14) can be correlated with the number of vacant states in the d bands of the materials, i.e. valence state.
- Temperature measurement using plasmon peak shift [29].
- Pressure measurement from plasmon peak shift [30].

We can use EELS to fingerprint different chemical environments for identification of compounds. EELS is routinely used for in-situ monitoring of chemical state of the sample as function of temperature and environment (liquid or gas) as explained in Chapters 3,4, 6, and 7.

1.5.4.2 EFTEM

EELS spectrometer can also be used to obtain images or DPs using the electron within a selected energy window and filtering out the rest, generally known as EFTEM. There are two types of filters commercially available: (i) in column filter (Omega or Wein) and (ii) post projector filter (Gatan Imaging Filter, GIF). Each system has slits to choose the windows between 5 and 100 eV for image/DP formation. Zero-loss images/DPs are obtained by choosing between 5 and 15 eV window, centered at the zero-loss peak, i.e. using only elastically scattered electrons. This method is used to obtain high-contrast images from thick samples. DPs obtained from elastically scattered electrons are used for structure determination as the intensities can be directly used to obtain structure factors (for more information, see Zuo and Spence [5]).

Images obtained using specific ionization intensities are used to obtain a distribution of certain element within the image or an elemental map of the sample. In order to obtain quantitative information, the contribution from the background must be subtracted. This can be achieved by following two ways:

(a) **Three window maps:** For quantitative elemental mapping, we select two pre-edge energy windows to form pre-edge images and another window centered at the elemental peak to form post-edge image (Figure 1.9). If the two pre-edge energy windows are adjacent to each other, they can be used to evaluate the background parameters, A and r, and background is calculated assuming the background has a form AE^{-r}. The background contribution thus can be subtracted from the post-edge image to obtain the image that contains the contribution only from the ionization edge intensities, i.e. elemental map [27].

(b) **Jump ratio maps:** It is simple method to obtain an elemental map and is particularly useful when pre-edge section is not wide enough to select two windows for collecting pre-edge images, as one of them might be overlapping with ionization edge of another element. Jump ratio maps, as the name implies, are obtained by recording imaged from (i) one pre-edge window for background and (ii) one for the ionization edge window of a specific element. A new image is then generated by dividing post-edge image by pre-edge image. This method yields a jump ratio map that is insensitive to the specimen thickness and diffracting conditions [27]. Both methods are routinely used.

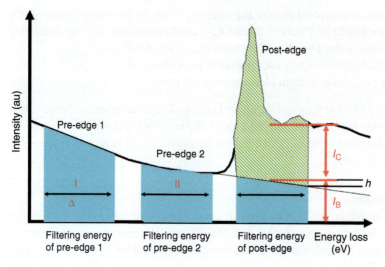

Figure 1.9 Schematic showing the process of choosing windows for collecting filtered images from two pre-edge and the ionization windows used to obtain an elemental map. The number of counts in background and edge spectra are marked as I_g and I_c, respectively. For jump ratio map, images from one background and ionization edge window are collected. Source: Egerton et al. [27]/Springer Nature.

1.5.4.3 Spectrum Imaging (SI)

Spectrum imaging (SI) is another method to use analytical data to obtain elemental maps. It is performed in STEM mode to collect either EDS or EELS signal from each pixel of the image. These scans provide 3-D data sets that are then used to obtain elemental distribution map at atomic scale [5, 31, 32]. Just as for the STEM imaging, the spatial resolution for SI is defined by the probe size; therefore, atomic-resolution elemental maps can be obtained using a monochromatic electron beam combined with C_s-corrected probe forming lens. The resultant image is like a normal TEM image, except each image pixel contains entire EELS or EDS data. Thus, the 3-D data cube (Figure 1.10) obtained can then be sliced to get 2-D information in any direction and allows us to perform quantitative and statistical analysis.

> God runs electromagnetics on Monday, Wednesday, and Friday by the wave theory, and the devil runs it by quantum theory on Tuesday, Thursday, and Saturday
>
> – Sir Lawrence Brag

1.6 Other Techniques

1.6.1 Lorentz Microscopy

Behavior of magnetic materials, just like others, is controlled by their nanoscale structure. Therefore, we need to determine a structure–property relationship, both for understanding fundamental science as well as for their potential applications in magnetic logic and memory devices. Lorentz TEM is ideally suited to find structure

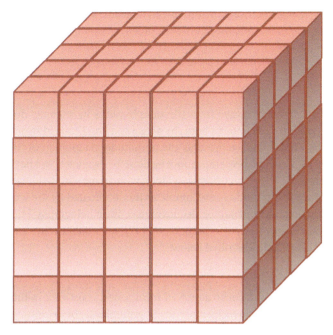

Figure 1.10 Schematic of a "data cube" obtained by collecting EELS data from each pixel in STEM mode. An elemental map can be produced using EELS spectrum thus collected.

and magnetic property relationship by determining the crystal and magnetic domain structures in correlation with their physical properties. For magnetic materials, the electron beam is deflected due to Lorentz interaction, arising from in-plane magnetic induction perpendicular to the electron beam. Therefore, the part of the sample, magnetized in-plane, can be imaged easily, and the part with perpendicular magnetization will need to be tilted to produce the same amount of deflection. The angular deflection, known as β, is given by $\beta = e\lambda B_s t/h$, where B_s is saturation induction, t is thickness of the sample, e is the magnitude of the electronic charge, λ is the wavelength of the electrons, and h is Planck's constant. We find that Lorentz microscopy can be used for both the qualitative and quantitative measures of the magnetic behavior of thin films [33, 34].

However, we must realize that in a TEM, the sample is immersed in high magnetic field of the objective lens that can alter or destroy the magnetic structure entirely. Therefore, we need a TEM with modified objective lens that can be switching off and use projector lenses for imaging. Other strategies such as adding a shielding around the objective lens to reduce the field or changing the pole pieces, etc. are also available [35].

Again, in the last few decades, we have moved from static imaging to in-situ investigation of effect of magnetic-field or temperature on the nanoscale configuration of magnetic components such vortices, skyrmions, bubbles, and phase transformation related to ferro- and ferrimagnetic properties [36, 37]. It is a developing field of research, and multiple imaging techniques such as Fresnel and Foucault modes, electron holography, and differential phase contrast (DPC) techniques are being applied [38, 39]. In-situ manipulation of magnetic domains, topological features, vortices, and skyrmions, using electrical current, magnetic fields, and temperature, has revealed novel phenomena that advance the development of nanodomain-based devices and spintronics [40–45].

1.6.2 Holography

As explained in Section 1.5, the exit wave from the sample varies in intensity due to electron–sample interaction and contains both the phase and amplitude information. The contrast in the TEM images represents the amplitude modulation of the exit wave, and phase shifts, although present, cannot be directly visualized. We need to find an alternate method to obtain phase images along with an amplitude image. However, retrieval of phase information is not simple as it is attenuated by the lens aberration, aperture function, etc. but can be deconvoluted using holography [46]. Cowley has identified 20 forms of electron holography [47], including in-line and off-axis holography that are briefly described below.

1.6.2.1 In-Line Holography

Two methods, Fresnel images and reconstruction of focal series, constitute in-line holography. Fresnel images are used to recover phases by minimizing the error function between simulated and experimental image intensities [48]. Image reconstruction from focal series to retrieve phase information is used to obtain 3-D shape of nanoparticles without the need to tilt the sample, unlike electron tomography [49]. The possibility to record atomic-resolution images using the aberration-corrected microscope extends our ability to obtain 3-D atomic-scale information by applying phase reconstruction algorithms to a focal series [46]. Another advantage of this method is that the dose rate for recording individual image can be adjusted to mitigate the electron-induced damage to the sample, i.e. image series can be acquired under low-dose conditions [50]. However, in-line holography requires tedious algorithms and computer time.

1.6.2.2 Off-Axis Holography

D. Gabor (1949) presented the idea to recover phase from electron microscope images using interreference methods such that the information lost by the lens imperfections can be restored [51]. However, it could not be fully developed due to the incoherent electron sources used at that time. After the advent of FEG electron source becoming commercially available, and development of electrostatic biprism, two crucial requirements for off-axis electron holography, the technique has been further developed [52–55]. Briefly, the method constitutes obtaining an image, after applying positive voltage to the biprism, where exit wave from the object superimposes an unscattered (reference) wave passing through the vacuum at an angle. The phase shift experienced by the electrons traveling through the sample can be retrieved from the image that is a direct measure of electrostatic potential and in-plane magnetic component in the sample. Therefore, off-axis holography, in combination with Lorentz microscopy, is frequently employed to study the strain in the sample and the magnetic properties of the materials [56–59].

1.6.3 UEM and DTEM

The main motivation behind time-resolved in-situ TEM observations is to reveal intermediate steps and/or transient phases that lead to the final product. Whereas TEM-based imaging, diffraction, and spectroscopy techniques are ideally suited to obtain 2D and/or 3D information at high spatial resolution, the time (temporal) resolution (fourth dimension) is limited to the acquisition rate of the recording media, currently reaching 1 ms by using a direct electron camera (e.g. Gatan's K3). Another way to overcome this limit is to reduce the electron emission time interval, used for recording diffraction, images, or chemical composition, such that recording media time becomes obsolete. Thus, we can uncover intermediate steps of the processes and visualize transient states, such as diffusion less phase transformation, dislocation motion, nucleation events, etc. that occur at a lightning speed. In past couple of decades, pump-probe approach has resulted in the development of ultrafast electron microscopes (UEM) and dynamic transmission electron microscope (DTEM) that has improved the time resolution to the order of femto seconds [60–64].

An UEM, first developed in Ziweil's group at California Institute of Technology, the thermionic source for electron, is replaced by a photo electron emission from a cathode (Figure 1.11a). The electron emission is achieved by using femtosecond laser pulses, a stroboscopic approach that generates single or few electron packets, used for imaging, diffraction, or spectroscopy. The temporal resolution is limited

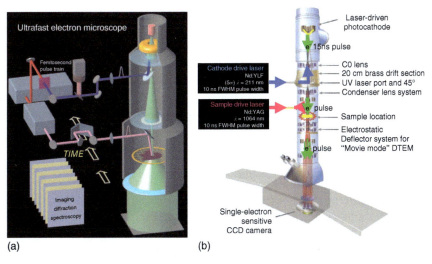

Figure 1.11 (a) The ultrafast electron microscope. Shown are the basic components, which involve interfacing the TEM with a train of femtosecond optical pulses to generate the on-axis electron beam in ultrafast packets of about one electron per pulse. The other optical beam delivers, after a well-defined delay time, initiating pulses at the specimen, thus defining the zero of time. Source: Lobastov et al. [65]/National Academy of Science. (b) Schematic of the DTEM from Lawrence Livermore National Laboratory (LLNL showing the modifications needed to convert a TEM into DTEM. Note that both technologies are based on the basic principle of pump probe but differ in the electron pulse rate and design philosophy. Source: LaGrange et al. [66]/With permission of Elsevier.

by the laser pulse rates used for electron emission, but the spatial resolution is compromised due to low signal-to-noise ratio. An improvement in spatial resolution is achievable only by recording a reversible phenomenon multiple times and averaging the images to obtain SNR with enough contrast to distinguish specific features. However, we need more electrons in a single shot to observe irreversible processes that is achieved using "pump-probe" method in a DTEM (Figure 1.11b). Here, the sample is subjected external stimuli (pump), and the effect is recorded (probe) with a suitable time interval [67]. Currently, two separate lasers are used, one as an external stimulus to bring the sample to a transient state and the other to generate photo electrons. As is obvious, the technology requires major modification to the conventional TEM column [68]. Recently, DOE and NIST team has eliminated laser excitation for electron pulse generation by placing RF cavity combs as tunable electron pulser between electron source and condenser lenses and have achieved picosecond resolution by generating electron pulses at GHz rate [69, 70]. These are developing techniques, and the TEM is modified in-house by the researchers, which has limited their applications to a small group of researchers.

1.7 Introduction to Different Stimuli Used for In-Situ TEM

During last few decades, a large number of TEM holders, as well as dedicated TEM for a specific application, have been developed to follow nanoscale changes in a material under different stimuli. In this book, we will explain the development, functioning, and applications of following stimuli:

1.7.1 Heating (Chapter 3)

Effect of thermal stimuli in vacuum, to decipher phase change, nucleation, dislocation formation and motion, decomposition behavior, etc.

1.7.2 Cooling (Cryo TEM – Chapter 4)

Low-temperature behavior of materials is obtained using time- and temperature-resolved imaging, diffraction, and spectroscopy, especially for magnetic phase transition, ice formation, glass transition, nucleation and precipitation, and other reactions in liquids using plunge-freeze technology, etc.

1.7.3 Interactions with Liquid/Electrochemistry (Chapter 6)

It is a growing field of research as it impacts several real-world problems such as corrosion, functioning of battery materials, and water splitting to produce hydrogen and oxygen.

1.7.4 Interaction with Gas Environment/Catalysis (Chapter 7)

This is another growing field with similar impact with applications including catalysis, redox processes, and corrosion.

1.7.5 Other Stimuli Not Included in this Book

1.7.5.1 Mechanical Testing

Applications of metals, alloys, and ceramics are dependent on their mechanical properties (plasticity and elasticity) that in turn are controlled by the atomic structure, grain size, grain boundaries, and defects, such as dislocations. Therefore, understanding the structure–property relationship is important for defining their scope and generating practical guidelines for their applications. We expect that understanding fundamental principles that define the structure–property relationship will result in designing materials with improved properties. Generally, the structure–property relationship is established by making static measurements, e.g. by finding the fracture point of a material and imaging it before and after the load (stress or strain) was applied. However, we cannot establish the structural changes that lead to the fracture by static characterization and dynamic in-situ measurements are needed [71].

> Breaking something is easy…understanding how it breaks is a different story
>
> – Andrew Minor, UC Berkeley

In past couple of decades, nano-indenters have been developed by individual research groups as well as by commercial companies [72–75]. Basic design of a nano-indenter is shown in Figure 1.12, where a pyramid-shaped diamond intender with a tip diameter of about 100 nm is attached to a shaft with a micrometer and piezo-drive capable of moving the intender tip in all three (x, y, and z) directions. The micrometer is used for rough alignments of the intender tip on to the sample, while piezo drive is used for fine-tuning and applying force on the sample. These advances in the design and functioning of nano-intender TEM holders, along with easy access to aberration corrected TEM, have made it possible to observe and measure the nucleation and motion of dislocations as a function of loading [76–79]. Here the sample preparation is also crucial for making successful measurements as the sample must be electron transparent, accessible to the intender tip, and mechanically stable such that force applied effects only the region of interest and does not bend the entire sample. The applications include, but are not limited to, measurement of elasticity/plasticity, nucleation and motion of dislocations, nucleation and annihilation of grain boundaries, etc. [59, 65, 66, 80, 81].

1.7.5.2 Ion Radiation/Implantation

Ion radiation damage to materials, such as metals, alloys, and ceramics, in specific environments alters their properties and applicability. Understanding

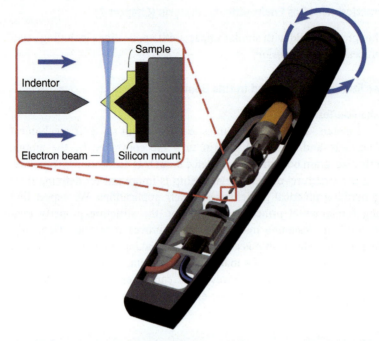

Figure 1.12 Schematic showing the tip of a nano-indenter holder with magnified region from the red square showing the sample mount and indenter location. Blue region represents the electron beam.

the phenomenon is especially crucial for the materials to be used in space and nuclear plants. Controlled irradiation, coupled with TEM, can be used to simulate the structural transformations caused by α- and/or β-decay, neutron, or fission fragments. As new nuclear reactor technologies are being developed, damage to the materials used over time has become more and more important. Intentional ion radiation is an expedited way to understand and measure the damage that may take years to happen in real reactor environment. Here, in-situ TEM observations provide unprecedented atomic-scale knowledge about the segregation within the material, formation of precipitates, defects and/or dislocations, etc. under the ion beam radiation [82, 83]. There are two main advantages of this exercise: (i) the life expectancy of a reactor material can be evaluated in short period by increasing the radiation dose to the levels expected for a longer period in the reactor, and (ii) a priori knowledge can be used to design new materials with improved properties.

Although small ion guns have been directly mounted on the TEM column, most of the facilities utilize free-standing ion accelerators. As expected, incorporation of an ion beam source in the TEM column is neither simple nor easy as ion accelerators are quite bulky and TEM column must be modified for ion beam to access the sample. Figure 1.13 shows layout of a dual-beam (medium and low energy) ion accelerator and its connectivity to the TEM, installed at University of Huddersfield, UK [84]. This system, named as MIAMI-2, occupies a two-story building, specifically built to house it. Similar systems, with up to triple-beam accelerators, are also available at few other places in the world, such as Xiamen

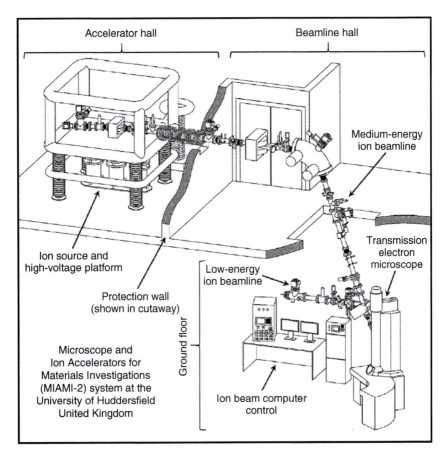

Figure 1.13 Overview of the layout of the MIAMI-2 system, which incorporates two ion beams that can be combined before entering the microscope. Source: Greaves et al. [84]/Elsevier/CC BY-4.0.

University, China [85]; Orsay, France [86]; University of Michigan, USA [87]; Sandia National Laboratory, USA [88], to name a few.

Lian et al. have combined a host of TEM-based techniques, such as TEM and Z-contrast imaging, EELS, and EFTEM, to decipher the effect of ion and electron radiation on complex ceramic materials [89]. Other similar studies include, but are not limited to, investigating the effect of ion radiation on ferric materials [90], formation of dislocation loop [91], nanoparticles of oxide dispersion strengthened (ODS) steels and austenitic steel [86], boron carbide [86], amorphization of an equiatomic martensitic NiTi alloys [92], effect of ion-radiation-induced amorphization of ABO_3 compounds on recrystallization [93], synthesis of nanosized bicrystalline (Pb, Cd) inclusions in Al by sequential ion implantation [94].

1.7.5.3 Biasing

Biasing the sample with known variable electric current is used to (i) generate magnetic/electric fields in the sample [95–97], (ii) trigger electromigration in sample [98], (iii) measure the I–V relationship [99], and (iv) conductance measurement

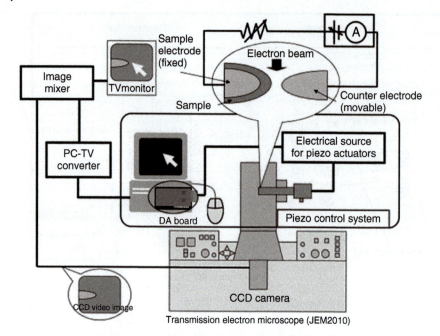

Figure 1.14 A schematic showing the entire system used for I–V measurements, where the modified holder with specimen and counter electrode (magnified image shown in balloon) that is aligned with respect to the sample using a piezo control system. Images are recorded the CCD camera of TEM. Source: Fujii et al. [99]/AIP Publishing LLC.

of nanostructures [100, 101], that reveal the structural modifications of electrode materials under operating conditions [102, 103]. In-situ TEM observations and measurements are performed using modified holders that are commercially available. Both the sample geometry and holder modifications depend on the type of measurements to be performed. For example, I–V curve measurements are done by scanning a flat surface, nanowire, or nanotube using an AFM/STM tip incorporated in the modified holder [102, 104]. A schematic of experimental plan used by Fujii et al. to make I–V measurements of NiO nanoregions is shown in Figure 1.14 [99]. They also developed a special holder with three main functions: First is to position the electrode with sub-nanometer accuracy with respect to the sample region; the second is to measure the current in nano-amp range using a built-in amplifier that can be switched off to measure larger currents; the third to measure the load using a semiconductor sensor [99]. By applying a voltage between the probe-tip and epitaxial bottom electrode, the TEM-STM holders can also be used to induce electric field along the film normal [105].

1.7.5.4 Magnetization

Magnetic nanostructures and thin films are of materials of interest due to their technological applications as random-access memory devices, magnetic recording media, ferroelectric field transistors, spintronics, etc. [38, 106]. The properties of these ever-shrinking structures are controlled by their morphology, shape, length, and distribution of magnetic domains within the nanostructure [43].

Nanoscale characterization of magnetic components within thin films and/or other nanostructures can be achieved using Lorentz microscopy [41, 106], or holography [107], or a combination of the two [108]. However, since their structure and properties change as a function of magnetic field and temperature, in-situ TEM observations and measurements are needed to establish a correlation between the structure and applied magnetic field or with temperature.

Controlled magnetic fields around the sample can be achieved in various ways such as by controlling the current in the objective lens [109, 110] or using a modified specimen holder [108, 111]. By exciting the standard objective lens to reduced levels, combined with sample tilt, it can be used manipulate the micromagnetic structure. This method allows us to design magnetization with other stimuli such as heating/cooling using appropriate holder or under different environments. However, the extent of magnetization by using this method is restricted by the inherent properties of the objective lens, and the perpendicular component of the objective can lead to distortions due to the sample geometry [112]. Alternatively, special magnetization holders have been built by incorporating a pair of electromagnets or coils [113]. Domain wall motion and switching [44, 109, 112], magnetization and demagnetization behavior [38, 114], hysteresis loops [109, 110, 115], formation and movement of vortices [111, 116], and skyrmions [117–119] are a few examples where in-situ TEM has played a crucial role in deciphering the magnetic behavior of nanostructures.

1.8 Potential Limitations and Cautions

As described above, last two decades have seen an unprecedented growth in the application of various in-situ techniques to explore and unveil atomic-level mechanisms involved in synthesis and functioning of nanomaterials, establish structure and property relationship, due to unprecedented developments in instrumentation and techniques. However, the more we have learnt about the applications, the more we have realized that the just like any other method, in-situ and operando measurements have their limitations. Some of the prominent ones, currently faced, are given below, whereas specific for different stimuli are explained in relevant chapters.

- **Reaction conditions:** Thermodynamic conditions, temperature, and gas pressure, achievable for in-situ TEM observations, are limited due to the instrumentation constraint. Similar restrictions are also applicable for other stimuli, such mechanical stress or electrical biasing.
- **Reaction time (not too fast nor too slow):** For us to observe and measure changes in the structure, chemistry, or property as a result of an external stimuli is very much dependent on the kinetics of the reaction. Therefore, we need a recording media that matches the reaction rate; otherwise, we will miss intermediate steps that may be important in controlling reaction. Gatan's K3 and other direct electron cameras have capability of image acquisition rate in the ms range, but the contrast in the images is very poor. However, we can obtain images with reasonable contrast with a time resolution of 0.025 seconds by averaging drift-corrected images acquired at higher rate.

Both EELS and DP can be acquired with higher temporal resolution and may be used instead of imaging if possible. UTEM and DTEM are other two techniques in development stages and currently limited to heating only.

What if the reaction is too slow? It is not practical to occupy the TEM for days to follow a reaction and may not be possible to monitor unless we can find a way to expedite it.

- **Reaction kinetics:** It is very tempting to measure the reaction rates and derive activation energies at nanoscale and compare them with theoretical or other experimental values. However, it can be elusive as the exact temperature of the observed area is questionable. The temperature measurement is further discussed in Chapters 3 and kinetics measurements in 2.
- Surface effects may dominate the process such that reaction mechanism of thin samples is not representative of bulk process. We can use STEM and/or tomography for 3D information at the cost of temporal resolution. We can also use bulk characterization techniques, such as X-ray diffraction to validate the TEM results (see Chapter 8).
- Last but not the least, limitation is due to ever-present electron beam effects. We should keep in mind that the images, DPs, spectroscopy data are all generated due to the interaction of high-energy electrons with the sample. Also, electrons interact with environment and ionize gas or liquid molecules, all of which can have negative consequences, especially making it difficult to distinguish between the electron beam affect and external stimuli.

> The rule of thumb is that electrons are affecting your sample unless proven otherwise
>
> – Peter Crozier, Arizona State University

However, all is not lost, as we can easily observe and even quantify the effect electron irradiation by (i) imaging the same area with "beam on" and "beam off" conditions, (ii) quickly moving to a previously unirradiated part to check if reaction starts again, and/or imaging unirradiated regions of the sample after external stimuli have been removed, i.e. the sample is cooled down. As we will see in Chapters 3 and 6, we can employ the electron energy to initiate certain reactions, which may be to our advantage as multiple sets of observations can be made from the same sample under same condition by just moving the beam to new areas.

Also, we have the possibility to mitigate the electron beam affects by reducing (i) the operating voltage and/or (ii) the electron dose. We can perform experiments to systematically measure the conditions affecting the sample and find a condition where the beam effects are eliminated or minimized. Keep in mind that this may not always be possible, and we may not be able to follow certain reactions using in-situ TEM techniques.

> The fact that we can "see" the electron-beam-induced changes makes it possible to determine the effect and find ways to mitigate it.

1.9 Take-Home Messages

- Motivation to make in-situ TEM/STEM observations arises from the shrinking of the technology world to nanosize.
- Most of the TEM-based techniques, combined with appropriate stimuli, can be used to decipher the morphological, structural, and chemical changes occurring during the synthesis and/or functioning of nanomaterials.
- Modification to either the TEM column or the specimen holders or both has enabled us to formulate structure–property relationships at nanoscale.
- The methodology has its limitations that should be considered.

References

Vendors Websites

Microscopes

1. Thermo Fischer: https://www.thermofisher.com/us/en/home/electron-microscopy/products/transmission-electron-microscopes.html
2. Hitachi: https://www.hitachi-hightech.com/global/science/products/microscopes/electron-microscope/tem
3. JEOL: https://www.jeolusa.com/PRODUCTS/Transmission-Electron-Microscopes-TEM
4. NION: http://www.nion.com/products.html

Specimen Holders

1. DENSsolutions: https://denssolutions.com
2. Gatan: https://www.gatan.com/products/tem-specimen-holders
3. Hummingbird: https://hummingbirdscientific.com
4. Protochips: https://www.protochips.com
5. Fischionne: https://www.fischione.com/products/holders
6. Mel-Build: https://melbuild.com

Articles

1 Knoll, M. and Ruska, E. (1932). Das Elektronenmikroskop. *Zeitschrift für Physik* 78: 318–339.
2 Williams, D.B. and Carter, C.B. (2009). *Transmission Electron Microscopy*, 2e, vol. 1–4. New York: Springer.
3 Muller, D.A. and Grazul, J. (2001). Optimizing the environment for sub-0.2 nm scanning transmission electron microscopy. *Journal of Electron Microscopy* 50 (3): 219–226.
4 Tiemeijer, P.C., Bischoff, M., Freitag, B., and Kisielowski, C. (2012). Using a monochromator to improve the resolution in TEM to below 0.5 Å. Part I: creating highly coherent monochromated illumination. *Ultramicroscopy* 114: 72–81.

5 Pennycook, S.J. and Colliex, C. (2012). Spectroscopic imaging in electron microscopy. *MRS Bulletin* 37 (1): 13–18.

6 Muller, D.A., Kourkoutis, L.F., Murfitt, M. et al. (2008). Atomic-scale chemical imaging of composition and bonding by aberration-corrected microscopy. *Science* 319 (5866): 1073–1076.

7 Rose, H.H. (1990). Electrostatic energy filter as monochromator of a highly coherent electron source. *Optik* 86: 95–98.

8 Krivanek, O.L., Ursin, J.P., Bacon, N.J. et al. (2009). High-energy-resolution monochromator for aberration-corrected scanning transmission electron microscopy/electron energy-loss spectroscopy. *Philosophical Transactions of the Royal Society A: Mathematical, Physical and Engineering Sciences* 367 (1903): 3683–3697.

9 Freitag, B., Kujawa, S., Mul, P.M. et al. (2005). Breaking the spherical and chromatic aberration barrier in transmission electron microscopy. *Ultramicroscopy* 102 (3): 209–214.

10 Kabius, B., Hartel, P., Haider, M. et al. (2009). First application of C_c-corrected imaging for high-resolution and energy-filtered TEM. *Journal of Electron Microscopy* 58 (3): 147–155.

11 Levin, B.D.A. (2021). Direct detectors and their applications in electron microscopy for materials science. *Journal of Physics: Materials* 4 (4): 042005.

12 Li, X., Mooney, P., Zheng, S. et al. (2013). Electron counting and beam-induced motion correction enable near-atomic-resolution single-particle cryo-EM. *Nature Methods* 10 (6): 584–590.

13 McMullan, G., Faruqi, A.R., Clare, D., and Henderson, R. (2014). Comparison of optimal performance at 300 keV of three direct electron detectors for use in low dose electron microscopy. *Ultramicroscopy* 147: 156–163.

14 Tietz, H.R. (2008). Design and characterization of 64 MegaPixel Fiber optic coupled CMOS detector for transmission electron microscopy. *Microscopy and Microanalysis* 14 (S2): 804–805.

15 Tate, M.W., Purohit, P., Chamberlain, D. et al. (2016). High dynamic range pixel array detector for scanning transmission electron microscopy. *Microscopy and Microanalysis* 22 (1): 237–249.

16 Zhu, W., Winterstein, J.P., Yang, W.-C.D. et al. (2017). In situ atomic-scale probing of the reduction dynamics of two-dimensional Fe_2O_3 nanostructures. *ACS Nano* 11 (1): 656–664.

17 Armigliato, A., Balboni, R., Carnevale, G.P. et al. (2003). Application of convergent beam electron diffraction to two-dimensional strain mapping in silicon devices. *Applied Physics Letters* 82 (13): 2172–2174.

18 Ozdol, V.B., Gammer, C., Jin, X.G. et al. (2015). Strain mapping at nanometer resolution using advanced nano-beam electron diffraction. *Applied Physics Letters* 106 (25): 253107.

19 Pekin, T.C., Gammer, C., Ciston, J. et al. (2017). Optimizing disk registration algorithms for nanobeam electron diffraction strain mapping. *Ultramicroscopy* 176: 170–176.

20 Zuo, J.M. and Spence, J.C.H. (1992). *Electron Microdiffraction*. New York: Springer US: 358 pages.

21 Zuo, J.M. and Spence, J.C.H. (1993). Coherent electron nanodiffraction from perfect and imperfect crystals. *Philosophical Magazine A* 68 (5): 1055–1078.

22 Zuo, J.M., Vartanyants, I., Gao, M. et al. (2003). Atomic resolution imaging of a carbon nanotube from diffraction intensities. *Science* 300 (5624): 1419–1421.

23 Schlossmacher, P., Burke, M.G., Haigh, S.J., and Kulzick, M.A. (2010). Enhanced detection sensitivity with a new windowless XEDS system for AEM based on silicon drift detector technology. *Microscopy Today* 18 (4): 14–20.

24 Zaluzec, N.J., Burke, M.G., Haigh, S.J., and Kulzick, M.A. (2014). X-ray energy-dispersive spectrometry during in situ liquid cell studies using an analytical electron microscope. *Microscopy and Microanalysis* 20 (2): 323–329.

25 Rez, P., Bruley, J., Brohan, P. et al. (1995). Review of methods for calculating near-edge structure. *Ultramicroscopy* 59: 159–167.

26 Sharma, R., Crozier, P.A., Kang, Z.C., and Eyring, L. (2004). Observation of dynamic nanostructural and nanochemical changes in ceria-based catalysts during *in situ* reduction. *Philosophical Magazine* 84: 2731–2747.

27 Egerton, R.F. (2014). *Electron Energy-Loss Spectroscopy in the Electron Microscope*, 3e. Boston, MA: Springer: 503 pages.

28 Varela, M., Oxley, M.P., Luo, W. et al. (2009). Atomic-resolution imaging of oxidation states in manganites. *Physical Review B* 79 (8): 085117.

29 Mecklenburg, M., Hubbard, W.A., White, E. et al. (2015). Nanoscale temperature mapping in operating microelectronic devices. *Science* 347 (6222): 629–632.

30 Taverna, D., Kociak, M., Stéphan, O. et al. (2008). Probing physical properties of confined fluids within individual nanobubbles. *Physical Review Letters* 100 (3): 035301.

31 Jeanguillaume, C. and Colliex, C. (1989). Spectrum-image: the next step in EELS digital acquisition and processing. *Ultramicroscopy* 28 (1): 252–257.

32 Hunt, J.A. and Williams, D.B. (1991). Electron energy-loss spectrum-imaging. *Ultramicroscopy* 38 (1): 47–73.

33 McVitie, S., McGrouther, D., McFadzean, S. et al. (2015). Aberration corrected Lorentz scanning transmission electron microscopy. *Ultramicroscopy* 152: 57–62.

34 Phatak, C., Petford-Long, A.K., and De Graef, M. (2016). Recent advances in Lorentz microscopy. *Current Opinion in Solid State and Materials Science* 20 (2): 107–114.

35 Zwek, J. (2012). Lorentz microscopy. In: *In-Situ Electron Microscopy: Applications in Physics, Chemistry and Materials Science* (ed. G. Dehm, J.M. Howe and J. Zweck), 347–369. Federal Republic of Germany: Wiley-VCH.

36 Gao, H., Zhang, T., Zhang, Y. et al. (2020). Ellipsoidal magnetite nanoparticles: a new member of the magnetic-vortex nanoparticles family for efficient magnetic hyperthermia. *Journal of Materials Chemistry B* 8 (3): 515–522.

37 Yu, X., JP, D.G., Hara, Y. et al. (2013). Observation of the magnetic skyrmion lattice in a MnSi nanowire by Lorentz TEM. *Nano Letters* 13 (8): 3755–3759.

38 Tanase, M. and Petford-Long, A.K. (2009). In situ TEM observation of magnetic materials. *Microscopy Research and Technique* 72 (3): 187–196.

39 Yu, X. (2019). Imaging magnetic vortices including skyrmions by Lorentz TEM and differential phase-contrast microscopy. *Microscopy and Microanalysis* 25 (S2): 28–29.

40 Peng, L.-c., Zhang, Y., Zuo, S.-L. et al. (2018). Lorentz transmission electron microscopy studies on topological magnetic domains. *Chinese Physics B* 27 (6): 066802.

41 Petford-Long, A.K. and Chapman, J.N. Lorentz microscopy. In: *Magnetic Microscopy of Nanostructures* (ed. H. Hopster and H.P. Oepen), 67–85. Berlin, Heidelberg: Springer-Verlag.

42 Graef, M.D., Willard, M.A., McHenry, M.E. et al. (2001). In-situ Lorentz TEM cooling study of magnetic domain configurations in Ni_2MnGa. *IEEE Transactions on Magnetics* 37 (4): 2663–2665.

43 Budruk, A., Phatak, C., Petford-Long, A.K. et al. (2011). In situ Lorentz TEM magnetization studies on a Fe–Pd–Co martensitic alloy. *Acta Materialia* 59 (17): 6646–6657.

44 Zak, A.M. and Dudzinski, W. (2020). Microstructural and in situ Lorentz TEM domain characterization of as-quenched and γ′-precipitated $Co_{49}Ni_{30}Ga_{21}$ monocrystal. *Crystals* 10 (3): 153.

45 Kim, J.J., Park, H.S., Shindo, D. et al. (2006). *In situ* observations of magnetization process in alnico magnets by electron holography and Lorentz microscopy. *Materials Transactions* 47 (3): 907–912.

46 Wang, A., Chen, F.R., Van Aert, S., and Van Dyck, D. (2010). Direct structure inversion from exit waves: Part I: Theory and simulations. *Ultramicroscopy* 110 (5): 527–534.

47 Cowley, J.M. (1992). Twenty forms of electron holography. *Ultramicroscopy* 41: 335–348.

48 Vincent, R. (2002). Phase retrieval in TEM using Fresnel images. *Ultramicroscopy* 90 (2): 135–151.

49 Chen, F.R., Van Dyck, D., and Kisielowski, C. (2016). In-line three-dimensional holography of nanocrystalline objects at atomic resolution. *Nature Communications* 7 (1): 10603.

50 Yu, Y., Zhang, D., Kisielowski, C. et al. (2016). Atomic resolution imaging of halide perovskites. *Nano Letters* 16 (12): 7530–7535.

51 Gabor, D. and Bragg, W.L. (1949). Microscopy by reconstructed wave-fronts. *Proceedings of the Royal Society of London. Series A. Mathematical and Physical Sciences* 197 (1051): 454–487.

52 Lichte, H. and Lehmann, M. (2007). Electron holography—basics and applications. *Reports on Progress in Physics* 71 (1): 016102.

53 McCartney, M.R., Dunin-Borkowski, R.E., and Smith, D.J. (2005). Electron holography of magnetic nanostructures. In: *Magnetic Microscopy of Nanostructures* (ed. H. Hopster and H.P. Oepen), 87–107. Berlin, Heidelberg: Springer-Verlag.

54 Lehmann, M. and Lichte, H. (2002). Tutorial on off-axis electron holography. *Microscopy and Microanalysis* 8 (6): 447–466.

55 Midgley, P.A. (2001). An introduction to off-axis electron holography. *Micron* 32 (2): 167–184.
56 Jia, C.-J., Sun, L.-D., Luo, F. et al. (2008). Large-scale synthesis of single-crystalline iron oxide magnetic nanorings. *Journal of the American Chemical Society* 130 (50): 16968–16977.
57 Hÿtch, M., Houdellier, F., Hüe, F. et al. (2008). Nanoscale holographic interferometry for strain measurements in electronic devices. *Nature* 453 (7198): 1086–1089.
58 Dunin-Borkowski, R.E., MR, M.C., Frankel, R.B. et al. (1998). Magnetic microstructure of magnetotactic bacteria by electron holography. *Science* 282 (5395): 1868–1870.
59 Harrison, R.J., Dunin-Borkowski, R.E., and Putnis, A. (2002). Direct imaging of nanoscale magnetic interactions in minerals. *Proceedings of the National Academy of Sciences* 99 (26): 16556.
60 Zewail, A.H. (2006). 4D ultrafast electron diffraction, crystallography, and microscopy. *Annual Review of Physical Chemistry* 57 (1): 65–103.
61 Barwick, B., Park Hyun, S., Kwon, O.-H., and Baskin, J.S. (2008). 4D imaging of transient structures and morphologies in ultrafast electron microscopy. *Science* 322 (5905): 1227–1231.
62 LaGrange, T., Armstrong, M.R., Boyden, K. et al. (2006). Single-shot dynamic transmission electron microscopy. *Applied Physics Letters* 89 (4): 044105.
63 King, W.E., Campbell, G.H., Frank, A. et al. (2005). Ultrafast electron microscopy in materials science, biology, and chemistry. *Journal of Applied Physics* 97 (11): 111101.
64 Montgomery, E., Leonhardt, D., and Roehling, J. (2021). Ultrafast transmission electron microscopy: techniques and applications. *Microscopy Today* 29 (5): 46–54.
65 Lobastov, V.A., Srinivasan, R., and Zewail, A.H. (2005). Four-dimensional ultrafast electron microscopy. *Proceedings of the National Academy of Sciences of the United States of America* 102 (20): 7069.
66 LaGrange, T., Campbell, G.H., Reed, B.W. et al. (2008). Nanosecond time-resolved investigations using the in situ of dynamic transmission electron microscope (DTEM). *Ultramicroscopy* 108 (11): 1441–1449.
67 Taheri, M., Lagrange, T., Reed, B. et al. (2009). Laser-based in situ techniques: novel methods for generating extreme conditions in TEM samples. *Microscopy Research and Technique* 72 (3): 122–130.
68 Browning, N.D., Bonds, M.A., Campbell, G.H. et al. (2012). Recent developments in dynamic transmission electron microscopy. *Current Opinion in Solid State and Materials Science* 16 (1): 23–30.
69 Jing, C., Zhu, Y., Liu, A. et al. (2019). Tunable electron beam pulser for picoseconds stroboscopic microscopy in transmission electron microscopes. *Ultramicroscopy* 207: 112829.
70 Lau, J.W., Schliep, K.B., Katz, M.B. et al. (2020). Laser-free GHz stroboscopic transmission electron microscope: components, system integration,

and practical considerations for pump–probe measurements. *Review of Scientific Instruments* 91 (2): 021301.

71 Soer, W.A. and De Hosson, J.T. (2008). In-situ transmission electron microscopy: nanointendation and straining experiments. In: *In-Situ Electron Microscopy and High Resolution* (ed. F. Banhart), 115–160. Singapore: World Scientific.

72 Spiecker, E., Oh, S.H., Shan, Z.-W. et al. (2019). Insights into fundamental deformation processes from advanced in situ transmission electron microscopy. *MRS Bulletin* 44 (6): 443–449.

73 Yu, Q., Legros, M., and Minor, A.M. (2015). In situ TEM nanomechanics. *MRS Bulletin* 40 (1): 62–70.

74 Zhang, J., Ishizuka, K., Tomitori, M. et al. (2020). Atomic scale mechanics explored by in situ transmission electron microscopy with a quartz length-extension resonator as a force sensor. *Nanotechnology* 31 (20): 205706.

75 Bobji, M.S., Pethica, J.B., and Inkson, B.J. (2005). Indentation mechanics of Cu–Be quantified by an in situ transmission electron microscopy mechanical probe. *Journal of Materials Research* 20: 2726–2732.

76 Dehm, G., Legros, M., and Kiener, D. (2012). In-situ TEM straining experiments: recent Progress in stages and small scale mechanics. In: *In-Situ Electron Microscopy: Applications in Physics, Chemistry and Materials Science* (ed. G. Dehm, J.M. Howe and J. Zweck), 227–254. Federal Republic of Germany: Wiley-VCH.

77 Gouldstone, A., Chollacoop, N., Dao, M. et al. (2007). Indentation across size scales and disciplines: recent developments in experimentation and modeling. *Acta Materialia* 55 (12): 4015–4039.

78 Minor, A.M. (2012). In-situ nanoindentation in the transmission electron microscope. In: *In-Situ Electron Microscopy: Applications in Physics, Chemistry and Materials Science* (ed. G. Dehm, J.M. Howe and J. Zweck), 255–277. Federal Republic of Germany: Wiley-VCH.

79 Tanji, T. (2005). Imaging magnetic structures using TEM. In: *Handbook of Microscopy for Nanotechnology II* (ed. N. Yao and Z.L. Wang), 361–394. Boston: Kluwer Academic Publishers.

80 Ishizuka, K., Tomitori, M., Arai, T., and Oshima, Y. (2020). Mechanical analysis of gold nanocontacts during stretching using an in-situ transmission electron microscope equipped with a force sensor. *Applied Physics Express* 13 (2): 025001.

81 Zheng, H., Wang, J., Huang, J.Y. et al. (2012). In situ visualization of birth and annihilation of grain boundaries in an Au nanocrystal. *Physical Review Letters* 109 (22): 225501.

82 Hinks, J.A. (2009). A review of transmission electron microscopes with in situ ion irradiation. *Nuclear Instruments and Methods in Physics Research Section B: Beam Interactions with Materials and Atoms* 267 (23): 3652–3662.

83 Birtcher, R.C., Kirk, M.A., Furuya, K. et al. (2005). In situ transmission electron microscopy investigation of radiation effects. *Journal of Materials Research* 20 (7): 1654–1683.

84 Greaves, G., Mir, A.H., Harrison, R.W. et al. (2019). New microscope and ion accelerators for materials investigations (MIAMI-2) system at the University of Huddersfield. *Nuclear Instruments and Methods in Physics Research Section A: Accelerators, Spectrometers, Detectors and Associated Equipment* 931: 37–43.

85 Tang, B., Zhang, J., Ma, R. et al. (2014). A triple beam *in-situ* facility at Xiamen University. *Materials Transactions* 55 (3): 410–412.

86 Gentils, A. and Cabet, C. (2019). Investigating radiation damage in nuclear energy materials using JANNuS multiple ion beams. *Nuclear Instruments and Methods in Physics Research Section B: Beam Interactions with Materials and Atoms* 447: 107–112.

87 Toader, O., Naab, F., Uberseder, E. et al. (2017). Technical aspects of delivering simultaneous dual and triple ion beams to a target at the Michigan Ion Beam laboratory. *Physics Procedia* 90: 385–390.

88 Hattar, K., Bufford, D.C., and Buller, D.L. (2014). Concurrent in situ ion irradiation transmission electron microscope. *Nuclear Instruments and Methods in Physics Research Section B: Beam Interactions with Materials and Atoms* 338: 56–65.

89 Lian, J., Wang, L.M., Sun, K., and Ewing, R.C. (2009). In situ TEM of radiation effects in complex ceramics. *Microscopy Research and Technique* 72 (3): 165–181.

90 Kirk, M., Baldo, P., Liu, A.Y. et al. (2009). In situ transmission electron microscopy and ion irradiation of ferritic materials. *Microscopy Research and Technique* 72 (3): 182–186.

91 Schäublin, R., Décamps, B., Prokhodtseva, A., and Löffler, J.F. (2017). On the origin of primary ½ a_0 <111> and a_0 <100> loops in irradiated Fe(Cr) alloys. *Acta Materialia* 133: 427–439.

92 Moine, P., Rivieri, J.P., Ruault, M.O. et al. (1985). In situ TEM study of martensitic NiTi amorphization by Ni ion implantation. *Nuclear Instruments and Methods in Physics Research Section B: Beam Interactions with Materials and Atoms* 7-8: 20–25.

93 Meldrum, A., Boatner, L.A., Weber, W.J., and Ewing, R.C. (2002). Amorphization and recrystallization of the ABO_3 oxides. *Journal of Nuclear Materials* 300 (2): 242–254.

94 Johnson, E., Touboltsev, V.S., Johansen, A. et al. (1997). TEM and RBS/channelling of nanosized bicrystalline (Pb, Cd) inclusions in Al made by sequential ion implantation. *Nuclear Instruments and Methods in Physics Research Section B: Beam Interactions with Materials and Atoms* 127–128: 727–733.

95 Liu, L.Z.-Y., McAleese, C., Sridhara Rao, D.V. et al. (2012). Electron holography of an in-situ biased GaN-based LED. *Physica Status Solidi C* 9 (3-4): 704–707.

96 Tan, X. (2012). In-situ TEM with electrical bias on ferroelectric oxides. In: *In-Situ Electron Microscopy: Applications in Physics, Chemistry and Materials Science* (ed. G. Dehm, J.M. Howe and J. Zweck), 321–346. Federal Republic of Germany: Wiley-VCH.

97 Möller, M., Gaida, J.H., Schäfer, S., and Ropers, C. (2020). Few-nm tracking of current-driven magnetic vortex orbits using ultrafast Lorentz microscopy. *Communications on Physics* 3 (1): 36.

98 Spolenak, R. (2012). Current-induced transport: electromigration. In: *In-Situ Electron Microscopy: Applications in Physics, Chemistry and Materials Science* (ed. G. Dehm, J.M. Howe and J. Zweck), 281–301. Federal Republic of Germany: Wiley-VCH.

99 Fujii, T., Arita, M., Hamada, K. et al. (2011). I–V measurement of NiO nanoregion during observation by transmission electron microscopy. *Journal of Applied Physics* 109 (5): 053702.

100 Hirose, R., Arita, M., Hamada, K. et al. (2005). In situ conductance measurement of a limited number of nanoparticles during transmission electron microscopy observation. *Japanese Journal of Applied Physics* 44 (24): L790–L792.

101 Takahashi, Y., Kudo M, Fujiwara I et al. (2015). Visualization of conductive filament during write and erase cycles on nanometer-scale ReRAM achieved by in-situ TEM. *2015 IEEE International Memory Workshop (IMW)*, Monterey, CA (17-20 May 2015). https://ieeexplore.ieee.org/document/7150312.

102 Lu, X., Adkins, E.R., He, Y. et al. (2016). Germanium as a sodium ion battery material: in situ TEM reveals fast sodiation kinetics with high capacity. *Chemistry of Materials* 28 (4): 1236–1242.

103 Cheng, Y., Zhang, L., Zhang, Q. et al. (2021). Understanding all solid-state lithium batteries through in situ transmission electron microscopy. *Materials Today* 42: 137–161.

104 Xu, T.T., Ning, Z.Y., Shi, T.W. et al. (2014). A platform for in-situ multi-probe electronic measurements and modification of nanodevices inside a transmission electron microscope. *Nanotechnology* 25 (22): 225702.

105 Li, L., Jokisaari, J.R., and Pan, X. (2015). In situ electron microscopy of ferroelectric domains. *MRS Bulletin* 40 (1): 53–61.

106 Budruk, A., Phatak, C., Petford-Long, A.K., and De Graef, M. (2011). In situ Lorentz TEM magnetization study of a Ni–Mn–Ga ferromagnetic shape memory alloy. *Acta Materialia* 59 (12): 4895–4906.

107 Dunin-Borkowski, R.E., MR, M.C., Kardynal, B. et al. (2000). Off-axis electron holography of patterned magnetic nanostructures. *Journal of Microscopy* 200 (Pt 3): 187–205.

108 Yano, T., Murakami, Y., Kainuma, R., and Shindo, D. (2007). Interaction between magnetic domain walls and antiphase boundaries in $Ni_2Mn(Al,Ga)$ studied by electron holography and Lorentz microscopy. *Materials Transactions* 48 (10): 2636–2641.

109 Volkov, V.V. and Zhu, Y. (2000). Dynamic magnetization observations and reversal mechanisms of sintered and die-upset Nd–Fe–B magnets. *Journal of Magnetism and Magnetic Materials* 214 (3): 204–216.

110 Rodríguez, L.A., Magén, C., Snoeck, E. et al. (2013). Quantitative in situ magnetization reversal studies in Lorentz microscopy and electron holography. *Ultramicroscopy* 134: 144–154.

111 Arita, M., Tokuda, R., Hamada, K., and Takahashi, Y. (2014). Development of TEM holder generating in-plane magnetic field used for in-situ TEM observation. *Materials Transactions* 55 (3): 403–409.

112 Dietrich, C., Hertel, R., Huber, M. et al. (2008). Influence of perpendicular magnetic fields on the domain structure of permalloy microstructures grown on thin membranes. *Physical Review B* 77 (17): 174427.

113 Inoue, M., Tomita, T., Naruse, M. et al. (2006). Development of a magnetizing stage for in situ observations with electron holography and Lorentz microscopy. *Journal of Electron Microscopy* 54 (6): 509–513.

114 Masseboeuf, A., Gatel, C., Bayle-Guillemaud, P. et al. (2009). The use of Lorentz microscopy for the determination of magnetic reversal mechanism of exchange-biased $Co_{30}Fe_{70}$/NiMn bilayer. *Journal of Magnetism and Magnetic Materials* 321 (19): 3080–3083.

115 Kryshtal, A., Mielczarek, M., and Pawlak, J. (2022). Effect of electron beam irradiation on the temperature of single AuGe nanoparticles in a TEM. *Ultramicroscopy* 233: 113459.

116 Zheng, H. and Zhu, Y. (2017). Perspectives on in situ electron microscopy. *Ultramicroscopy* 180: 188–196.

117 Peng, L., Zhang, Y., Ke, L. et al. (2018). Relaxation dynamics of zero-field skyrmions over a wide temperature range. *Nano Letters* 18 (12): 7777–7783.

118 Ding, B., Li, Z., Xu, G. et al. (2020). Observation of magnetic skyrmion bubbles in a van der Waals ferromagnet Fe_3GeTe_2. *Nano Letters* 20 (2): 868–873.

119 Jiang, W., Zhang, S., Wang, X. et al. (2019). Quantifying chiral exchange interaction for Néel-type skyrmions via Lorentz transmission electron microscopy. *Physical Review B* 99 (10): 104402.

2

Experiment Design Philosophy

> *Failure is the opportunity to begin again more intelligently*
> – Henry Ford

Currently, the TEM column is invariably being used as a nano laboratory, where we perform in-situ observations of materials in vacuum, gas, or liquid environment while subjected to different stimuli. We are performing a wide range of experiments, including, but not limited to, chemical synthesis, electrical and magnetic property measurements, mechanical testing, etc. Needless to say, we are using the TEM column as a furnace, flow reactor, chemical vapor deposition chamber, nano-calorimetry cell, electrochemical cell, to name a few [1]. Therefore, we must pay the same considerations to plan an in-situ TEM experiment as we do for planning any experiment in the laboratory. Successful operation of a TEM is dependent on its location and environment, and we assume that it is located in a room/building that fulfills the vendors' requirements to achieve best performance and passed all required safety regulations. Now we need to pay attention as to how to utilize the TEM most effectively for in-situ experimental observations/measurements. In this chapter, we will discuss the general philosophy, pitfalls, and remedies for in-situ experimental planning. Some of the topics specific to certain environment or stimuli will be further discussed in relevant chapters.

> Almost every TEM examination of a new material reveals something unexpected. This is even more true for in-situ TEM observations. However, it is making sure that the new observations represent the real material behavior, which is the most difficult, but most critical aspect
> – Robert Sinclair, Stanford University

2.1 General

In-situ TEM experiments reveal a wealth of fundamental as well as practical information but can be tedious to conduct and have limitations as pointed at the end of Chapter 1. However, they require careful planning, are time-consuming, not simple,

In-Situ Transmission Electron Microscopy Experiments: Design and Practice, First Edition. Renu Sharma.
© 2023 WILEY-VCH GmbH. Published 2023 by WILEY-VCH GmbH.

and are not always needed. Therefore, before planning an in-situ TEM experiment, we should ask following questions (Figure 2.1):

1. **What is the missing data that is needed to answer my scientific question?**
2. **Are there any other techniques available that can be used to get the same set of information?**
   For example, overall structural and chemical transformations can be examined using time-resolved X-ray techniques, such as diffraction, XAS, EXAFS, XPS, etc. On the other hand, nanoparticle size–dependent behavior, structure of isolated defects, long-range order, structure of grain boundary, and dislocation, as well as mechanisms of their formation, require TEM-based techniques.
3. **If not, which of the TEM-based techniques or a combination of is most appropriate?**
   For example, we may choose electron diffraction, imaging, or spectroscopy, or a combination, depending on the answer to our first question. We can learn more about making these choices from the information each technique can provide, as described in Chapter 1 and summarized in Table 2.1.
4. **Do I know enough about the experimental conditions required to obtain the desired information?**
   Ideally, we don't want to use the in-situ TEM measurements to determine the thermodynamic conditions or limits as it is time-consuming and may not provide any unequivocal results, especially as the temperature of the particle or area under observation cannot be measured accurately. We should use ex-situ or bulk measurement methods, such as TGA, DTA, DSC, etc., to establish ballpark numbers of reaction parameters required for the process we want to make in-situ time-resolved measurements.
   Most profitable application of in-situ TEM is to reveal, atomic-scale mechanisms, metastable phases, or transient states and not to determine the processing conditions.
5. **Can I achieve the experimental conditions required in TEM?**
   This is one of the most important considerations, for example, we cannot study a sample that is beam-sensitive if we do not have low-dose techniques or fast-recording media available to us. Or we want to study the structural and chemical changes in a nanoparticle for a reaction whose thermodynamic conditions cannot be achieved by current technology. Or we want to study the dislocation formation above the level of mechanical force or temperature that can be applied using currently available holders. Some of these limits also described in Section 1.7.

Once we have established that in-situ TEM-based techniques are the only option for us to uncover the missing information or settle an ongoing debate in the literature, we should move forward. In Sections 2.2–2.5, we will discuss the strategic steps to perform successful in-situ experiments because execution of in-situ TEM/STEM experiments is not trivial and requires careful planning.

> I am going to make a name for myself. If I fail, you will never hear of me again.
>
> – Eadweard Muybridge

Table 2.1 Choosing the TEM/STEM/Holder for specific applications.

Type of external stimuli	Goal	TEM/STEM open system	Type of holder-open or closed system	Characterization technique
Heat	Phase transformation, crystallization, nucleation and growth of nanostructures, strain relaxation, dislocation formation and movement, vacancy ordering, charge order etc.	TEM/STEM, EELS	Furnace or MEMS based heating holder	Imaging, diffraction, and EELS
Cryo	Phase transformation, crystallization in glasses, beam sensitive materials, reactions in liquids	TEM/STEM, EELS/EDS	Liquid N_2 or liquid He TEM holder	Imaging, diffraction, EELS, EDS
Liquid environment	Corrosion, electrochemistry, nucleation and growth from solutions	TEM/STEM, EELS, EDS	Open and closed cell (latter recommended)	Imaging, diffraction and spectroscopy
Gaseous environment	Catalysis, corrosion, redox, hydration, de-hydration, hydrogenation, nanomaterial synthesis, CVD, MOCVD	TEM/STEM, EELS (CL in open cell)	Open and closed cell	Imaging, diffraction, EELS, EDS, CL
Radiation	Electron beam and, ion radiation effects	TEM/STEM	Open system	Imaging, diffraction, EELS, EDS
Electrical bias	Structure, electrochemistry, I–V curves	TEM/STEM	Biasing, or biasing and heating or biasing and liquid holder	Low- and high-resolution imaging, combined with electrical measurement
Mechanical	Effect of stress on structure, fracture, density of dislocations	TEM/STEM	Straining, or straining and heating holder	Low- and high-resolution imaging, diffraction
Magnetic	Identification of magnetic domains, skyrmions, vortex formation and movement	Lorentz microscopy and holography	Modified imaging system/biprism	Low- and high-resolution imaging, diffraction, EDS, EELS

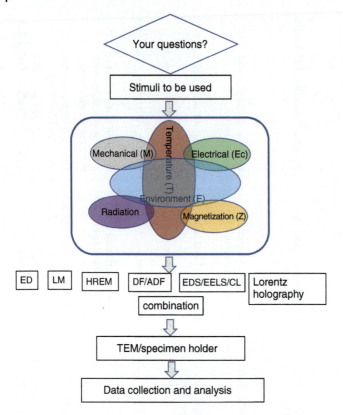

Figure 2.1 Block diagram describing the strategic elements of in-situ TEM experiments. Note that either temperature or environment or both may be combined with other stimuli such as electrical, mechanical, radiation, and magnetization.

2.2 Choice of Technique and the Microscope

2.2.1 Stimulus and Technique Selection

After we have answered the questions identified in Section 2.1, we need to define the stimulus we want to employ to establish synthesis–structure–property relationship. We can choose an appropriate stimulus by revisiting the detailed description given in Section 1.6. Next, we need to make sure that appropriate instrumentation for our choice of stimulus is readily available, and if not, we have resources to make modifications or build a new one, such as funding, access to machine shop, etc. Some of the ideas for designing and building TEM holders are provided in Chapter 5.

We can also choose one or multiple TEM-based techniques, as described in Section 1.4, which will provide us the answers we are looking for. We should keep in mind that HREM or HAADF images are not the only methods to provide answers. Morphological or structural changes can be easily identified using low-magnification images and/or electron diffraction. For example, mechanisms of nanowire growth as a function of temperature and oxygen partial pressure were first revealed by low-magnification images [2–6]. Similarly, electron diffraction has been successfully employed to decipher Cu oxidation steps [7, 8], defect formation [9], phase transition in perovskite thin films [10]. Hydrogen adsorption dynamics and conversion of Pd nanoparticles to PdH_x using dark-field imaging are yet another example of low-magnification imaging [11]. However, we should try to combine as many TEM-based techniques as possible, e.g. combine electron diffraction with HREM images to obtain micro-scale and atomic-scale structural information or combine diffraction, imaging, and spectroscopy to determine the phase transformations, etc. Moreover, we should always employ ex-situ or an in-situ technique other than TEM such as X-ray diffraction, XPS, Raman, theoretical simulations, etc., to corroborate our findings (see Chapter 8).

2.2.2 Microscope Selection

In this section, we discuss the selection of a TEM/STEM instrument suitable for in-situ experiments that is relevant if you are in market to buy a new microscope, or you have possibility to select one within your lab or collaborate with someone outside your lab (e.g. National lab or another institute). As mentioned earlier, we use the TEM sample area or sample stage/holder for conducting in-situ or operando experiments. Therefore, certain features of the microscope should be considered at the time of the purchase or when retrofitting an existing TEM with specific stimuli. Following is a short list of helpful criteria.

> The modern-day HREM is capable of atomic-scale resolution on a routine basis, BUT there are many potential pitfalls along the way which may affect microscope operation and preclude correct image interpretation.
> – David J. Smith, Arizona State University

2.2.2.1 Operating Voltage

Currently, we have an option to choose a TEM that can be operated from 60 to 300 keV. While higher operating voltage is desirable to obtain atomic-resolution images and chemical maps, we should expect an increase in the probability of electron beam effects (damage) from high-energy electrons [12]. As we will show later, high-energy electrons can also be used as a stimulus to initiate some reactions, such as crystallization, nucleation, and growth of nanoparticles from liquid solutions or excite localized surface plasmon resonance [13–15]. Moreover, a modern microscope, equipped with a monochromated FEG and aberration corrected lenses, can be operated at a voltage as low as 60 keV to obtain atomic-resolution images. But low-voltage electrons may not reduce the electron beam damage entirely as they can heat the sample (especially ceramics) to initiate reactions without applying heat as a stimulus [16, 17]. As will be explained in Chapters 6 and 7, radiolysis of liquids and ionization of gases should also be considered [18]. We should also keep in mind that while spatial resolution increases with increasing voltage, the energy resolution increases with decreasing voltage such that changes in the low-loss region due to external stimuli are easy to decipher, for example, change in phonon resonances due to heating [19]. As discussed in Section 1.8, electron beam effects are omnipresent, and we will continue to discuss their effects and mitigation possibility in almost every chapter of this book.

2.2.2.2 TEM/STEM and Pole-Piece Gap

Dedicated TEM, STEM, or even ATEM may have slightly different configurations, for example, dedicated STEM instruments have an FEG source with better stability, monochromator, aberration-corrected condenser lens, top entry stage, and ultrahigh vacuum column for best performance. These attributes help in forming a small probe with high intensity, reduce sample drift, and provide high image and energy resolutions. However, modifying a top entry stage to incorporate other stimuli such as heating, electrical biasing, liquid, or gas flow is not easy (although it has been done) and adversely affects their performance. Moreover, STEM is rarely used for in-situ observations due to high data acquisition time. Note that most of the modern microscopes, with a few exceptions, accept side entry TEM holders.

The sample is located between the two pole-pieces, also known as objective lens, which is the most critical component that produces an axially symmetric magnetic field. For in-situ experiments, we need this gap to be as large as possible. Larger pole-piece gap in a dedicated ATEM makes it easier to use side entry TEM holders with thicker tip, needed to incorporate external stimuli, and are well suited for in-situ experiments (see Chapter 10 for detailed discussion).

In the past, large pole-piece gap was avoided for high-resolution imaging as it increased the C_s of the objective lens; however, a modern microscope with aberration correctors makes it a moot point [20, 21]. The gap between the pole pieces is crucial for in-situ experiments such that:

o Sample can be tilted (i) to align along one of the crystalline zone axes parallel to the incident beam, (ii) to collect highest X-ray signal for chemical analysis, and (iii) to obtain 3D images using tomography.
o Have space to insert objective aperture in the back focal plane of the sample, light source, and/or a cathodoluminescence (CL) detector.
o Have space to insert liquid and gas cell holders with other stimuli and thicker tip (see Chapter 5 for details).
o Other stimuli such as ion source and detectors can be incorporated.

In practice, we find that dedicated TEMs for one functionality are now a thing of the past (with a few exceptions). Most of the modern microscopes can operate in both TEM and STEM modes and are equipped with EDS and EELS detector with mid-size pole-piece gap, and further modifications are in the works (see Chapter 10).

2.2.2.3 Image Acquisition System and Detectors

As described in Section 1.4.4, several cameras and detectors are available to record images, DPs, or spectroscopy data (EELS or CL, etc.). Time (temporal) resolution is an important aspect of in-situ characterization such that all reaction steps are acquired by the camera, or the detector used. In recent years, direct electron detectors have improved the image acquisition speeds up to 1600 frames per second (fps) or 0.006 25 seconds. However, individual images acquired at this rate have very low SNR. On the other hand, directly acquiring images at a rate of 400 fps (0.0025 seconds) or summing the images recorded with 0.006 25 resolution after drift correction can improve the SNR. Moreover, these cameras are ideally suited for imaging beam-sensitive materials at lower electron dose to mitigate the beam damage. The major drawbacks of these cameras are that they are quite expensive and produce large datasets (see Chapter 9 on data management).

CCD cameras are commonly used for recording images, DPs, and EELS signal. Whereas the camera to record images and DPs is placed under the viewing screen, EELS detector is located at the end of the spectrometer, and EDS detector is mounted in the sample chamber (see Section 1.4.3 for further details).

2.2.3 Development or Modification of New Tool

In-situ and operando TEM/STEM are growing fields of research and new technologies and tools are constantly being developed. We have two options, (i) to build our own TEM holder for specific projects or (ii) to contact a vendor and discuss your project with their scientific staff to have them build one for you. Most of the vendors, especially for specialized TEM holders, are open to collaborate with the researchers to modify or incorporate a new feature in their existing systems. Both options are currently being pursued [22, 23], also see Chapter 5 and references therein.

As we have options, it is important to choose the right TEM, specimen holder, and characterization technique, keeping following criteria in mind:

1. We should use at least two, if not more, characterization methods to obtain unambiguous data, for example, we can combine atomic-resolution imaging and chemical analysis for structural characterization. We can also combine electron diffraction if possible.
2. Always, double check the electron beam effects by collecting the data under beam-on and beam-off conditions. Also, employ another technique, such as in-situ X-ray diffraction, XPS, EXAFS, or Raman spectroscopy, to confirm TEM observations.
3. Your choice should be to start from an easy to difficult set of experimental plan and technique.

Table 2.1 provides a quick guide to help you select a microscope, holder, and technique.

> Only those who dare to fail greatly can ever achieve greatly.
> – Robert F. Kennedy

2.3 TEM Holder Design and Selection

The TEM specimen holder to be used depends on the nature of the stimuli to be applied for the in-situ experiments. As explained in Sections 2.1 and 2.2, we need to make sure our experimental conditions are compatible with the holder design, material, and achievable experimental conditions. Currently, TEM holders for heating, cooling, biasing, in vacuum, gas, or liquid environments are readily available, and a list of vendors is provided in the reference sections of Chapters 1 and 5. Similarly, mechanical testing holders, combined with heating, are also available. We should choose a holder appropriate for achieving experimental requirements as explained above. Nuances of holder design and performance are discussed in detail in Chapter 5.

> The story of civilization is, in a sense, the story of engineering – that long and arduous struggle to make the forces of nature work for man's good.
> – Lyon Sprague DeCamp

2.4 Specimen Design and Preparation

As we have learnt in the Chapter 1, high-energy electrons can only transmit through very thin samples; 500 nm or thinner; depending on the characterization technique to be used. For example, samples need to be thinner than 5 nm for atomic-resolution imaging but could be thicker for diffraction and spectroscopy data collection. However, most of the as-synthesized materials of interest are much thicker. Therefore, preparing TEM-compatible specimens/samples is challenging, time-consuming, but most important step. We will briefly review the commonly used methods for TEM sample preparation as they also apply for preparing samples for in-situ observations. The choice of specimen preparation method depends on several factors, such the morphology (bulk, powder, single crystal), amount of the material available, compatibility with instrumentation to be used, location and size of region of interest, information to be gained, etc. [24, 25]. Specific requirements, if any, for in-situ observations using a specific stimulus, will be further discussed in the relevant chapters.

2.4.1 Direct Dispersion on a TEM Grid

Measurement of particle size distribution, structure, and chemistry of nanoparticles or single crystals (that can be crushed into powder) can be achieved by dispersing them directly on support films such as amorphous carbon, SiO_2, or SiN_x on 3 mm metal grids or microchips (Figure 2.2a). Commercially available film supports can be continuous with varying thickness or perforated (also called "holey" or "lacy"), and the grid martial can be Cu, Ni, Au, Al, Mo, Si, etc. Holey support films are preferred as the sample region extending into the hole can be examined without any electron scattering contribution from the support. There are two ways to disperse powder samples on the TEM grids. First is by making a suspension of the powder (crushed) sample by sonicating a mixture of powder in organic solvent, such as iso-propanol, drop casting a known volume of the suspension on the grid and let it dry. Water can also be used to make a suspension but should be avoided as many materials are soluble in water, and it takes much longer to dry than an organic solvent. Acetone should always be avoided for sample preparation as it is not clean and can be a source of contamination during electron radiation.

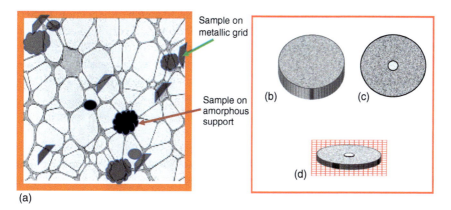

Figure 2.2 (a) Powder sample loaded, using dry or wet technique, on a holey (perforated) electron transparent membrane supported on a metal grid. Note the distribution of different size and shape of particle on support as well as anchored to the grid. (b) Powder sample is pressed in a pallet, (c) a hole is milled in the center using electropolishing (Section 2.4.4) or mechanical thinning (Section 2.4.5) techniques and (d) placed on 400 or 500 mesh grid to keep any loose particles from falling in the column. Source: Miler et al. [26]/from Elsevier.

Second method is to simply role the TEM grid, with or without thin film support, in the powder, gently shake the grids to remove any extra or loose powder, and load it in the specimen holder. The powder sticks to the grid bars or film due to static interaction and is most simple procedure (Figure 2.2a). Bare grids may be preferred for in-situ observations if the support films get destroyed by the external stimuli or environment. For example, carbon films will burn off in oxygen environment. Direct dispersion is a simple and efficient method, but unfortunately, it has limited applicability as it unsuitable for bulk materials.

> Sample preparation for electrical measurements is especially difficult. We need to make sure that we have sufficient electrical contact to current collectors and the contact is not lost due to thermal expansion during heating. Sufficient background resistance and capacitance is important and will depend on the chip geometry. Also, beware of leak currents.
> – Soren B. Simonsen, Technical University of Denmark (DTU)

2.4.2 Sintering Pallets

Making pallets from powder or nano-crystalline sample may be needed as a first step before thinning the sample for TEM visualization. For example, when a specific sample geometry is needed for making mechanical or electrical measurements or to make thin sections of needle-shaped crystals perpendicular to the long axis direction that cannot be achieved by crushing. However, the method can only be used if large amount of sample is available. Here, we mix the powder with an appropriate binding material that has similar mechanical properties, is high vacuum-compatible, and will not degrade under external stimuli to be used. The mixture is then pressed into 3 mm pallets or sintered at high temperature with glass fibers (Figure 2.2b), followed by thinning at the center (Figure 2.2c) using an ion mill (Section 2.4.5) or by preparing thin slices using ultramicrotomy (Section 2.4.3) or FIB (Section 2.4.6) and loaded on a 400 or higher mesh grid to keep any debris from falling in the TEM column during observations (Figure 2.2d). This method was successfully employed to increase the amount of reactants to produce enough gaseous products that could be measured using EELS during a catalytic reaction [26].

2.4.3 Ultramicrotomy

It is one of the most common methods to make biological samples where the tissues or muscles are sectioned into thin slices in any desired direction and thickness using a diamond saw. However, this method can also be used to make thin sections of powders, nanocrystals, polymers, or ceramics. As expected, slice thickness can be controlled more easily for soft materials than for hard materials. It is most suitable for crystals that are prone to cleave only in one specific crystallographic direction, restricting the alignment of a crystalline particle in all orientations with the limited tilt angles available for HREM imaging or when small amount of sample is available, for example, needle-shaped crystals or debris, respectively [27]. These samples are embedded in a raisin of similar mechanical properties or pressed into pallets and sliced in desired directions using a diamond saw [27]. Main disadvantage of this method is the introduction of mechanical deformation such that the shape of the material in microtomed slices may be entirely different than original sample.

2.4.4 Electropolishing

The methods mentioned above are not suitable to examine intrinsic structural defects in metals, alloys, or ceramics as grinding them may alter the bulk structure. Similarly, grinding epitaxial thin films deposited on substrate can alter their epitaxial properties, structure, and composition. Therefore, other methods of preparing thin samples have been developed over the years. For most of them, the first step is preparing a 3 mm disc or large strips, few mm to 1 cm thick that are easy to handle, using diamond or wire saw. Further thinning, usually concentrated to the central section of the disc or strip, can be achieved by using electropolishing where a potential difference is applied between sample and flowing electrolyte [28, 29]. Anodic dissolution of the samples results in localized thinning that is monitored by shining LED light from the backside. The method is suitable for making metal and or alloy samples, and the process is controlled by the composition, temperature, flow rate of the electrolyte, and applied voltage. The LED assembly can be used for automatically shut-off after a perforation in the sample is reached [30]. The perforated area can extend up to 50 µm around the hole with thinner sections at the edge of the hole.

2.4.5 Mechanical and Ion Milling

It is one of the most common methods employed to prepare TEM samples of electronic and semiconductor materials [25]. Most of the semiconducting devices

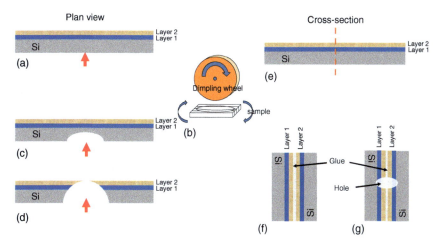

Figure 2.3 Mechanical preparation of epitaxially grown thin films on a substrate (Si chosen as an example here) for making TEM ready samples. (a) A strip of thin film and substrate is cut from the bulk sample using diamond saw or scribe. For making a plan view sample: (b) the dimpling wheel is used to (c) thin the polished strip from the backside of the substrate until Si is almost transparent and (d) is ion milled to create a small hole in the center. To make a cross-section sample, (e) the strip is cut in half and (f) the two pieces are glued together with thin film side facing each other. (g) The sample is thinned using the dimpling wheel (b) in the same way as for the plan-view sample and ion milled as final step.

are prepared by depositing multilayer components on a substrate. Examination of the quality, thickness, and epitaxial relationships between substrate and the layers using TEM requires least destructive sample preparation technique. First of all, the sample preparation is divided in two categories, plan view and cross section [31, 32]. As the name implies, plan view samples are thinned from the top of substrate (Figure 2.3a,c). This technique is best suited to interrogate density and structure of defects and/or dislocations in bulk material or the nature of the deposited film [33].

Cross-section samples are needed to view the epitaxial nature of the film with the substrate and each other. In the case of devices with multilayered deposited films, cross-section samples also allow us to characterize the thickness, intra-layer epitaxy, structure, and quality of each film. For cross-section samples, two strips are cut from the substrate and glued together with deposited films facing each other such that deposited films are perpendicular to the electron beam direction (Figure 2.3b). Making either of these samples is multistep process [34–36], with few deviations based on the nature of the material as given below.

1. **Cutting strips:** This step is like the one described in the electropolishing or plan view sample preparation. Si wafers are often used as substrate for making semiconductor devices. Therefore, the first step is to make two thin strips, often 0.5 mm × 1.5 mm (Figure 2.3a), or make two identical size strips by cutting one strip in half for cross-section sample (Figure 2.3e).
2. **Gluing:** This step is needed for cross-section samples only, where two identical strips are glued together with deposited layers facing each other using

vacuum-compatible and temperature-resistant glue (Figure 2.3f). The strips are then mounted on a 3 mm TEM grid.

3. **Polishing/Grinding:** Both plan view and cross-section samples are polished/grinded using special equipment under running water to keep them from getting hot.
4. **Chemical/Reactive ion etching:** Electron transparent samples can be prepared by chemical etching or reactive ion etching, depending on the material. Alternatively, we continue with the next steps.
5. **Dimpling:** The center of the single strip or glued two strips can be further thinned in the middle using diamond paste and a dimpling wheel (Figure 2.3b). This step reduces the ion milling power and time but is avoided if low-angle milling is to be employed as the edges of partially thinned central section will shadow low-angle ion beam to reach the center [34].
6. **Ion milling:** This is the last step, where low-energy ions are used to create a small hole in the center of the plan view or cross-section strips (Figure 2.3d,g). Thin sections around the hole can thus be used for TEM observations. For more than 20 years, before the advent of FIB instruments, it was the most popular method and constant improvements were made to the instrumentation to reduce the damage and artifact generation during milling, [12, 37–39] such as low-power, low-angle ion beam, liquid N_2 cooling of sample stage to reduce the heat, etc. [40].

Making either the plan view or the cross-section sample, using mechanical and ion milling methods, is time-consuming and requires a lot of practice and patience. Also, a number of artifacts have been identified associated with ion milling technique at almost every step [25]. For example, brittle samples can break during mechanical grinding, polishing, dimpling steps; epoxy bonding may peel off; preferential milling of different layers; sample heating; ion implantation; chemical composition change; and beam damage during the ion milling are often present [24, 34, 41]. Careful procedures or alternative methods, as explained in the references given in this section, should be followed to avoid these problems.

> Failure isn't fatal, but failure to change might be
> – John Wooden

2.4.6 Focused Ion Beam (FIB)

In the last couple of decades, FIB has replaced the regular Ar ion beam milling as it can be used to make site-specific sample (within microns) of nearly uniform thickness at a faster rate [42–44]. A FIB instrument is very similar to SEM, where ion source, usually Ga, can be used for imaging as well as milling. A dual-beam FIB combines both electron and ions sources and is often equipped with an EDS detector [45]. Imaging with electrons reduces the ion beam damage, and EDS aids in precisely locating the area of interest (site specific) if it has different chemical composition than the matrix [46]. Pre-FIB sample preparation may be needed and follows

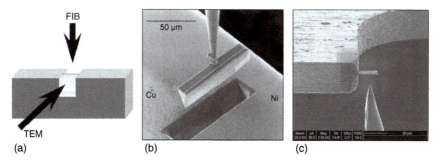

Figure 2.4 (a) Schematic illustration of the H-bar FIB technique. Material on opposite sides of a region of interest is FIB-milled until it is nearly electron-transparent. (b) An 80-mm-long piece of material is mounted to the nanomanipulator needle by Pt deposition. (c) Shorter lamellae are mounted from the side to the central posts on special half-grids. Source: Mayer et al. [47]/from Springer Nature.

the same steps of cutting, polishing, dimpling, etc., as described above (Figure 2.3). Although a FIB instrument is more expensive than an ion mill, their multipurpose applications, such as SEM, SIM, EBID, 3-D imaging, etc., make them an integral part of a materials science laboratory.

In a FIB, we generally start with a polished sample block, mounted on a stub or holder, and loaded in the FIB column. Basic steps include making a thin lamella by milling trenches on both sides of the region of interest to be extracted for TEM observation (Figure 2.4a), normally referred as H-bar due to its shape [47]. Ga is commonly used as ion source, although other sources are available. Pt is usually deposited as a stop layer to define region to be milled. The lamella is then lifted out by pining it to a micromanipulator by Pt deposition using the EBID followed by IBID capability of the FIB (Figure 2.4b) and mounted to a special half grid (Figure 2.4c). Carbon instead of Pt has also been used as protective layer. The sample can be thinned further using low-angle thinning by tilting the lamella to appropriate angles [47]. Further details of procedures and limitations, including transfer to MEMS chips [48, 49], can be found in Refs. [50–56]. YouTube videos are also available for making FIB sample for a regular TEM grid (https://www.youtube.com/watch?v=YaZ9lyzLIB8) and for MEMS chips (https://www.youtube.com/watch?v=uXJI5GmdGCw). Moreover, specific modification for making samples for heating [57], cryo TEM [53], thermal and electric experiments [58] have also been reported.

Major drawbacks of this method are the deposition of Ga in the sample, redeposition of sputtered materials, and ion beam damage resulting in the amorphization of the sample. Depositing a metal, usually Pt or carbon, as protective layer and using low-angle and low-power ion beam are some of the techniques applied to mitigate these issues [42, 51]. In-situ heating experiments have also been used to demonstrate the thermal recovery of damaged FIB lamella of Cu single crystals [59]. Advantages and disadvantages of FIB sample preparation methods, their mitigation, and existing concerns have been reported [47, 60, 61].

2.4.7 Tripod Polishing

To avoid the artifacts generated during ion milling process, a mechanical process to thin the samples is a welcome alternative. Tripod polishing is a thinning method to prepare wedge-shaped samples and is also referred as wedge polishing method, where the thin edge of the sample is used for TEM observations. Here, a strip of the sample is fixed in place by wax mounting and a tripod polisher equipped with 3 µm screws to adjust the polishing height and direction is used to create a wedge [62, 63]. The polisher is set at a desired angle using micrometer screws, and the sample is polished from one side by using progressively finer grit diamond lapping films. It is then flipped and polished from the other side using the same procedure. The sample is then dismounted from the polisher and bonded to a 3 mm TEM grid with glue such that the thinnest part of the wedge is at the center. Sample can be further cleaned to remove any surface damage using low-power ion milling for a short time. Recently, a quadripod, instead of tripod, has also been employed to prepare thin samples from specific area of interest [64]. This method avoids or minimizes the ion milling, and the thickness of the sample can be estimated from the user-defined wedge angle.

2.4.8 Cryo Sample Preparation

Like ultramicrotomy, plunge freezing has been routinely used to prepare biological TEM samples such that the sample is embedded in a thin layer of vitrified ice and remains hydrated during TEM observations. Making samples covered with vitrified ice is the "holy grail" for biological cryo TEM. However, after the Nobel prize in Chemistry was granted to Dubochet, Frank, and Henderson for their cryo-TEM work in 2017, its application to solve materials science problems has generated a lot of interest (also see Chapter 4). Cryo-TEM is being explored as a new way to follow the reactions in liquid phase, characterize electron beam sensitive and soft materials, such as soft polymers, metal organic frameworks, certain zeolites, batteries, colloids, heterogenous catalysts, functional molecules, quantum materials, to name a few [65–68]. Although cryogenic specimen preparation methods are essential for dedicated cryo-TEM instruments, they are also applicable to prepare samples for low-temperature observations using liquid N_2 or liquid He TEM holders for materials science applications [12].

Cryo samples are prepared by the plunge-freezing technique where a thin layer of sample suspension is drop caste on a TEM grid and plunged quickly in liquid ethane or propane to vitrify the suspension [69]. The liquid ethane or propane is kept cold by liquid N_2 in a surrounding container. Low vapor pressure of ethane or propane mitigates the formation of a vapor layer between sample and the liquid [70]. It is important to avoid crystalline ice formation during plunge freezing and keep the vitrified layer as thin as possible, the latter is controlled by the thickness of suspension on the grid, quality of the support film, temperature of the cryogen, plunge speed, etc. [71–73].

> Beware, cryo sample needs to stay frozen, starting from insertion to FIB, making lamella, storage, to loading in the holder through TEM observation.
>
> – Michael Zachman, Oak Ridge National Laboratory

Cryo-FIB techniques, developed to prepare frozen hydrated sample, are also being used to make samples from solid suspensions or solid–liquid interfaces for cryo TEM observations (see Section 4.3) [74, 75]. Special lift-out and sample transfer techniques, such as design of cryogenically cooled manipulator probe for subsequent attachment to the TEM grid; avoid condensation or ice formation on the sample during cryo-FIB lamella preparation; environmentally protect the sample transfer from FIB to TEM under cryogenic conditions, have been developed and retrofitted on an FIB to offer possibility to lift out a targeted site-specific area of interest from the frozen bulk [46, 76].

2.5 Guidelines for Experimental Setup

Some of the general guidelines for making TEM observations, such as keeping the samples and holders clean to avoid contamination, are always applicable.

> Source of contamination is invariably your sample and/or sample holder. Make sure to plasma clean the holder and heat the TEM grid loaded with sample under a heat lamp (or plasma clean) immediately before loading it in the column is a good practice to keep the TEM operation contamination-free.

Additionally, the TEM alignments, especially electron beam tilt, must be kept the same throughout the experimental observations as it affects the structural information such as lattice spacing and symmetry [77]. Here we will explain the settings that affect in-situ experiments in general. A quick reminder that for most of the in-situ experiments, covered in this book, we are using TEM sample chamber as a reactor. Therefore, we need to pay the same attention to our choice of TEM sample grid and holder material as for setting up an ex-situ reactor. We will cover some general guideline here and come back to them for specific experiments in chapters associated with a specific stimulus.

2.5.1 Electron Beam Effects

For TEM characterization, high-energy electrons are necessary evils as the information is obtained due to their interaction with the sample, which can also damage or alter the material under observation. Electron beam effects vary with materials system and experimental condition and must be evaluated using control experiments. We will continue to discuss them throughout this book to emphasize their importance.

> Design of experiment is critical. Understanding the potential and limitations of electron beam techniques and small sample size requirements will help the design of specific experiment.
> – JM Zuo, University of Illinois, Urbana-Champaign

Table 2.2 Relevant properties of grid and support materials.[a]

Grid material	MP/Tamman temperature (K)	Known reactivity	Support	Stability/reactivity
Cu	1356/678 (407 °C)	Oxidizes	Carbon	Fragile/burns in O_2/H_2O
Ni	1725/863 (518 °C)	Oxidizes	SiO_2	Possible charging/stale
Au	1336/668 (401 °C)	N/A	PolySi	Stable
Mo	2883/1442 (865 °C)	Oxidizes	SiN_x	Stable/mostly no reactive
Pt	2028/1014 (608 °C)	N/A		
Ta	3290/1645 (1372 °C)			
W	3695/1847 (1574 °C)	Oxidizes		
Al	933/466.5 (193.5)	Not suitable for heating		

a) Si grids not included in the table as they are considered safe for in-situ experiments.

2.5.2 Choice of TEM Grid and Support Material

First and foremost, we need to define out experimental parameters, such as temperature, electrical bias, nature of the environment, and check the compatibility of the experimental setup with the grid and sample holder materiel as well as other internal parts of the TEM column. Materials Safety Data Sheets (MSDS) should always be consulted, especially when working with liquids and gases. They are readily available on the web, and the vendors always include them when shipping the chemicals. We know that we must consider the nature and reactivity of a solvent with the container while mixing solutions (a very simple process) or the sample during heating. Similarly, we will choose the container (grid) material based on the temperature and reactivity of our sample with it. For example, Pt does not react with most of the materials and is safe to use below its Tamman temperature (half the melting temperature in Kelvin) (Table 2.2), silicon may be used as an alternate for high-temperature experiments. Mo grids are safe for heating in vacuum, and Be grids are generally used for EDS data collection to avoid peaks from the grid material (Figure 1.7) but should be avoided in oxidizing environments. A few examples are described in Sections 2.5.2.1–2.5.2.5. Chapters 3, 4, 6, and 7 will further illustrate the experimental planning for specific stimulus or a combination of stimuli.

2.5.2.1 Reactivity of Sample with Grid and/or Support Material

Generally, continuous/holey carbon or graphene film supports on metal grids are used to load the samples for TEM observations. They are especially suitable to load powder samples as they provide support but have negligible influence on information derived from the images, DP, EDS, or EELS data collected. However, the carbon films are prone to break under electron beam irradiation or upon heating as they "burn-off" even in low-vacuum environment or break due to thermal expansion. As a result, the area under investigation may be lost and the experiment will fail to provide conclusive results. **It is important to remember that small amount of oxygen partial pressure (chemical potential) is always present in the TEM column due to the residual water content even in 10^{-4} Pa vacuum.** Moreover, metal nanoparticles, such as Ni, have been reported to graphitize carbon thin films at 600 °C due to strong metal–support interaction [78, 79].

A detailed study of interaction of amorphous carbon with some other transition metals (Ti, Cr, Fe, Co, Ni, and Cu) during heating up to 800 °C in vacuum revealed that while Ni and Co graphitized carbon, it was accompanied by metastable carbide formation for Fe and Cr and no activity was observed with Ti and Cu [80]. Silica (SiO_x) thin films may be used as an alternative support, especially to study catalytic processes at high temperature or in gaseous environment. The powder samples containing nanoparticles may also be directly loaded on the metal grids, thereby avoid a support film. In this case, the sample will attach to the grid bars due to electrostatic forces and the particles sticking out in the vacuum, away from the grid bars will be suitable for in-situ observations (Figure 2.2a). However, in this case the chemical nature of the grid becomes important and should be selected after careful considerations as explained below.

2.5.2.2 Reactivity of TEM Grids Upon Heating

Cu is the most commonly used grid material, but other metal grids, such as Ni, Al, Mo, Pt, Au, and stainless steel, are also available. The melting point of the grid material must be considered for heating samples to temperatures above 300 °C. It has been shown that the metal atoms (or molecules in a solid) become mobile and start to diffuse at a temperature that is half of their melting point in Kelvin (Tamman temperature), which is around 400 °C for Cu [81]. Zhang et al. have investigated the behavior of Cu, Ni, Mo, and Au TEM grids, coated with ultrathin films of amorphous C or SiO_x, in the temperature range of 500–850 °C [81]. They reported that although different metals and supports behaved differently, most of the grid materials adversely changed the structure of the support material above 600 °C. For example, Cu particles were observed to nucleate (Figure 2.5a) on amorphous C film at 600 °C either due to evaporation and re-deposition or due to surface mobility above the Tamman temperature (≈ 400 °C).

Similarly, diffusion of Au atoms from the grid to the Ni/SiO_2 catalyst particles improved the catalytic activity of Ni for the growth of carbon nanotubes (Figure 2.5b,c) at 520 °C using C_2H_2 as precursor. Ni–Au catalyst particles of various composition, formed by co-depositing Au and Ni thin films, were then used to systematically investigate the effect of Au incorporation in Ni nanoparticles on CNT synthesis. HAADF images and EDS maps acquired after CNT growth confirmed formation of nanoparticles with Au rich surface and tip (Figure 2.5d,e) with Ni rich core. Detailed analysis and DFT calculations confirmed that small amount ($\approx 8\%$) of Au in Ni improved the yield of CNTs [82]. Metal mobility also depends on the gas environment, i.e. the reactivity of metal to form volatile compounds such as Mo to form $Mo(OH)_2$ in water vapor rich vacuum, can be source of metal deposition on the thin film support or on the sample.

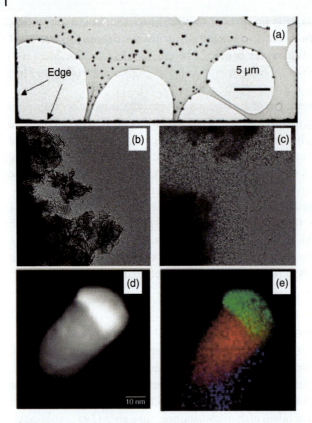

Figure 2.5 (a) TEM image of amorphous carbon film supported on Cu grid after in-situ heating at 600 °C for 0.2 hours showing growth of Cu particles on carbon film. Source: Zhang et al. [81]/from Europe PMC. Low-magnification images of CNTs grown at 520 °C in 70 Pa of C_2H_2 using Ni/SiO$_2$ catalyst loaded on (b) Ni grids and (c) Au grids. Note comparatively high density of CNTs formed when Au grids were used. Source: Sharma et al. [82]/from American Chemical Society. (d) Dark-field STEM images from 0.2 mol fraction of Au in Ni sample, recorded ex-situ after CNT growth, (e) a color overlay emphasizes the spatial extent of Ni-rich, Au-rich, and surrounding carbon nanostructure (blue). Source: Sharma et al. [82]/from American Chemical Society.

2.5.2.3 Reactivity of TEM Grids in Gaseous Environment

Gaseous environment does not adversely affect the grid or support material at room temperature but can disintegrate carbon support films upon electron beam radiation if water vapor is present in the column or adsorbed on the sample. Moreover, oxidation of certain metallic grids/or heating wires is also possible at high temperatures that can be used to our advantage. For example, growth mechanisms of CuO nanowires have been postulated by observing their growth directly from Cu grids in O_2 environment at 400 °C using an ETEM [83]. $W_{18}O_{49}$ nanowires have been reported to grow directly from the tungsten wire heating holder at 600 °C at 10^{-4} Pa [84]. Here, both the nature of gaseous environment and temperature play a role and reactivity of grid material can also result in generating contaminants or artifacts. Therefore, suitable inert grid material with respect to the reaction temperature and gas environment under investigation must be used. Si grids with SiO_2, SiN_x or polycrystalline Si support with holes provide a comparatively inert environment.

2.5.2.4 Reactivity of Liquids with the Windows

Hydrophobic or hydrophilic nature of the membranes is important for the static or flow type liquid cells. Solid particles can attach to the windows that will stop their motion and change the reaction kinetics. However, it is easier to obtain high-resolution images from particle stuck on the windows than from the floating particles. We can alter the hydrophilicity of the window materials by plasma cleaning or heating under UV lamp (see Section 6.5.4). Although the SiN_x windows, generally used for liquid cells, do not react with water and most of the organic solvent, they can be damaged by corrosive solutions. For example, some of the salt solutions may be acidic or basic, depending upon the concentration of the salt in solution. **Best practice is to leave a sacrificial window chip in the liquid to be used overnight and make sure it is undamaged before performing experiments.**

2.5.2.5 Reactivity of Gases/Liquids with the TEM Holder Parts

For in-situ observations of gas–solid interactions, reactivity of gases with the other parts of the TEM holder must also be considered. Note that a TEM holder or column is very expensive, and we don't want either of them to be destroyed on top of a failed experiment. Best first step should be to consult with the scientific advisor/applications engineer from the vendor's (TEM or specimen holder) team and explain your experimental plan in detail. They may not disclose the materials used in their instrument but will let you know if there is any danger of you harming the TEM column or holder. Most vendors will be disposed to sign a non-disclosure agreement (NDA) with your team if you are worried about your research idea becoming public knowledge or hijacked.

Most of the MEMS-based holders use heating elements embedded in a SiN_x membrane supported on Si that can tolerate mild oxidizing and reducing environments. Also, both conductive and radiative heat transfer to the sample or environment from miniature heating elements is negligible (Figure 3.4). Moreover, the exposed connections to the sample housing do not heat up appreciably, still care should be taken to keep the connections unharmed when loading the sample or during experiment.

Although MEMS-based holders have become more popular, we are still using furnace-based holders, or heating wires for direct heating. In first case, the furnace as well as connections materials to the heating unit and thermocouple (if present) can be damaged in gaseous environment or at high temperature. For example, Gatan heating holders are available with two types of furnace bodies: Ta and Inconel. Following is a list of some of the known issues with these two materials:

1. Ta is easily oxidized in low vacuum or O_2 environment upon heating.
2. Certain compositions of Inconel are safe to heat the samples in oxidizing environment but not in disilane or digermane. Also, CO leaches out Ni from Inconel to form nickel carbonyl that decomposes in electron beam to deposit

Ni nanoparticles on the sample (Peter Crozier, Arizona State University, private communication).
3. Disilane will also corrode Pt in thermocouple wire at high temperatures (above 700 °C).

There are two ways to check the compatibility of holder parts with your sample or environment: (i) check the phase diagram and/or (ii) use washer of the furnace holder to check its reactivity ex-situ. The first method is also applicable to all of our tangible experiments, but should be aware that the reaction temperature or melting point for nanomaterials may deviate from the bulk by as much as 100 °C [85, 86]. For example, while Si nanowires are reported to grow via vapor–liquid–solid (VLS) mechanism, they were observed to nucleate from Au–Si melt slightly above the eutectic [2]; on the other hand, nucleation and growth of Si nanowires using Pd catalyst or Ge nanowire were observed to start below the eutectic temperature [87, 88].

2.5.3 Purity of Gases

It is obvious but often forgotten that impurities, present in the commercial gas cylinders, can adversely affect the reaction. For example, presence of water vapor in H_2 will change the oxygen potential in the reactant volume and thereby diminish the reduction rate for a given reaction or require higher reduction temperature. Therefore, the gas cylinders with six nines purity (99.9999) are highly recommended. If the gas we want to use is not available with such high purity (e.g. C_2H_2), then we need to pay attention to the impurity component and their possible effect on our reaction. Another issue to keep in mind is the reactivity of the gas with the container and delivery lines, for example, CO reacts with alloy contents of stainless-steel cylinder and/or gas delivery lines to form metal carbonyls [89]. These carbonyls decompose easily by electron beam to deposit contaminants on the sample. Figure 2.6a shows a TEM image of Au nanoparticle covered by contamination after being exposed 100% CO at 100 Pa for 10 minutes at RT. C, Fe, and Ni peaks are present in the EDS (Figure 2.6b) acquired from the position marked by X in the TEM image. On the other hand, no contamination was observed in the TEM image of Au nanoparticle from the same sample when a gas purifier was used before introducing the same amount of CO for the same time (Figure 2.6c) in the TEM column, confirming that the contaminants are carried by CO. Absence of C, Fe, and Ni peaks in EDS data (Figure 2.6d) from the particle in Figure 2.6c further confirmed that the purifier has removed the metal carbonyls from the CO source. Therefore, we should use CO in Al cylinder or use an active coke filter to remove carbonyl vapors before introducing the gas into the reaction cell.

The ability to mix gases and change composition of the mixture is often desired to explore the changes in the structure, morphology, and composition occurring in nanomaterials under synthesis or reactor conditions. However, mixing gases is not a straightforward process and depends on the diffusivity of gases involved. Most of the ETEMs and some of the gas-reaction holders are equipped with residual gas analyzers (RGAs) or mass spectrometer (MS). However, their sensitivity to

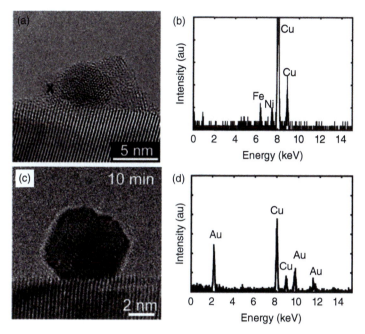

Figure 2.6 (a) TEM image of an Au nanoparticle covered by a contamination layer after ETEM observation in 100% CO gas. (b) An EDX spectrum obtained from the point indicated in the TEM image showing the presence of Fe and Ni in the contaminant layer. (c) A TEM image of an AuNP recorded after 10 minutes of exposure to 100% CO that was introduced in TEM column using a gas purifier. (d) An EDX spectrum obtained from the AuNP in (c) in vacuum after the ETEM observation. Note the absence of C, Fe, and Ni peaks, the Cu peak in EDS spectrum (b) and (d) come from the Cu grid. Source: Uchiyama et al. [89]/from Oxford University Press.

distinguish between some of the gases, e.g. N_2 (28) and CO (28), in small volumes is low. Crozier et al. have used low-loss EELS to determine the time it takes to obtain the desired gas composition of the mixtures in a mixing tank. The system was calibrated by acquiring spectra after filling the sample chamber (reaction cell) with individual gases (H_2 and O_2) at known pressure, one at a time. Then a gas mixing tank was charged with O_2 and H_2, in 2:1 ratio, adding O_2 first and then H_2. Evolution of changes in the spectrum was monitored with time after introducing this gas mixture (pressure ≈ 330 Pa) into the specimen area (reaction cell) of an ETEM (Figure 2.7). Low-loss spectrum acquired immediately after adding the H_2 to the mixing tank contained only O_2 plasma peaks (Figure 2.7a), the peak located at 12.5 started to increase in intensity after 30 minutes (Figure 2.7b) and increased to a value similar to the main H_2 plasma peak after 60 minutes (Figure 2.7c). Their report includes a detailed procedure and application of their measurement technique to other gas mixtures [90].

Experiments in liquids require similar considerations. SiN_x windows or graphene, used to encapsulate liquids, is chemically robust, still highly acidic or basic solutions should be avoided. Also, the nanoparticles may stick to the windows depending on their hydrophilicity, and its control is further discussed in Section 6.5.4.

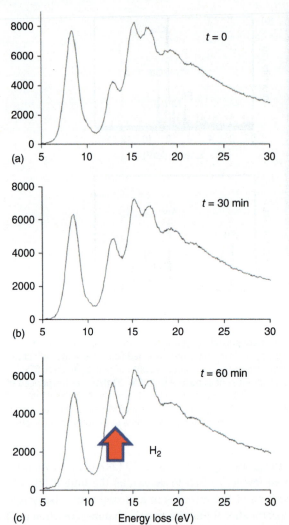

Figure 2.7 Evolution of energy loss spectrum in the specimen chamber with time after charging the mixing tank with $2O_2 + H_2$, adding O_2 to the mixing tank first. (a) Immediately after changing gas mixture in tank, (b) after 30 minutes and (c) after 60 minutes (cell pressure ≈ 330 Pa). The intensity H_2 plasma peak (marked by red arrow) increased with time.

2.5.4 Liquid Cell Experiments

Windows of a liquid cell, irrespective of the material, are the weakest point of the cell as they are fragile and can have holes generated during manufacturing. Thereby, loading the liquid sample on the chips is critical for executing successful experiments. Most of the available liquid cell holders have two chips containing windows, top and bottom, that are sealed together after loading the sample. First, a visual inspection of individual windows under high-magnification optical microscope is highly recommended to make sure that the windows are completely intact. For sealed liquid cells, the liquid should be drop casted using microliter pipettes with precisely known amounts as recommended by the window or chip manufacturer. Sudden drop or weight of liquid can break the windows. After sealing the two chips, containing the liquid, the entire assembly should be leak-tested, using a vacuum chamber, before loading in the microscope. The holder assembly should also be vacuum-tested for a liquid flow cell for possible leaks.

Apart from the resolution limits for imaging and EELS, specifications such as the thickness of windows, maximum flow rate, drift rate, etc. should be considered when buying or making a liquid cell holder. The electron dose needs to be carefully monitored to avoid radiolysis effects as discussed in detail in Chapter 6.

2.5.5 Experiments Using Other Stimuli

Other stimuli such as nanoindentation or biasing also require planning strategies. For example, the sample size, shape, and orientation with respect to nano indenter will play an important role in our ability to observe the effects. The sample should be loaded at a height that is within the translation limits of the indenter such that they can be aligned in the TEM. Also, orientation of the sample should be in the direction perpendicular to the indenter to image the formation or defects and dislocations as a function of applied stress. Similarly, the sample size and orientation need to be considered to visualize the electromigration in the sample, initiated as a function of biasing current.

Similar step-by-step process can be orchestrated for other in-situ experiments independent of the stimuli used or questions to be answered.

2.6 Practical Example of Designing In-Situ TEM Experiment

2.6.1 Growth of GaN Nanowires Using ETEM

Here, we go over the experimental planning steps for the in-situ growth of GaN nanowires including challenges and limitations. This example is chosen as it required most of the strategic steps described above. Our goal was to determine the growth mechanism of GaN nanowires from Au–Ga solid solution and the growth parameters for the formation of selective nanowire structures. In-situ TEM observations have been reported to answer similar question for Si and Ge nanowire growth [2, 4, 88]. GaN nanowires are reported to form via VLS mechanism using a catalyst (Au, Ni, Fe, or In) that dissolves Ga deposited from a metalorganic source (trimethyl Ga, i.e. TMG) that in turn is nitrided using NH_3 or N_2 upon heating [91]. However, most of the MOCVD chambers are not compatible to using metal catalysts. So, the need to use in-situ TEM to answer our questions was clear, and we proceeded with planning our experiments using Au as catalyst as described below:

1. **Ex-situ determination of reaction parameters:** Reaction temperature and gas pressures were determined from literature [92]. We inferred that catalytic growth of GaN nanowires is a two-step process, first TMG decomposes above 450 °C to deposit Ga on pre-patterned Au catalyst and forms liquid droplets upon heating. The nitridation reaction occurs above NH_3 decomposition temperature (≈ 800 °C).

2. Next step was to make sure that our specimen heating holder and other TEM components were compatible with TMG and NH_3. The latter has been used in our ETEM to observe NH_3 intercalation mechanisms in dichalcogenides [93], so it was safe to be used. We had access to Ta or Inconel furnace-type heating holders only. So, we dipped washers from both furnaces and heated to the reaction temperature in a mixture of TMG an NH_3 for an hour and examined them using an SEM. We found that inconel washer did not corrode and was safe to use.
3. We then designed the plumbing to introduce TMG vapor from one inlet to the specimen chamber and another for NH_3 without mixing the two precursors to avoid gas phase reaction between them [94]. We started with Au nanoparticles deposited on poly-Si thin film and heated to 480 °C (above the TMG decomposition temperature). We then opened the valve to TMG to deposit Ga on Au and make liquid solution of the two.

> Success is not final, failure is not fatal: it is the courage to continue that counts.
>
> – Winston Churchill

4. We found that Ga starts to evaporate above 500 °C, but the nitridation reaction occurs above 800 °C. From the Au–Ga phase diagram (Figure 2.8a), we note that a solid solution of Au–Ga is formed below 20% Ga in Au and is stable up to quite high temperatures. However, GaN nanowire growth will occur only from the liquid Au–Ga droplets requiring us to carefully control the amount to Ga deposited on the grids to form as many liquid droplets as possible that will be stable until NH_3 is introduced in the reaction chamber (Figure 2.8b). As expected, we found that while liquid droplets nucleated GaN nanowires after NH_3 was introduced in the chamber, solid particles remained inactive [6]. Furthermore, deactivation of a liquid particle, within the red circle in Figure 2.8b, was also observed after coalescing with remanent of thin Au particle on the substrate (Figure 2.8c), and becoming solid (Figure 2.8d) and remained inactive while liquid particles continued to grow GaN nanowires.

Growth mechanisms and reaction parameter for other III–V nanowires, such as multi-directional growth GaN [95], phase switching during growth of GaAs [96], hybrid structure of GaP-Si nanowires [97], have also been reported. Hetherington et al. have recently reported a gas handling system specially designed for in-situ TEM observations of MOCVD growth of III–V semiconducting nanowires [98].

2.6.2 Applications of Quantitative Data

We are employing in-situ TEM to follow changes in materials as a function of external stimuli, such as temperature, electrical, mechanical, magnetic, gas, and/or liquid. Above and beyond revealing atomic scale mechanisms, we can make qualitative or quantitative measurements of these changes and use them to either fine-tune the process parameters or obtain kinetic parameters such as reaction rates and/or activation energy. For example, by measuring the growth rate of Si nanowires with and

2.6 Practical Example of Designing In-Situ TEM Experiment | 65

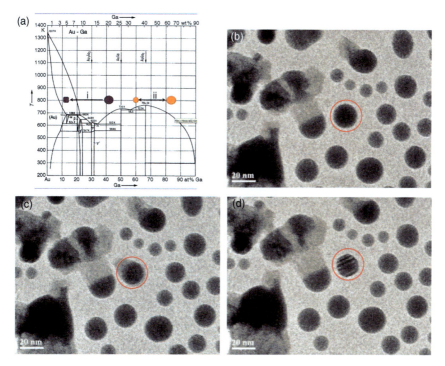

Figure 2.8 (a) Au–Ga phase diagram showing existence of a solid phase in the Au-rich region, marked by purple solid square, and at 800 °C liquid phase is stable beyond 20 at% Ga. (b) A TEM image extracted from the GaN growth video showing nucleation of GaN nanowires and Au–Ga liquid droplets. (b, c) Note the liquid particle, enclosed in red circle, sucking Au from the substrate and becoming solid, evident from faceting contrast in (d). Source: Diaz et al. [6]/from Elsevier.

without oxygen, it was established that addition of small amounts of oxygen with disilane increased the growth period resulting in the formation of long wires [3]. Similarly, temperature-resolved observations established the temperature regime to grow straight or kinked Si nanowires [5].

Growth rates for nanowires (Cu, Si, III–V) [99–103] and redox reaction rates (e.g. ceria, CuO, Fe_2O_3) [104–106] have been measured under reaction conditions, mostly to quantify the reaction mechanism that can then be verified using DFT calculations. Also, distinction between coalescence and Ostwald ripening can be made by measuring the growth rate of individual particles over time, where the former is controlled by particle diffusion on the surface [107], and latter by vapor phase diffusion [108], i.e. the particles move on the surface and coalesce upon contact, or some particles will slowly decrease in size and disappear while others will slowly grow.

> Conduct your own quantifications and check to be sure that you are consistence with your measurements and calibrations to obtain quantitative data. Clearly identify if the data presented is qualitative or quantitative and make sure to report the beam current and dose estimations.
>
> – Katherine Jungjohann, NREL, Denver

On the other hand, we need to measure following parameters to calculate reaction rate and/or activation energy:

1. Concentration of reactants and/or products.
2. Change in physical state or chemistry, i.e. morphology, surface area, chemistry as a function of time under controlled conditions, i.e. with or without a catalyst.
3. Sample temperature.
4. Nature and properties of the catalyst used, if any.

Our ability to measure some of these parameters as function of time and temperature have motivated researchers to use quantitively measured rates of dynamic changes to obtain kinetic parameters. We may or may not be able to measure reactants, but we can often measure the products or relative change in the concentrations of reactants and/or products. On the other hand, we do not to need these parameters to measure the functionalities of material, such as electrical conductivity or mechanical strength. This discussion is to bring home the point that the possibility and requirements to measure the kinetics of a process under observation are different and should be treated accordingly. We can use following broad distinctions to apply quantitative measurements made during the experiments in gases and liquids.

2.6.2.1 Physical and Materials Science

For most of the experimental measurements, relevant to this broad category, the reactants and the products or at least one of them is solid, and a catalyst may or may not be used. Therefore, rate of dynamic change in size, structure, or chemistry, will be proportional to the reaction rate, and temperature-resolved measurements may be used to calculate activation energy of the process. Baker et al. have reported comprehensive measurements of growth rate of carbon nanofibers nucleating from various transition metal nanoparticles (Ni, Co, Fe) in acetylene environment as a function of temperature [109]. They calculated activation energies for carbon fiber growth using Arrhenius equation ($k = Ae^{-E_a/RT}$), where k is rate constant, A is pre-exponential factor, E_a is activation energy, R is universal gas constant, and T is absolute temperature in Kelvin and compared them with the activation energies for carbon diffusion in respective metals and found to be similar. Therefore, they concluded that carbon diffusion in metal nanoparticles is the rate-limiting step and proposed a vapor solid liquid growth mechanism that was accepted for carbon nanotube growth until in-situ TEM observations proved it to be vapor–solid–solid mechanism (see Section 7.5.2).

Ko and Sinclair have calculated activation energy for the formation of an amorphous layer at the Pt/GaAs interface during heating by measuring growth rate of interlayer formed as a function of temperature [110]. They also applied Arrhenius equation, and the measured activation energy of 1.2 ± 0.1 eV, which could be assigned as the activation energy for diffusion of Pt through an amorphous Pt-GaAs ternary mixture.

On the other hand, Avrami equation can be used to measure the kinetics of the crystallization process, where the degree of crystallization can be measured from TEM images. In-situ heating experiments have been used to measure the nucleation and growth rate of crystalline particles in NiTi amorphous alloy films to obtain kinetic parameters for the process using Johnson–Mehl–Avrami–Kolmogorov (JMAK) theory (modified Avrami equation) [111]. Similar experimental measurements and analysis for the precipitation of ternary $Al_{11}(Cr,Mn)_2$ phase in Al matrix during heating of powder-processed Al–2.5Cr–1.35Mn–0.25Co–0.31Zr alloy gave an activation energy of 277 kJ mol^{-1}, which is consistent with Cr diffusion in the Al matrix and identified as the rate-limiting process for the precipitation [112]. Grain growth as a function of temperature has also been used to calculate the activation energies [113].

These examples show that careful measurements of physical changes in the structure and morphology can be used to understand the dynamic process or measure the kinetic parameters under similar conditions as used in these reports.

2.6.2.2 Catalysis

As mentioned in Sections 2.6.2 and 2.6.2.1, we need to measure the reaction rate under controlled reaction environment that was achieved in the examples provided in previous section but is almost impossible when following catalytic reactions. In-situ TEM, especially ETEM or gas-cell nanoreactor (see Chapter 7), have been extensively used to observe catalytic processes during last couple of decades. The atomic-resolution imaging and spectroscopy have provided unprecedented information about the morphological, structural, and chemical changes in the catalytic nanoparticles at elevated temperatures and in gaseous environments. There have been efforts to measure reaction rates by measuring the product gas composition [114, 115], but they may not be used to obtain kinetic parameters. One of the limitations is the "pressure gap," i.e. it is not possible for us to achieve the reactor conditions. Also, the nature and property of the catalyst, such as the surface area, total concentration of catalyst loaded on a TEM grid, are unknown. Other parameters as temperature distribution across the gas cell, heat, and mass transport may not be quantified.

However, phase transformation, nanomaterials synthesis, dislocation formation, and motion described in Section 2.6.2.1 don't have such stringent requirements. While in-situ TEM observations provide unprecedented information about the reaction mechanisms at atomic scale, retrieving kinetic parameters is not trivial.

2.7 Review

- For in-situ observations and measurements, we are extending the TEM sample chamber into a nano laboratory.
- In-situ experiments require careful planning and a thorough understanding of functioning of the TEM, the physical and chemical behavior of the sample and its interactions with the specimen holder and the environment.
- The nature of the stimuli, instrumental constraints, effect of electron beam, specification of the TEM/STEM, characterization techniques, all should be a part of strategic experimental plan.

- While nucleation and growth rates of nanowires, solid interfaces, nanotubes, precipitates, etc. can be measured to obtain kinetic parameters, chemical kinetics for catalytic reactions remains elusive, mainly because the achievable conditions for in-situ TEM measurements are far from reactor conditions. However, we can decipher the atomic-scale mechanisms and measure the product.

References

1 Robertson, I.M., Schuh, C.A., Vetrano, J.S. et al. (2011). Towards an integrated materials characterization toolbox. *Journal of Materials Research* 26 (11): 1341–1383.
2 Kodambaka, S., Tersoff, J., Reuter, M.C., and Ross, F.M. (2006). Diameter-independent kinetics in the vapor–liquid–solid growth of Si nanowires. *Physical Review Letters* 96 (9): 096105.
3 Kodambaka, S., Hannon, J.B., Tromp, R.M., and Ross, F.M. (2006). Control of Si nanowire growth by oxygen. *Nano Letters* 6 (6): 1292–1296.
4 Kodambaka, S., Tersoff, J., Reuter, M.C., and Ross, F.M. (2007). Germanium nanowire growth below the eutectic temperature. *Science* 316 (5825): 729–732.
5 Madras, P., Dailey, E., and Drucker, J. (2009). Kinetically induced kinking of vapor–liquid–solid grown epitaxial Si nanowires. *Nano Letters* 9 (11): 3826–3830.
6 Diaz, R.E., Sharma, R., Jarvis, K., and Mahajan, S. (2012). Direct observation of nucleation and early stages of growth of GaN nanowires. *Journal of Crystal Growth* 341 (1): 1–6.
7 Yang, J.C., Kolasa, B., Gibson, J.M., and Yeadon, M. (1998). Self-limiting oxidation of copper. *Applied Physics Letters* 73 (19): 2841–2843.
8 Yang, J.C., Bharadwaj, M.D., Zhou, G., and Tropia, L. (2001). Surface kinetics of copper oxidation investigated by in situ ultra-high vacuum transmission electron microscopy. *Microscopy and Microanalysis* 7 (6): 486–493.
9 Bourgeois, J., Hervieu, M., Poienar, M. et al. (2012). Evidence of oxygen-dependent modulation in $LuFe_2O_4$. *Physical Review B* 85 (6): 064102.
10 Meyer, T., Kressdorf, B., Roddatis, V. et al. (2021). Phase transitions in a perovskite thin film studied by environmental in situ heating nano-beam electron diffraction. *Small Methods* 5 (9): 2100464.
11 Narayan, T.C., Hayee, F., Baldi, A. et al. (2017). Direct visualization of hydrogen absorption dynamics in individual palladium nanoparticles. *Nature Communications* 8: 14020.
12 Russo, C.J. and Egerton, R.F. (2019). Damage in electron cryomicroscopy: lessons from biology for materials science. *MRS Bulletin* 44 (12): 935–941.
13 Zheng, H.M., Smith, R.K., Jun, Y.W. et al. (2009). Observation of single colloidal platinum nanocrystal growth trajectories. *Science* 324 (5932): 1309–1312.
14 Baldi, A., Narayan, T.C., Koh, A.L., and Dionne, J.A. (2014). In situ detection of hydrogen-induced phase transitions in individual palladium nanocrystals. *Nature Materials* 13 (12): 1143–1148.

15 Yang, W.-C.D., Wang, C., Fredin, L.A. et al. (2019). Site-selective CO disproportionation mediated by localized surface plasmon resonance excited by electron beam. *Nature Materials* 18 (6): 614–619.
16 Sinclair, R., Yamashita, T., Parker, M.A. et al. (1988). The development of in situ high-resolution electron microscopy. *Acta Crystallographica Section A* 44 (6): 965–975.
17 Hinode, H., Sharma, R., and Eyring, L. (1990). A study of the decomposition of neodymium hydroxy carbonate and neodymium carbonate hydrate. *Journal of Solid State Chemistry* 84 (1): 102–117.
18 Egerton, R.F. (2013). Control of radiation damage in the TEM. *Ultramicroscopy* 127: 100–108.
19 Yan, X., Liu, C., Gadre, C.A. et al. (2019). Unexpected strong thermally induced phonon energy shift for mapping local temperature. *Nano Letters* 19 (10): 7494–7502.
20 Taheri, M.L., Stach, E.A., Arslan, I. et al. (2016). Current status and future directions for in situ transmission electron microscopy. *Ultramicroscopy* 170: 86–95.
21 Kabius, B., Hartel, P., Haider, M. et al. (2009). First application of Cc-corrected imaging for high-resolution and energy-filtered TEM. *Journal of Electron Microscopy* 58 (3): 147–155.
22 Kishita, K., Sakai, H., Tanaka, H. et al. (2009). Development of an analytical environmental TEM system and its application. *Journal of Electron Microscopy* 58 (6): 331–339.
23 Schilling, S., Janssen, A., Zaluzec, N.J. et al. (2017). Practical aspects of electrochemical corrosion measurements during in situ analytical transmission electron microscopy (TEM) of austenitic stainless steel in aqueous media. *Microscopy and Microanalysis* 23 (4): 741–750.
24 Roberts, S., McCaffrey, J., Giannuzzi, L. et al. (2000). Advanced techniques in TEM specimen preparation. In: *Progress in Transmission Electron Microscopy 1* (ed. X.-F. Zhang and Z. Zhang), 301–352. Springer: Tsinghua University Press.
25 Rao, D.S., Muraleedharan, K., and Humphreys, C. (2010). TEM specimen preparation techniques. In: *Microscopy: Science, Technology, Applications and Education* (ed. A.D.J. Mendez-Vilas), 1232–1244. Badajoz: Formatex Research Center.
26 Miller, B., Barker, T., and Crozier, P. (2015). Novel sample preparation for operando TEM of catalysts. *Ultramicroscopy* 156: 18–22.
27 de Wit, E., Walet, M.A.M., and Celis, J.-P. (1998). Ultramicrotomy on fretting wear debris. *Microscopy Research and Technique* 40 (6): 492–494.
28 Kestel, B.J. (1986). Non-acid electrolyte thins many materials for TEM without causing hydride formation. *Ultramicroscopy* 19 (2): 205–211.
29 LePore, J.J. (1980). An improved technique for selective etching of GaAs and $Ga_{1-x}Al_x$As. *Journal of Applied Physics* 51 (12): 6441–6442.
30 French, P.J., Nagao, M., and Esashi, M. (1996). Electrochemical etch-stop in TMAH without externally applied bias. *Sensors and Actuators A: Physical* 56 (3): 279–280.

31 Romano, A., Vanhellemont, J., Bender, H., and Morante, J.R. (1989). A fast preparation technique for high-quality plan view and cross-section TEM specimens of semiconducting materials. *Ultramicroscopy* 31: 183–192.

32 Sáfrán, G., Szász, N., and Sáfrán, E. (2015). Two-in-one sample preparation for plan-VIew TEM. *Microscopy Research and Technique* 78 (7): 599–602.

33 Weaver, L. (1997). Plan view TEM sample preparation for non-continuous and delaminating thin films. *Microscopy Research and Technique* 36 (5): 378–379.

34 McCaffrey, J.P. and Barna, A. (1997). Preparation of cross-sectional TEM samples for low-angle ion milling. *Microscopy Research and Technique* 36 (5): 362–367.

35 Ma, G.-H.M. and Chevacharoenkul, S. (1990). A technique for preparing "two in one" cross-sectional TEM specimens. *MRS Online Proceedings Library* 199 (1): 235–242.

36 Bravman, J.C. and Sinclair, R. (1984). The preparation of cross-section specimens for transmission electron microscopy. *Journal of Electron Microscopy Technique* 1 (1): 53–61.

37 Madsen, L.D., Weaver, L., and Jacobsen, S.N. (1997). Influence of material properties on TEM specimen preparation of thin films. *Microscopy Research and Technique* 36 (5): 354–361.

38 Park, Y.M., Ko, D.S., Yi, K.W. et al. (2007). Measurement and estimation of temperature rise in TEM sample during ion milling. *Ultramicroscopy* 107 (8): 663–668.

39 McCaffrey, J.P., Phaneuf, M.W., and Madsen, L.D. (2001). Surface damage formation during ion-beam thinning of samples for transmission electron microscopy. *Ultramicroscopy* 87 (3): 97–104.

40 Süess, M.J., Mueller, E., and Wepf, R. (2011). Minimization of amorphous layer in Ar+ ion milling for UHR-EM. *Ultramicroscopy* 111 (8): 1224–1232.

41 Srot, V., Gec, M., van Aken, P. et al. (2014). Influence of TEM specimen preparation on chemical composition of $Pb(Mg_{1/3}Nb_{2/3})O_3$–$PbTiO_3$ single crystals. *Micron* 62: 37–42.

42 Ishitani, T. and Yaguchi, T. (1996). Cross-sectional sample preparation by focused ion beam: a review of ion-sample interaction. *Microscopy Research and Technique* 35 (4): 320–333.

43 Li, C., Habler, G., Baldwin, L.C., and Abart, R. (2018). An improved FIB sample preparation technique for site-specific plan-view specimens: a new cutting geometry. *Ultramicroscopy* 184: 310–317.

44 Giannuzzi, L.A., Drown, J.L., Brown, S.R. et al. (1998). Applications of the FIB lift-out technique for TEM specimen preparation. *Microscopy Research and Technique* 41 (4): 285–290.

45 Nakahara, S. (2003). Recent development in a TEM specimen preparation technique using FIB for semiconductor devices. *Surface and Coatings Technology* 169–170: 721–727.

46 Zachman, M.J., Asenath-Smith, E., Estroff, L.A., and Kourkoutis, L.F. (2016). Site-specific preparation of intact solid–liquid interfaces by label-free in situ localization and cryo-focused ion beam lift-out. *Microscopy and Microanalysis* 22 (6): 1338–1349.

47 Mayer, J., Giannuzzi, L.A., Kamino, T., and Michael, J. (2007). TEM sample preparation and FIB-induced damage. *MRS Bulletin* 32 (5): 400–407.

48 Duchamp, M., Xu, Q., and Dunin-Borkowski, R.E. (2014). Convenient preparation of high-quality specimens for annealing experiments in the transmission electron microscope. *Microscopy and Microanalysis* 20 (6): 1638–1645.

49 Lee, H., OFN, O., Kim, G.Y. et al. (2021). TEM sample preparation using micro-manipulator for in-situ MEMS experiment. *Applied Microscopy* 51 (1): 8.

50 Li, J., Malis, T., and Dionne, S. (2006). Recent advances in FIB–TEM specimen preparation techniques. *Materials Characterization* 57 (1): 64–70.

51 Giannuzzi, L.A. and Stevie, F.A. (1999). A review of focused ion beam milling techniques for TEM specimen preparation. *Micron* 30 (3): 197–204.

52 Sugiyama, M. and Sigesato, G. (2004). A review of focused ion beam technology and its applications in transmission electron microscopy. *Journal of Electron Microscopy (Tokyo)* 53 (5): 527–536.

53 Parmenter, C.D. and Nizamudeen, Z.A. (2021). Cryo-FIB-lift-out: practically impossible to practical reality. *Journal of Microscopy* 281 (2): 157–174.

54 Gasser, P., Klotz, U.E., Khalid, F.A., and Beffort, O. (2004). Site-specific specimen preparation by focused ion beam milling for transmission electron microscopy of metal matrix composites. *Microscopy and Microanalysis* 10 (2): 311–316.

55 Jublot, M. and Texier, M. (2014). Sample preparation by focused ion beam micromachining for transmission electron microscopy imaging in front-view. *Micron* 56: 63–67.

56 Krueger, R. (1999). Dual-column (FIB–SEM) wafer applications. *Micron* 30 (3): 221–226.

57 Vijayan, S., Jinschek, J.R., Kujawa, S. et al. (2017). Focused ion beam preparation of specimens for micro-electro-mechanical system-based transmission electron microscopy heating experiments. *Microscopy and Microanalysis* 23 (4): 708–716.

58 Radić, D., Peterlechner, M., and Bracht, H. (2021). Focused ion beam sample preparation for in situ thermal and electrical transmission electron microscopy. *Microscopy and Microanalysis* 27 (4): 828–834.

59 Kiener, D., Zhang, Z., Šturm, S. et al. (2012). Advanced nanomechanics in the TEM: effects of thermal annealing on FIB prepared Cu samples. *Philosophical Magazine* 92 (25–27): 3269–3289.

60 Engelmann, H.J. (2003). Advantages and disadvantages of TEM sample preparation using the FIB technique; Vor- und Nachteile der TEM-Probenpraeparation mittels FIB. *Practical Metallography*, https://doi.org/10.1515/pm-2003-400403.

61 Kim, S., Jeong Park, M., Balsara, N.P. et al. (2011). Minimization of focused ion beam damage in nanostructured polymer thin films. *Ultramicroscopy* 111 (3): 191–199.

62 Benedict, J., Anderson, R., and Klepeis, S.J. (1991). Recent developments in the use of the Tripod Polisher for TEM specimen preparation. *MRS Proceedings* 254: 121.

63 Cha, H.-W., Kang, M.-C., Shin, K. et al. (2016). Transmission electron microscopy specimen preparation of delicate materials using tripod polisher. *Applied Microscopy* 46 (2): 110–115.

64 Li, H. and Salamanca-Riba, L. (2001). The concept of high angle wedge polishing and thickness monitoring in TEM sample preparation. *Ultramicroscopy* 88 (3): 171–178.

65 Li, Y., Huang, W., Li, Y. et al. (2020). Opportunities for cryogenic electron microscopy in materials science and nanoscience. *ACS Nano* 14 (8): 9263–9276.

66 Patterson, J.P., Xu, Y., Moradi, M.-A. et al. (2017). CryoTEM as an advanced analytical tool for materials chemists. *Accounts of Chemical Research* 50 (7): 1495–1501.

67 Sheng, O., Zheng, J., Ju, Z. et al. (2020). In situ construction of a LiF-enriched interface for stable all-solid-state batteries and its origin revealed by cryo-TEM. *Advanced Materials* 32 (34): 2000223.

68 Bianco, E. and Kourkoutis, L.F. (2021). Atomic-resolution cryogenic scanning transmission electron microscopy for quantum materials. *Accounts of Chemical Research* 54 (17): 3277–3287.

69 Dobro, M.J., Melanson, L.A., Jensen, G.J. et al. (2010). Plunge freezing for electron cryomicroscopy, Chapter 3. In: *Methods in Enzymology* (ed. G.J. Jensen), 63–82. Academic Press.

70 Galway, M.E., Heckman, J.W., Hyde, G.J. et al. (1995). Advances in high-pressure and plunge-freeze fixation, Chapter 1. In: *Methods in Cell Biology* (ed. D.W. Galbraith, H.J. Bohnert and D.P. Bourque), 3–19. Academic Press.

71 Cabra, V. and Samsó, M. (2015). Do's and don'ts of cryo-electron microscopy: a primer on sample preparation and high quality data collection for macromolecular 3D reconstruction. *Journal of Visualized Experiments* 95: e52311.

72 Tivol, W.F., Briegel, A., and Jensen, G.J. (2008). An improved cryogen for plunge freezing. *Microscopy and Microanalysis* 14 (5): 375–379.

73 Dubochet, J. and McDowall, A.W. (1981). Vitrification of pure water for electron microscopy. *Journal of Microscopy* 124 (3): 3–4.

74 Schreiber, D.K., Perea, D.E., Ryan, J.V. et al. (2018). A method for site-specific and cryogenic specimen fabrication of liquid/solid interfaces for atom probe tomography. *Ultramicroscopy* 194: 89–99.

75 Kuba, J., Mitchels, J., Hovorka, M. et al. (2021). Advanced cryo-tomography workflow developments – correlative microscopy, milling automation and cryo-lift-out. *Journal of Microscopy* 281 (2): 112–124.

76 Zachman, M., De Jonge, N., Fischer, R. et al. (2019). Cryogenic specimens for nanoscale characterization of solid–liquid interfaces. *MRS Bulletin* 44 (12): 949–955.

77 Smith, D.J., de Jonge, N., Fischer, R. et al. (1983). The importance of beam alignment and crystal tilt in high resolution electron microscopy. *Ultramicroscopy* 11 (4): 263–281.

78 Anton, R. (2008). On the reaction kinetics of Ni with amorphous carbon. *Carbon* 46 (4): 656–662.

79 Aikawa, S., Kizu, T., and Nishikawa, E. (2010). Catalytic graphitization of an amorphous carbon film under focused electron beam irradiation due to the presence of sputtered nickel metal particles. *Carbon* 48 (10): 2997–2999.

80 Sinclair, R., Itoh, T., and Chin, R. (2002). In situ TEM studies of metal–carbon reactions. *Microscopy and Microanalysis* 8 (4): 288–304.

81 Zhang, Z.L. and Su, D.S. (2009). Behaviour of TEM metal grids during in-situ heating experiments. *Ultramicroscopy* 109 (6): 766–774.

82 Sharma, R., Chee, S.W., Herzing, A. et al. (2011). Evaluation of the role of Au in improving catalytic activity of Ni nanoparticles for the formation of one-dimensional carbon nanostructures. *Nano Letters* 11 (6): 2464–2471.

83 Sun, X., Zhu, W., Wu, D. et al. (2020). Atomic-scale mechanism of unidirectional oxide growth. *Advanced Functional Materials* 30 (4): 1906504.

84 Chen, C.L. and Mori, H. (2009). In situ TEM observation of the growth and decomposition of monoclinic $W_{18}O_{49}$ nanowires. *Nanotechnology* 20 (28): 285604.

85 Gamalski, A.D., Tersoff, J., Sharma, R. et al. (2012). Metastable crystalline AuGe catalysts formed during isothermal germanium nanowire growth. *Physical Review Letters* 108 (25): 255702.

86 Zhang, M., Efremov, M.Y., Schiettekatte, F. et al. (2000). Size-dependent melting point depression of nanostructures: nanocalorimetric measurements. *Physical Review B* 62 (15): 10548–10557.

87 Ross, F.M. (2010). Controlling nanowire structures through real time growth studies. *Reports on Progress in Physics* 73 (11): 114501.

88 Hofmann, S., Sharma, R., Wirth, C.T. et al. (2008). Ledge-flow controlled catalyst interface dynamics during Si nanowire growth. *Nature Materials* 7 (5): 372–375.

89 Uchiyama, T., Yoshida, H., Kamiuchi, N. et al. (2016). Revealing the heterogeneous contamination process in metal nanoparticulate catalysts in CO gas without purification by in situ environmental transmission electron microscopy. *Microscopy* 65 (6): 522–526.

90 Crozier, P.A. and Chenna, S. (2011). In situ analysis of gas composition by electron energy-loss spectroscopy for environmental transmission electron microscopy. *Ultramicroscopy* 111 (3): 177–185.

91 Hersee, S.D., Sun, X., and Wang, X. (2006). The controlled growth of GaN nanowires. *Nano Letters* 6 (8): 1808–1811.

92 Parikh, R.P. and Adomaitis, R.A. (2006). An overview of gallium nitride growth chemistry and its effect on reactor design: application to a planetary radial-flow CVD system. *Journal of Crystal Growth* 286 (2): 259–278.

93 McKelvy, M.J., Sharma, R., and Glaunsinger, W.S. (1993). Atomic-level imaging of intercalation reactions by DHRTEM. *Solid State Ionics* 63–65: 369–377.

94 Thon, A. and Kuech, T.F. (1996). High temperature adduct formation of trimethylgallium and ammonia. *Applied Physics Letters* 69 (1): 55–57.

95 Gamalski, A.D., Tersoff, J., and Stach, E.A. (2016). Atomic resolution in situ imaging of a double-bilayer multistep growth mode in gallium nitride nanowires. *Nano Letters* 16 (4): 2283–2288.

96 Jacobsson, D., Panciera, F., Tersoff, J. et al. (2016). Interface dynamics and crystal phase switching in GaAs nanowires. *Nature* 531 (7594): 317–322.

97 Hillerich, K., Dick, K.A., Wen, C.-Y. et al. (2013). Strategies to control morphology in hybrid group III–V/group IV heterostructure nanowires. *Nano Letters* 13 (3): 903–908.

98 Hetherington, C., Jacobsson, D., Dick, K., and Wallenberg, L. (2020). In situ metal-organic chemical vapour deposition growth of III–V semiconductor nanowires in the Lund environmental transmission electron microscope. *Semiconductor Science and Technology* 35 (3): 034004.

99 Sun, X., Wu, D., Zhu, W. et al. (2021). Atomic origin of the autocatalytic reduction of monoclinic CuO in a hydrogen atmosphere. *The Journal of Physical Chemistry Letters* 12 (39): 9547–9556.

100 Kikkawa, J., Ohno, Y., and Takeda, S. (2005). Growth rate of silicon nanowires. *Applied Physics Letters* 86 (12): 123109.

101 Rackauskas, S., Jiang, H., Wagner, J.B. et al. (2014). In situ study of noncatalytic metal oxide nanowire growth. *Nano Letters* 14 (10): 5810–5813.

102 Chou, Y.-C., Hillerich, K., Tersoff, J. et al. (2014). Atomic-scale variability and control of III–V nanowire growth kinetics. *Science* 343 (6168): 281–284.

103 Sharma, R. (2008). Kinetic measurements from in situ TEM observations. *Microscopy Research and Technique* 72 (3): 144–152.

104 Sharma, R., Crozier, P.A., Kang, Z.C., and Eyring, L. (2004). Observation of dynamic nanostructural and nanochemical changes in ceria-based catalysts during *in situ* reduction. *Philosophical Magazine* 84: 2731–2747.

105 Zhu, W., Winterstein, J.P., W-CD, Y. et al. (2017). In situ atomic-scale probing of the reduction dynamics of two-dimensional Fe_2O_3 nanostructures. *ACS Nano* 11 (1): 656–664.

106 Ding, Y., Choi, Y., Chen, Y. et al. (2020). Quantitative nanoscale tracking of oxygen vacancy diffusion inside single ceria grains by in situ transmission electron microscopy. *Materials Today* 38: 24–34.

107 Wang, Y.Q., Liang, W.S., and Geng, C.Y. (2009). Coalescence behavior of gold nanoparticles. *Nanoscale Research Letters* 4 (7): 684–688.

108 Ross, F.M., Tersoff, J., and Tromp, R.M. (1998). Ostwald ripening of self-assembled germanium islands on silicon(100). *Microscopy and Microanalysis* 4 (3): 254–263.

109 Baker, R.T.K., Harris, P.S., Thomas, R.B. et al. (1973). Formation of filamentous carbon from iron, cobalt and chromium catalyzed decomposition of acetylene. *Journal of Catalysis* 30 (1): 86–95.

110 Ko, D.-H. and Sinclair, R. (1994). In-situ dynamic high-resolution transmission electron microscopy: application to Pt/GaAs interfacial reactions. *Ultramicroscopy* 54 (2): 166–178.

111 Lee, H.-J., Ni, H., Wu, D.T., and Ramirez, A.G. (2005). Experimental determination of kinetic parameters for crystallizing amorphous NiTi thin films. *Applied Physics Letters* 87 (11): 114102.

112 Leonard, H.R., Rommel, S., Li, M.X. et al. (2021). Precipitation phenomena in a powder-processed quasicrystal-reinforced Al–Cr–Mn–Co–Zr alloy. *Materials Characterization* 178: 111239.

113 Dannenberg, R., Stach, E.A., Groza, J.R., and Dresser, B.J. (2000). In-situ TEM observations of abnormal grain growth, coarsening, and substrate de-wetting in nanocrystalline Ag thin films. *Thin Solid Films* 370 (1): 54–62.

114 Chenna, S. and Crozier, P.A. (2012). Operando transmission electron microscopy: a technique for detection of catalysis using electron energy-loss spectroscopy in the transmission electron microscope. *ACS Catalysis* 2 (11): 2395–2402.

115 Vendelbo, S.B., Elkjær, C.F., Falsig, H. et al. (2014). Visualization of oscillatory behaviour of Pt nanoparticles catalysing CO oxidation. *Nature Materials* 13 (9): 884–890.

3

In-Situ Heating

> *Heat, like gravity, penetrates every substance of the universe, its rays occupy all parts of space. The object of our work is to set forth the mathematical laws which this element obeys. The theory of heat will hereafter form one of the most important branches of general physics.*
> – Baron Jean-Baptiste-Joseph Fourier

Heat is one of the fundamental stimuli that affects both the physical and chemical properties of materials. It is invariably used for materials synthesis using solid–solid, solid–liquid, solid–gas, or any other combinations of interactions. We find that most of the chemical process and materials properties are a function of temperature. Moreover, most of the materials applications generate heat that may change the functioning of the entire system. Therefore, our interest in observing physical and chemical changes with temperature is important for both fundamental and applied points of view. Nucleation, phase transformation, sintering, melting, dislocation movement, etc. are a few examples that are fundamental steps for materials synthesis and may transpire during application. For example, phase transformation temperature is critical to form stable and/or metastable phases with different properties. Whereas we may use this information for synthesis of new phases with desired properties or inhibit the change to maintain the properties of the original phase. We also find that heating is often combined with other stimuli such as biasing, magnetic and mechanical force as well as in liquid and gas environments. In this chapter, we will concentrate on heating in vacuum with or without other stimuli, and the heating experiments in liquid and gaseous environments will be covered in Chapters 6 and 7, respectively. We will start with a brief history, examine design criteria for heating holders, sample preparation, experimental setup needs, temperature measurement at nanoscale, typical results, limitations, and future prospects.

3.1 History

As mentioned above, we can observe the effect of heat in our daily life – most simple example is melting of ice and vaporization of water or cooking food. Of course, with the development of TEM, we became interested in understanding the nanoscale processes such as nucleation and growth of nanoparticles or defects or dislocations, and phase transformation. Therefore, time and temperature-resolved imaging and electron diffraction are one of the earliest approaches for transforming the TEM sample chamber into a nano lab for in-situ experiments. Development of heating holders/stages, compatible with a specific TEM, started in the 1960s. Following are some of the considerations (valid even today) that were delibrated for selecting the heating holder:

1. Desired operating temperature limit, higher the better.
2. High heating rate.
3. Low thermal expansion coefficient.
4. Low reactivity with residual gas environment in TEM column.

Pt is ideally suited for heating up to 1000 °C as it has reasonably high melting temperature, is non-reactive, with high heat conductivity. Mo, Ni, and Ta are also suitable for high-temperature operation in high vacuum. Graphite or tungsten support is used to achieve temperatures above 1200 °C [1].

Early designs can be divided in to two categories: (i) direct heating and (ii) indirect heating using a furnace. Current holder designs are using the same basic principles (Section 3.2). Direct heating was achieved by loading the sample directly on a metallic strip or sandwiched between two metal grids that were resistively heated by the passage of electric current. First direct heating top-entry stage was designed by Wheeler for a Siemens ELMISKOP in 1960 [2]. A few other top-entry stages, both with single-tilt and double-tilt mechanisms, were also constructed during early 1960s [1, 3, 4]. However, the electrical leads, which supplied heating current in top-entry holders, were located too near to the electron beam and generated magnetic field that caused electron beam deflection during operation. Therefore, a side-entry direct heating stage was designed by Peter Swann (who later founded Gatan, Inc.), where sample grid was sandwiched between two perforated Pt-Rh ribbons with spring-loaded current leads that were supported over two quartz fibers [5]. The size of the holder body helped to keep the current supplying wires (for heating), and thermocouple (to measure the temperature), away from the electron beam, thus mitigating the electron beam deflection.

Later, indirect heating option was developed by incorporating a small furnace, with either a single-tilt or a double-tilt mechanism, for both the top-entry and side-entry specimen stages [6, 7]. A review of early heating holder designs can be found in Ref. [8].

3.2 Currently Available Heating Holders

Currently, we have several different types of heating holders commercially available, and some of them can be further modified to accommodate our experimental requirements. A brief description of their design, functioning, and limitations is given below.

3.2.1 Direct Heating Holder

Direct heating idea has been further improved by Saka's group in Japan for a side-entry holder compatible with Hitachi TEMs. Here, a heating coil is used for resistive (Joule) heating as well as for loading the powder sample. It is possible to place two heating coils across the tip of a holder (Figure 3.1) [9]. Advantage of the two-filament design is that two different samples can be loaded at the same time but observed individually, or one coil can be used as an evaporator and other as collector to follow solid-state reactions [10]. The heating coils, made of thin refractory metals, such as tungsten, are connected to a dry battery placed outside the TEM column that controls the current, measured by using an ampere meter. A DC power supply source, devoid of any instability or power fluctuation, is the key component for obtaining high-resolution images at very high (up to 1500 °C) temperature. A temperature calibration curve is obtained by measuring the temperature against current supplied using a pyrometer. The sample is dry loaded by rolling the heating element in powder before placing it on the holder. Advantages of this design with small thermal mass are (i) reduced sample drift during heating, (ii) reduced heating and stabilization time, (iii) temperature limit is dependent on the material used for heating coil, and up to 1500 °C is easily achieved using W filament, (iv) no large capacity heating power is needed, (v) the thickness of the holder tip can be less than 2 mm to fit in the small pole-piece gap.

This holder has been further modified to be used for heating in gas environments (Section 7.2.2.1). Although the holder is most suitable for powder samples, FIB lamella transfer to the heating wire is possible [11]. However, due the direct contact of the sample with heating coil, the reactivity between the two materials must be considered (Section 2.5.2), and the commercially available holder can only be used in Hitachi microscopes.

3.2.2 Indirect Heating Holders

Indirect heating holders, employing resistive (Joule) heating, are more popular than direct heating systems and are universally available. Furnace and MEMS-based are the two types of holders as described below.

3.2.2.1 Furnace Heating Holders

TEM holders equipped with a Joule heating furnace, in which the TEM sample grids can be loaded, are commercially available. The heating furnace consists of a cradle, miniature metal coil with two leads, and four-point connections; two for circulating the current through the coil and the other two to measure the temperature of the furnace body (Figure 3.2a,b). The furnace body is generally made of Ta or Inconel, but other metals can be used. The operating temperature depends on the thickness of the wire and number of coils, i.e. total resistance. The current flowing through the coil is kept low, in m-Amp range, to reduce the production of electrical or magnetic fields. All electrical leads run through the body of the holder and are connected to an outside control unit.

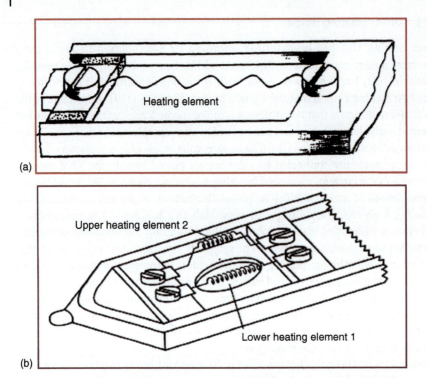

Figure 3.1 (a) A single wire heating element placed in the tip area of a side-entry TEM holder. (b) Double wire design to heat two materials independently or simultaneously. Current for resistive heating is supplied using dry battery that is essential to provide constant current without fluctuations for high-resolution imaging. Source: Adapted from Refs. [9, 10].

A few variations of the furnace holder described above became commercially available, from Philips (Figure 3.2c) and Gatan, Inc.(Figure 3.2a,b), in the 1980s that made it easier for researcher to concentrate on performing experiments and not spend time in designing and machining holders. Whereas the heating coil in the Philips furnace is coated with the insulating ceramic material (Figure 3.2d,e), it is embedded in a bed of ceramic that filled the furnace body in the holders sold by Gatan, Inc. (Figure 3.2a,b). Furnace body for Gatan holder can be made from either Ta or Inconel, providing flexibility for sample environment as explained in Section 2.5.2. Other difference between the two is: no cooling water was needed for the first and furnace can be easily repaired by opening the body to change the heating coil when it fails. On the other hand, the Gatan holder requires cooling water circulating within the holder body to keep it from becoming too hot and needs to be sent back to the factory for repairs. Unfortunately, Philips, which changed its name to FEI and now is part of Thermo-Fisher, no longer sells this type of holder, but it is possible to build similar holders. Hummingbird Scientific also markets a furnace holder with low drift rate without need for water flow (check their website for further details).

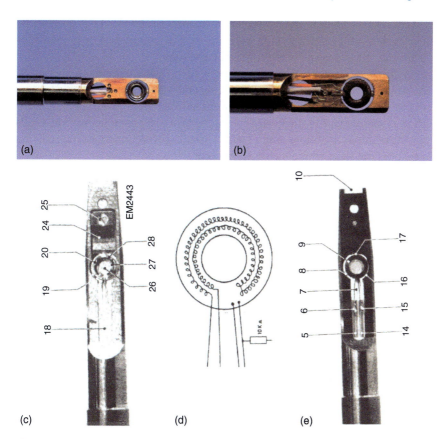

Figure 3.2 Photographs of the tip section of a Gatan heating holder: (a) top view showing the furnace with over 3 mm diameter well where the sample grids are loaded; (b) back view showing two sets of wires for heating and measuring temperature (thermocouple); (c) drawing of the top-view of the single tilt Philips heating holder showing similar space for sample loading as for the Gatan holder; (d) diagram of the heating coil coated with insulating ceramic to be enclosed in the furnace used for resistive heating; (e) bottom view of this holder showing the heating and thermocouple wires connected to the furnace body. Source: (c–e) are extracted from the User's Manual of Philips heating holder.

However, these furnace heating holders have following problems:

1. The heat transfer from the furnace body to sample requires a perfect contact with sample grid/support, but there is no way to check.
2. The heating rate is very slow due to the large mass of the furnace body.
3. The current required to increase temperature follows a parabolic curve and is difficult to predict or calibrate accurately.
4. Specimen drift is substantial that makes it quite difficult to keep the sample in viewing area during heating.
5. It takes long time for sample to stabilize and stop drifting after reaching the set temperature, thus making it impossible to record intermediate metastable phases during heating.

6. Gatan holders require water circulation through the holder body to keep the O-ring seal at low temperatures. However, the pump used for water flow generates vibrations and results in loss of resolution.
7. The wiring to the cradle is fragile and exposed to TEM environment making the holder susceptible to fail due to (i) swinging of the cradle during sample loading, (ii) connecting wire getting corroded from residual gases in TEM column, (iii) short circuiting within the heater coil as the insulating ceramic material cracks during multiple operations, and/or due to thinning of sections of heating coil that generates hot spots where melting may occur.

Despite these shortcomings, they have been successfully used to observe structural changes in thin metallic films, phase changes in alloys, nucleation and growth during recrystallization, etc. Further developments of this design have been made to include single- or double-tilt specimen holders, commercially available, that are still in use.

3.2.2.2 MEMS-Based Heating Holders

The drawbacks of the furnace heating holders described above and the advancements in cleanroom lithographic techniques have led to a major breakthrough in designing furnace heating holders using microelectromechanical systems (MEMS) [12]. Basic concept is to modify the sample loading section of a side-entry stage such that direct contact between the heater and the sample can be made by a user on the fly. Motivations behind the developments have been to (i) mitigate wire bonding to the furnace, (ii) decrease the thermal mass of the heater for rapid heating and quick stabilization, (iii) provide embedded connections for power input and system response output, and mitigate wire connections. Several holders with MEMS-based heaters in contact with sample support films have been developed during last couple of decades. Basic philosophy is to create electron-transparent thin films, mostly SiN_x, on Si with inbuilt electrical contacts for heating, called microchips or chips in short (see Chapter 5 for microchip fabrication).

A design of one such holder, with an eight-contact microchip, is shown in Figure 3.3. Part "a" is a baseplate, made of Al, on which the eight contacts, marked "b," are chemically etched using a CuBe alloy and coated with Au for corrosion-free low-contact resistance. Each set of contacts is fabricated as a single piece, and they are sandwiched between base plate "a" and Al frame "c" such that the contacts are held in place and guide alignment of the sensor "d" with contact pads (rectangular boxes on the top of the plate). A pair of spring clips "e" are used to hold the parts together. "f" designates the grooves machined to electrically clamp the wires (not shown) in place. These wires run through the body of the holder to an external connector, and part "g" attaches the assembly to the holder body. All of the parts are held together by nonmagnetic screws (not shown). Parts (a), (c), and (f) are all anodized Al, where the aluminum oxide coating provides electrical insulation [12]. This holder can accommodate multiple types of microchips, including commercially available SiN_x thin films supported on Si [12].

Design of the microchip, containing mini heaters, is fundamental to the performance of these holders and may have subtle variations (Figure 3.4). In one of the

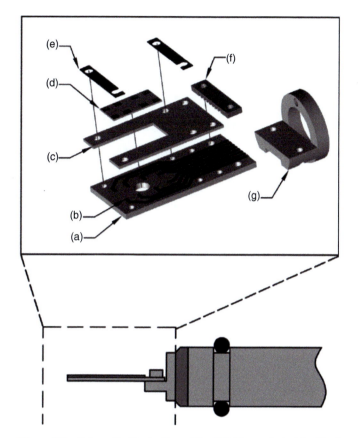

Figure 3.3 Schematic illustration of the MEMS sample holder. The bottom section shows a side view of the assembled holder. The top section shows the individual components: (a) baseplate, (b) electrical contacts, (c) alignment frame, (d) MEMS device, (e) spring clips to hold the device in place, (f) clamp to align and hold the contacts, and (g) connector to the holder body.

designs (Figure 3.4a), used by Protochips Inc., low-conducting ceramic membrane with holes, supported by Si, is used for loading the sample. The membrane is heated using electrical contacts etched in the thick section of the membrane [13]. Another microchip design, currently used by DENSsolutions, is shown in Figure 3.4b [14]. Here, a spiraled Pt resistive wire, embedded in the membrane, acts as a heater. The heated area is kept small (0.34 mm × 0.34 mm) to minimize the thermal mass and increase the heating rate. The resistance of the wire is monitored by a four-point measurement to determine the temperature. Custom-made amplifiers, controlled by a LabView program, are used to control the heating, and read out, which is recorded as a function of time.

Similar concept is used in the design of *NanoEx™-i/v MEMS microheater*, with heating element, shaped as a meander, embedded in a thin squared free-standing membrane made of SiN_x (Figure 3.4c). The heat is generated by Joule effect when current is forced through the spiral, and the entire device is fabricated

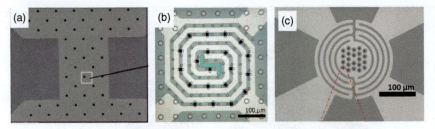

Figure 3.4 (a) Backscattered electron image showing the pattern of holes in a low-conductivity ceramic membrane that provides heating surface as well as sample support. Current is introduced through the membrane using an external constant current supply. Source: Allard et al. [13]/John Wiley & Sons. (b) Optical microscope image of membrane with Pt spiral (white). The ovaloids and circle are the electron-transparent SiN_x windows and spacers, respectively. Source: Creemer et al. [14]/IEEE. (c) The NanoEx™-i/v MEMS microheater with the meandered resistor with the heat spread in the center of chip. The circles in the middle are thin SiN_x windows with holes. Source: Mele et al. [15]/John Wiley & Sons.

with standard cleanroom silicon microfabrication technology (see Chapter 5 for details). Electron-transparent windows are etched in the locally heated area for TEM characterization. The resistance value of the spiral, accurately measured by four-probe local sensing, is used to determine the temperature and continuous monitoring of the resistance with a closed control loop that results in precise control on temperature stabilization [15]. MEMS-based heaters for in-situ TEM, with subtle differences and thermal property measurement capabilities, are being developed continuously [16–19], and a detailed review can be found in Ref. [17].

Most of the MEMS-based holders can combine multiple stimuli, for example, heating can be combined with biasing by using eight contact pads: four dedicated to heating and the other four are used for biasing experiments [15, 20]. In most holders, the heater is optimized to achieve uniform temperature distribution over the entire electron-transparent window area, but in practice, thermal gradients have been reported [21]. Also, currently available commercial holders may have different heating rate, drift rate, stabilization time, and/or temperature limit, and these parameters are available on their websites.

Microchip-based nanocalorimetry holders for thermal analysis of endothermic and exothermic reactions during high heating and cooling rates have also been developed [22]. As we will see in Chapters 4, 6 and 7 that MEMS-based heating chips are also well suited for combining heating with other stimuli, such as electrical, mechanical, liquid, and gas.

3.3 Experimental Considerations

3.3.1 General

We have discussed the general design philosophy for in-situ TEM experiments in Chapter 2, and we recommend you read it carefully. However, we will highlight

3.3 Experimental Considerations

specific considerations in each chapter dealing with a particular stimulus as the success to obtain unambiguous data is dependent on it. Moreover, you can save your instrument and/or time by not repeating mistakes that others have made. Specific considerations for heating experiments are listed below:

- **Sample:** First and foremost, you must know chemical and physical properties of your sample that will determine the safe operating temperature for your reaction. Keep in mind, this temperature should be below the temperature achievable by your TEM heating holder. Physical form of TEM sample, i.e. powder, plan-view, cross-sectional, or fibbed, will require different considerations as described in detail in Section 2.5.2.
- **Heating holder:** Your TEM heating device, direct, furnace, or microchip heater, will determine the (i) maximum achievable temperature, (ii) heating rate, (iii) sample drift rate, (iv) sample preparation method. For example, it is not easy (but possible) to load a thin film, plan view, or cross-sectioned sample on a MEMS chip or wire heaters, but furnace heating holders accept any kind or sample.
- **Grid and support material:** This is an important consideration when using a furnace-type heating holder. Holey carbon (or lacy carbon), SiN_x, or SiO_x films supported on metal grids are commonly used to load powder samples. As described in Section 2.5.2, powder samples can also be dry loaded directly on the metal grid where particles sticking away from the grid bars are ideally suited for observation as it avoids contribution of the support for imaging and/or spectroscopy and provides direct contact with the grid. However, the operating temperature must be well below the melting point and/or reactivity of the grid material to avoid misleading or false results (Figure 2.4) [23, 24]. The reactivity of wire material with powder sample is also important consideration. Also, operating (reaction) temperature should be well below the tolerance of the bonding glue for cross-sectional samples.
- **TEM vacuum:** High vacuum environment in a TEM is generally considered non-reactive. Unfortunately, it is not entirely true as the base pressure in TEM column is mostly due to water vapor that results in a slightly oxidizing environment and should be kept in mind.
- **Temperature rise due to electron beam:** Various types of interactions of high-energy electrons, which modify the structure and chemistry of the material, have been established and heating is one of them [25–27]. Watanabe et al. have used an Au-Pd micro-thermocouple to measure the temperature rise due to electron beam irradiation. Their systematic study reported a temperature increase up to a few hundred °C, depending on the irradiated area and electron dose at 80 keV [28].

Currently, microchips commonly use SiN_x on Si as support film that does not react with most of the materials at the operational temperature of 1000 °C to 1300 °C, and does not require the same considerations as furnace or direct heating holders. Still, temperature measurement at nanoscale is important and traditionally used methods, thermocouple or pyrometer, do not have the spatial resolution required for in-situ measurements. Therefore, in past couple of decades, researchers have reported other methods for nanoscale temperature measurements as described Section 3.3.3.

3.3.2 Electron Beam

Electron beam interaction with the sample is always present and should be considered when interpreting the experimental results. However, intentionally heating the samples by electron beam has been successfully used to observe atomic-scale mechanisms for grain boundary formation, dislocation motion, phase transformation, solid-state synthesis, crystallization, etc. It has been established that the heating temperature can be controlled by varying the electron beam current as well as the operating voltage of the microscope. Electron beam heating is considered advantageous as it allows us to observe multiple areas of the same sample by moving the electron beam to different location to confirm the reproducibility of the observed mechanisms.

In the 1980s, Eyring's group reported a number of electron-beam-induced reactions, including silicide formation [29], decomposition, and phase transformation related to rare-earth oxides [30]. His group also reported the collective motion of PbO_2 atoms on the surface [31]. Images, extracted from a video (Figure 3.5a–h), show a remarkable coordinated motion of scores of atoms on the surface of modified PbO_2. An island with 23 atoms and 5-unit cell (Figure 3.5a) in projection underwent subtle structural modification with time. For example, number of atoms in top layer decreased and bottom layer increased (Figure 3.5b) followed by reduction in number of layers (height) and number of atoms in top-layer but an increase in the projected width, i.e. number of atoms in bottom layer (Figure 3.5c,d). Later, top two layers separated as they reached the end of the island and moved down on the surface of the larger particle (Figure 3.5e–h) and another layer thereafter (Figure 3.5h). It is an

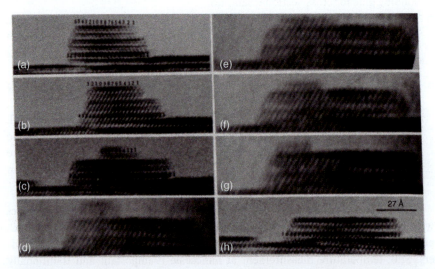

Figure 3.5 (a–h) Images showing cooperative motion of large numbers of atoms of modified β-PbO_2 during HREM observation. Source: Kang et al. [31]/from Elsevier.

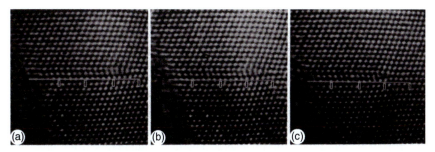

Figure 3.6 Photographs showing the gradual disappearance of an extrinsic stacking fault, by removal of successive atomic columns at the interior end of the fault. The camera exposure time was 0.5 second, and so each picture is a time average of 25 frames. The time intervals and number of image spots (corresponding to CdTe atomic columns) in the fault are as follows: (a) 00 second, 18 spots; (b) 28 seconds, 17 spots; (c) 48 seconds, 16 spots. The reference markings correspond to every fifth image spot. Source: Sinclair et al. [34]/from Springer Nature.

interesting example of the surface reduction and changes during redox reactions. In recent years, similar collective motion has been reported during Pt/GaAs interfacial reactions upon heating [32], on CuO surface during reduction restructuring [1], Au nanoparticle surface in reactive environment [33].

During the same period, Sinclair et al. reported direct near atomic resolution observation of defect annihilation in CdTe thin films [34]. A partial dislocation terminating a fault is marked by white line with marked dislocation cores in Figure 3.6a. The disappearance of an extra plane, with an average of 20 seconds time interval, occurred as one column pair of (CdTe) disappeared (Figure 3.6b,c). A mechanism for the defect annihilation was deduced based on the assumption that atomic motion along the final dislocation column results in the loss of atoms as they reach the surface and diffuse away [34]. These results were important as they conveyed the technological importance of continuous recording of atomic resolution images for direct observations of defects, dislocation, and atomic motion that set the stage for in-situ microscopy. His group later reported similar results for crystallization of amorphous Si using electron beam or heating holder [35].

However, the difference between electron beam and resistive heating can be monitored as reported by Kooi et al. [36]. They followed the crystallization of the 10, 40, and 70 nm thick amorphous $Ge_2Sb_2Te_5$ films by in-situ heating using TEM and found that crystallization started at RT in the areas irradiated by 400 keV electron beam [36]. The incubation time was found to be inversely proportional to the current density in the electron beam. Also, the crystallization started at 130 °C without electron beam exposure and between 70 °C and 130 °C when exposed to 200 keV electrons. In principle, electron irradiation always affected the crystallization kinetics, strongly promoting nucleation and not hampering the growth. Also, Wittig et al. used a combination of annealing and beam blanking to follow the cubic to tetragonal phase transformation of FePt magnetic nanoparticles at 500 °C [37].

Therefore, we need to pay attention to electron beam effects by making observations with and without the electron beam and checking the unirradiated areas often during heating. We should also explore the ways to reduce or mitigate the radiation effects by reducing the electron beam current or changing the operating voltage.

3.3.3 Sample Temperature at Nanoscale

Most of the heating holders are equipped with a method to calibrate or measure the temperature of the wire, furnace, or microchip but not at nanoscale. Precise temperature measurement is critical for establishing thermodynamic parameters of a metastable phase and/or to deduce kinetic parameters from time and temperature-resolved measurements (Section 1.6.2). Most of the furnace-based holders are outfitted with a thermocouple that measures the temperature of the furnace body quite accurately. However, for the metallic or ceramic samples made as discs, or powder samples loaded directly on a TEM grid, the actual temperature will depend on the following facts:

- How good is its contact with furnace body?
- What is samples' heat conductivity?
- How far is the observed area from the furnace body?

The situation for powder samples or nanoparticles loaded on a thin film supported by metal grid is far worse as the conductivity, and particle's contact to thin film and film to grid material also need to be considered. Moreover, above 600 °C, radiative heating from the furnace body, not directly measured by thermocouple, to the sample becomes significant and must be accounted for.

The temperature of most of the MEMS-based heating chips is calibrated and programmed by the vendor such that a desired temperature can be reached and maintained. However, by using in-situ Raman spectroscopy, a temperature gradient within a single chip was reported by Picher et al. [21]. Temperature gradient with a microchip device has also been observed using infrared thermal camera [38]. Currently, heating chips from DENSsolution are delivered with a temperature gradient map, and the temperature of the entire chip can be obtained from resistivity data. Still, measuring precise temperature at nanoscale spatial resolution is needed to correctly interpret the observations and obtain kinetics data.

Multiple methods have been developed and tested during last decade to address this deficiency. First couple of attempts were made by depositing thin film thermocouples directly on the TEM sample to measure the temperature rise in a sample due to electron beam [28, 39]. In recent years, temperature calibrations and nanoscale measurements have been achieved by using the known parameters for physical or chemical transformations, such as melting point, phase transformation, and bulk plasmon. Each of these methods has limited applications as described in detail in Ref. [40], but leads us in the right direction. Here we will describe some of the relevant methods for direct nanoscale temperature measurements and modeling the temperature distribution across the sample.

Kim et al. calibrated the thermocouple readings of heating holder by performing multiple sets of experiments for different reactions, such as solid-state epitaxial regrowth, melting point of pure Sn, Al-95 wt% Zn eutectic alloy, NiO/carbon nanotube composite, and pure Al. Their systematic measurements revealed the thermocouple reading to deviate by 4% on an average [41].

Isothermal sublimation of Ag nanocubes is yet another method to measure local temperature. Vijayan and Aindow have used this method for systematic measurements from various nanocubes loaded on MEMS microchip, at different locations on the same chip or different chips [42]. They report the variation in their measurements within the chip or from chip to chip to be $\pm 5\,°C$ and within $0.5\,°C$ of the mean of the calculated values using Kelvin equation. Interestingly, they also found measured values to be 5% to 7.5% lower than the externally calibrated set point for the holder, which could be due to imperfect contact of the sample with the support film and emphasize the importance of direct nanoscale measurement [42].

Although Raman spectroscopy provides a convenient method for direct temperature measurement under reaction conditions from $10\,\mu m$ area with 5% relative error or $38\,°C$ at $500\,°C$ [21], most of TEMs do not have this facility connected to the column. Other temperature calibration routines, such as relating thermal expansion of metal from electron diffraction, have been reported using Ag thin film [43]. This method was further improved by Niekiel et al. by using Au nanoparticles and parallel beam electron diffraction to measure the temperature within $\pm 20\,K$, and statistical precision as low as $2.8\,K$ [44]. On the other hand, Wehmeyer et al. have measured thermal diffuse scattering from STEM diffraction patterns to obtain a repeatable thermal coefficient that can be used to measure the local temperature [45].

Thermal expansion also changes the density of material that can be measured from the bulk plasmon peak shifts using low-loss EELS. Mecklenburg et al. have successfully related the bulk plasmon peak shift of 80 nm thick Al nanowires to the local temperature during Joule heating with 10% accuracy [46] and have been employed by other research groups [47, 48] (Figure 3.7a). For 2D materials, the plasmon peak shifts were found to depend on the temperature as well as on the number of layers due to quantum confinement effects. However, by measuring the sample thickness, in units of atomic layers, contribution from the layer thickness can be accounted for, making it possible to extract the local temperature and thermal expansion coefficients from the measured plasmon shifts [49].

High-energy-resolution EELS provides us an access to measure the vibrational modes [50] that are also temperature-dependent. For example, for SiC films, surface phonon polaritons (SPhP) energy, measured with electron beam placed away from the surface of the sample (aloof mode) shifts as function of temperature (Figure 3.7b), for both energy-loss and energy gain, but in opposite direction [51]. Similar but slightly larger shifts were also observed for bulk phonon peaks measured with electron beam placed on the sample (Figure 3.7c). Quantitatively, bulk phonon and SPhP modes show significant energy shifts with rates of -3.47 and $-5.10\,meV$ per $1000\,K$, respectively, with measured temperature of $1270\pm 10\,K$ for set-point of $1273\,K$ [51].

These experimental methods provide us a choice for measuring/mapping temperature of the sample at nanoscale, although not all of them are applicable to every experimental condition. Finite element-based analysis using programs like COMSOL can also be used to evaluate temperature distribution within the TEM grid

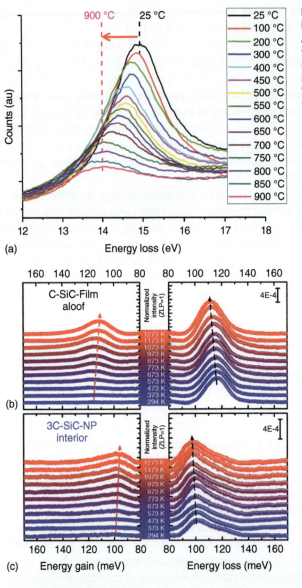

Figure 3.7 (a) Measured bulk plasmon peak shift of Al nanoparticles as a function of set temperature using a MEMS-based heating holder. (b, c) Temperature-dependent vibrational spectra of surface phonon polariton (SPhP) and transverse optical (TO) mode. (b) Electron energy-gain and energy-loss spectra of 3C-SiC-Film under aloof configuration with different temperatures. (c) Electron energy-gain and energy-loss regions of vibrational spectra of 3C-SiC-NP under interior configuration with different temperatures. Source: (b, c) Yan et al. [51].

or microchip [52]. For example, Pérez Garza et al. modeled a membrane heater (Figure 3.8a) as a thermal network and mapped the temperature distribution over the entire microchip (Figure 3.8b) and have used it to evaluate the design of a biasing and heating microchip [16, 53].

3.3.4 Specimen Design and Selection

Whereas most of the samples prepared using the techniques described in Section 2.3 can be loaded in the furnace holders, both the wire and microchips require special procedures to load samples other than powders. Although preparing a thin lamella

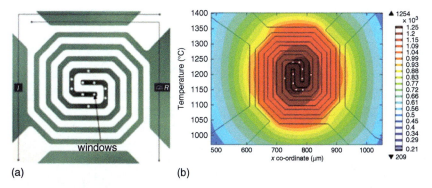

Figure 3.8 (a) Spiral-shaped microheater uses a four-point-probe technique to accurately determine the temperature (b) COMSOL simulations were used to define the position of the windows (white circles), to ensure that regardless of the selected temperature by the user, they will all be at the same temperature. Source: Pérez Garza et al. [53]/John Wiley & Sons.

from bulk using FIB is now routinely used, its transfer to heating chips requires special considerations. Therefore, significant efforts have been directed toward milling and transferring FIB lamella to the microchips or wire heater, including modification to the FIB stage or using optical microscope with a special nano-manipulator [11, 54–56].

A step-by-step procedure for FIB lamella preparation and transfer to a MEMS-based heating window can be found in Ref. [55] and shown in Figure 3.9. Briefly, a thick lamella is cut from the bulk sample as first step and protective Pt layer is deposited on the region of interest (also see Section 2.4.6). Next, the trenches are milled on either side of the protective layer and lamella is pre-thinned to $\approx 0.8\,\mu m$. The EasyLift nanomanipulator is then attached to the Pt cap to cut the lamella free from the bulk sample (Figure 3.9a) that is attached to the Cu grid on the flip stage (Figure 3.9b) for the final milling (Figure 3.9c). The ion beam voltage/current is iterated between 30 keV/0.23 pA and 5 keV/15 pA, respectively, to obtain an electron-transparent region with minimum ion beam damage (Figure 3.9d). The lamella is transferred to the MEMS heating chip by attaching the nanomanipulator probe to thick section on onside to cut it free from the Cu grid (Figure 3.9e,f). It is important that FIB stage is tilted to 52° so that probe axis is inclined to lamella plane for easy transfer.

3.3.5 Thermal Drift

Specimen drift upon heating is due to the thermal expansion of the support or heating unit and is a major problem for furnace-based heaters due to their large thermal mass. As a result, high magnification images or spatial-resolution spectroscopy data cannot be recorded until the temperature is stabilized and sample stops drifting. It not only compromises the time resolution as a function of temperature but also makes it difficult to obtain before and after data from the same area/nanoparticle. The drift rates are reduced in MEMS-based heating microchips due to small thermal

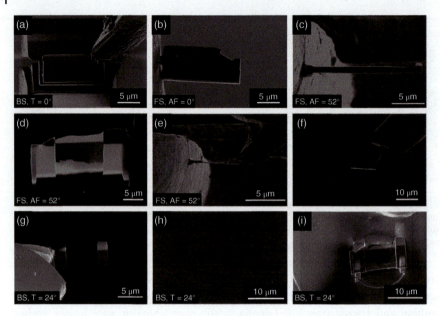

Figure 3.9 Sequence of operations: (a–c) SE ion beam images showing the transfer of the lamella from the bulk stage to the flip stage, (d) HAADF STEM image of the electron-transparent lamella after final milling, (e, f) SE ion beam images showing the nanomanipulator reattachment to the side of the specimen and separation from the flip stage, (g, h) alignment and attachment of the specimen to the MEMS chip, and (h, i) SE images of the final configuration obtained using the ion and electron beams, respectively. BS = bulk stage; FS = flip stage. Source: Vijayan et al. [55]/from Microscopy Society of America.

mass and can be negligible for some commercial holder (see specifications on the websites of respective companies provided in the reference section of Chapter 1).

3.4 Select Applications

Here, we include examples of in-situ heating experiments in vacuum using both electron beam heating and heating holders. It is interesting to note that some of the results obtained by electron beam heating could be directly related to thermal effects. The examples of heating in liquid or gas environment are covered in Chapters 6 and 7, respectively. Relevant references to combine heating with electrical, mechanical, and/or magnetic stimuli are included in Chapters 1 and 8.

In the early days, electron beam was used the main source of heating, which is still prevalent for certain experiments (see Sections 6.5.2 and 6.6.1). We can control the electron dose by controlling the condenser aperture size and thereby the temperature and/or heating rate. Structural modification of Au nanoparticles was among the first few observations reported under the electron beam radiation [57]. Authors reported that the rate of dynamic modifications was directly proportional to the current density of the incident electron beam, the degree of contact of the particle with the substrate, inversely proportional to the size of the particle but not on the nature of the support, i.e. carbon or silicon oxide [57].

However, its limitations were soon realized as (i) we could not measure the temperature of the viewing area, (ii) relationship between the electron dose and temperature could not be established, (iii) fine control of temperature or heating rates was not possible, and (iv) the electron beam effects are not limited to heating only; thus, heat may not be the reason for observed transformation. Later, as mentioned earlier, Joule heating stages/holders became commercially available and are used for controlled experiments. Following examples include specific mechanisms revealed using in-situ heating at nanoscale and should help you in planning similar experiments.

3.4.1 Dislocation Motion

Dislocations and grain boundaries are planar defects that are considered be a measure of stress in the material. Therefore, the formation and motion of dislocations have been related to mechanical properties (elasticity and plasticity) of thin films using in-situ TEM observations. One of the approaches has been measuring the mechanical properties as a function of temperature or strain or both combined with simultaneous imaging using specially modified TEM holders. The other way is to measure the effect of temperature on mechanical properties ex situ and perform TEM imaging under same thermal conditions for in-situ observations of formation and motion of dislocations. Atomic-scale mechanism of dislocations motion has also been reported using electron beam heating as shown in Figure 3.6 [34].

It is not possible to cover the large number of published reports on dislocation and grain boundary nucleation, motion, and annealing. Therefore, we briefly describe the dislocation formation and motion during multiple thermal cycles as function of film thickness and provide references of some of the other relevant reports. Balk et al. monitored the thermomechanical behavior of 50 nm to 2 µm Cu thin films by measuring the stress in the films during thermal cycles between RT and 500 °C [58]. They found that the stress flow increased with decreasing film thickness at RT but plateaued at 400 nm. They used this information to select a 200 nm Cu film for in-situ TEM imaging of the grains during thermal cycling.

At 500 °C, the grains were found to be dislocation free in agreement with the ex situ stress measurement showing that the stress reduced with increasing temperature and becoming almost stress free above 240 °C. Weak beam TEM imaging conditions were used such that dislocations appear as white lines in the images (Figure 3.10). First dislocation was observed to emerge from a triple junction at 355 °C during cooling (Figure 3.10a) followed by second just below the first (Figure 3.10b) and continued to move upward. New dislocations continued to emerge from the same location, pushing previously formed dislocations upward with time and decreasing temperature (Figure 3.10c–h). A total of 10 dislocations emerged; at fairly regular intervals as the sample was cooled down to 40 °C, note that the dislocations did not immediately glide across the entire the grain but only a fraction of its length. We can see that the new dislocations push the old ones into the grain that results in a pile up at the triple

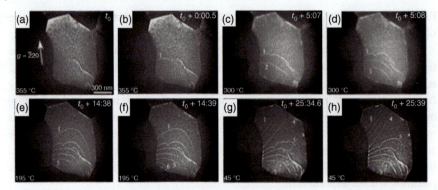

Figure 3.10 A sequence of weak-beam TEM video images recorded during cooling of a 200 nm Cu film. The TEM images were recorded in weak-beam mode and dislocations thus appear as white lines. Note that the time, in the upper right corner of each frame, is written in minutes, so that the time span of this sequence is less than 26 minutes (a–h). A total of 10 dislocations (numbered in white) started sequentially from the source at the lower left triple junction (a). Dislocations were pushed forward by later dislocations, which in turn were not able to glide as far into the grain as the earlier dislocations (compare Figure 3.3b,d,f,h). Based on their motion and on the grain geometry, dislocations must have undergone glide on the (111) plane parallel to the film/substrate interface. Source: Balk et al. [58]/from Elsevier.

junction (Figure 3.10g,h) at 40 °C, following threading mechanism [58]. Also note that the dislocations curve as they glide further into the interior of the grain.

Interestingly, the dislocation formation in a 270 nm film, where the stress flow is expected to be lower than for 200 nm thick film, the dislocations glide on a (111) plane parallel to the film/substrate interface, termed as parallel glide, completely replacing threading dislocation motion observed in films thinner than 200 nm. Moreover, dislocation formation and motion are completely reversible as shown in the images recorded at 500 °C at the beginning of third and fourth cycles (Figure 3.11a,c) and at 40 °C (Figure 3.11b,d). Here the dislocation arrangement and number are identical in these pairs of images, confirming the ex-situ stress measurements to be directly related to the dislocation behavior in the sample.

Some of the other examples reporting in-situ TEM observations of dislocation formation and motion, during thermal cycling include, but are not limited to, the effect of cluster formation on dislocation mobility and thereby ductility of Nb micro-alloyed steel [59], formation, evolution, and annihilation of large closed GB loop defects in graphene with atomic resolution [60], role of thermally activated dislocation motion in ultrafine grained Al–Zn–Mg alloys for heterogeneous nucleation of precipitates and their coarsening at dislocation cores [61], dynamic behavior of nanoscale prismatic dislocation loops in pure α-Fe sheets [62], cyclic temperature-resolved activity of dislocation motion in thermally strained Al nanowires [63].

3.4.2 Nucleation, Precipitation, and Crystallization

Nucleation, precipitation, and crystallization are the fundamental steps that lead to the formation of solids. Also, creation of precipitates in alloys alters their properties,

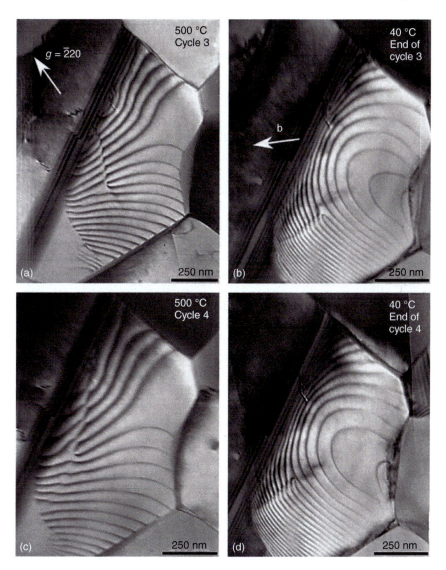

Figure 3.11 TEM images recoded from a 270 nm thick Cu film during cooling, after third heating and cooling cycle, at (a) 500 °C and (b) 40 °C, respectively. (c) Fourth thermal cycle 500 °C. (d) Fourth thermal cycle, 40 °C (during cooling). The dislocation configurations at 500 °C are virtually identical (a, c), as are those at 40 °C (b, d). The dislocation pile-up in (b) and (d) is very dense, with a separation of only 10 nm between dislocations in the lower end of the grain. Source: Balk et al. [58]/from Elsevier.

and therefore it is important to understand their nucleation and growth conditions and mechanisms. TEM/STEM imaging and spectroscopy has played an important role in achieving this goal [64]. Similarly, crystallization of amorphous material with temperature is important to determine the boundary conditions for their optimum applications. Electron radiation has been reported to trigger the crystallization at lower temperature and/or affect the growth kinetics. Therefore, in-situ observations should always be compared with ex situ characterization such as X-ray diffraction and/or differential scanning calorimetry (DSC) [65]. Despite this concern, in-situ heating experiments, using TEM/STEM imaging and spectroscopy, have been successfully employed to reveal atomic-scale mechanisms and kinetics of these processes as evidenced in the following examples.

Sinclair and his group have performed some of the earliest experiments such as single crystal regrowth from amorphous Si thin film, using a furnace heating holder [35]. As-deposited Si films on sapphire were polycrystalline and highly defective that became amorphous after ion implantation. They found that crystallization of these amorphous films started at 600 °C and atomic-resolution images show that upon annealing at 700 °C, the defect annihilation and single crystal growth followed similar mechanism as electron-beam-induced process.

The crystallization of amorphous NiTi thin films can be considered as a representative example of understanding and measuring the kinetics of nucleation extracted using in-situ transmission electron microscopy [66]. Here, the sample was heated to 490 °C using a single-tilt holder and kept at the temperature for several minutes while recording a video of bright-field TEM images (Figure 3.12). First nuclei were observed after 20 seconds (Figure 3.12a) and time-resolved images were extracted from the video shown in Figure 3.12b–i. The nucleation and the growth rate were extracted from these images by counting number of new grains within each image and the increase in their size. The growth rate follows the well-known S-curve (Figure 3.12j), with incubation period, almost linear growth period and plateau (steady state). Individual kinetic rates thus measured could be fitted to conventional Johnson–Mehl–Avrami–Kolmogorov (JMAK) equation (Figure 3.12k). This quantitative analysis provided a platform to study the development and control of microstructures in amorphous materials [67].

Atomic resolution STEM images, to monitor the evolution of $CuAl_x$ precipitates and their growth in Al alloy (Figure 3.13), have been reported by Liu et al. [68]. The low specimen drift and high stability of a MEMS-based heating holder enabled them to record time-resolved STEM videos under isothermal conditions at 140, 160, 180, and 200 °C, to evaluate the temperature dependence on the kinetics. They found that the nucleation rate and precipitate growth rate increased as a function of higher aging temperature, but the precipitate density decreased. Figure 3.13 shows time-resolved images, extracted from video recorded during isothermal heat at 160 °C that illustrate these observations. We note the existence of two precipitates, P-I and P-II (Cu is heavier than Al and thus has bright contrast in HAADF images), and faint bright spots indicating possible nucleation sites for other three precipitates. New precipitates, P-III, P-IV, and P-V, appear after 120 minutes of heating and their growth starts with time. For example, P-I continues to grow until it comes in contact with P-II and growth stops between 158 and 196 minute period, then P-II and P-V disappeared between 196 and 202 minute period. Next, P-I and P-III continue to grow, but no change is observed in the size of P-IV indicating the reason for reduced precipitation density at higher temperature. The process was affected by interaction of growing precipitate with matrix, dislocation, and other

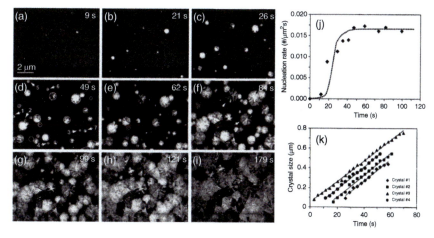

Figure 3.12 Video-captured TEM bright-field images recorded during in-situ heating show crystallization of the amorphous nickel–titanium (NiTi) films at 490 °C. The first nuclei formed after 20 seconds at which the clock was started. Each image was taken after the onset of nucleation at (a) 9 seconds; (b) 21 seconds; (c) 26 seconds; (d) 49 seconds; (e) 62 seconds; (f) 84 seconds; (g) 99 seconds; (h) 121 seconds; and (i) 179 seconds. The illustration in (d) describes how the kinetic details such as the number of nuclei, the crystal size change, and transformed fraction were estimated. (j) The nucleation rate as a function of time exhibits a sigmoidal or S-shaped behavior and reaches its steady state within 25 seconds, within approximately 15% of the total 180 seconds of crystallization time. (k) A plot of the crystal size with time for different crystals showing a linear growth behavior, indicating the growth rate is constant over time. Source: Lee et al. [66]/AIP Publishing LLC.

precipitates [68]. This study paves a way to understand the nucleation and growth of precipitates in alloys systems.

Similarly, Leonard et al. have followed the kinetics of the precipitation of $Al_{11}(Cr,Mn)_2$ phase in Al cellular matrix with time at different temperatures using STEM images [69]. The activation energy of the precipitation process (277 kJ mol^{-1}), calculated from Awrami equation using the kinetics determined from the STEM images, is consistent with Cr diffusion in the Al matrix and was determined to be the rate-limiting process [69]. Other examples include, but not limited to, carbide formation and α-Fe recrystallization in twinned martensite as function of heating [70], the crystallization kinetics of HfO_2 films grown by atomic layer deposition (ALD) on SiO_2 passivated Si wafers and its relationship to leakage current [71], quantitative assessment of the precipitate phases in the Al–Mg–Si–Cu alloy system [72], multistep crystallization and precipitation during heating of gold based metallic glass ($Au_{49}Ag_{5.5}Pd_{2.3}Cu_{26.9}Si_{16.3}$ [at%]) [73], crystallization and growth of Si nanoparticles during heating of AlSi10Mg alloy [74], crystallization and phase transformation of superionic Na_3PS_4 solid-state battery material [75], crystallization mechanism of amorphous iron particles [76], initial stages of nucleation and growth from amorphous to crystalline transformation in electron beam sensitive $Ge_2Sb_2Te_5$ thin films [77], size-dependent crystallization temperature of metallic glass nanorods [78], formation of the δ'-ZrO phase during the annealing of thin foils of both pure Zr and a Zr–Sn–Nb–Mo alloy at 700 °C [79], metallic nanoparticle

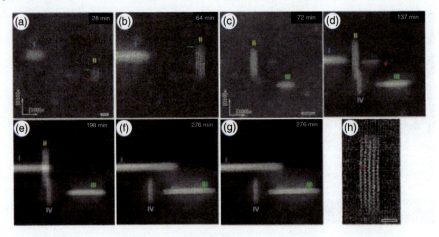

Figure 3.13 The growth of an initial-stage precipitate. (a, b) Images extracted from HR-STEM movie recorded after heating the quenched Al–Cu alloy at 160 °C for (a) 28 minutes and (b) 64 minutes. The blue arrow and red arrow point to the upper edge and lower edge of the precipitate, respectively. Scale bar is 2 nm. (c–g) The interaction of precipitates in close proximity to each other showing the features of precipitate morphology at various moments during continued aging time. Note that precipitate I grows further and faster after coming in contact with precipitates II and V, by absorbing them. Scale bars is 5 nm. (h) The atomic-scale ADF-STEM image of the precipitate in (b), scale bar is 2 nm. Source: Liu et al. [68]/from Springer Nature, CC BY 4.0.

formation and growth in Ag/TiO_2 and Au/TiO_2 thin films upon annealing at 400 °C and its relationship to the appearance of LSPR [80].

3.4.3 Sintering

Sintering of nanoparticles is commonly observed upon heating and is driven by the surface energy considerations, i.e. larger particles are more stable as they have lower surface energy. The sintering process can modify the materials properties in a desirable or an undesirable manner. For example, sintering of ceramic nanoparticles results in condensed material with improved mechanical properties. On the other hand, sintering of catalyst nanoparticles reduces the surface area and thereby catalytic efficacy. Therefore, understanding the sintering mechanism can help us to identify the conditions that promote or mitigate the sintering behavior. Sintering can proceed by coalescence, where two particles move near to each other and collapse into one or via Ostwald ripening where atomic-scale migration via surface diffusion or vapor transport results in growth of some particles at the cost of others [81]. The two can be easily distinguished during in-situ observations as particles movement and necking is observed during coalescence, while in the latter case, some particle reduce in size and disappear while others grow slowly [82]. A large number of studies revealing the sintering processes have been reported, and here a couple of examples are provided for you to get a flavor of types of experiments that are possible to perform using a heating holder.

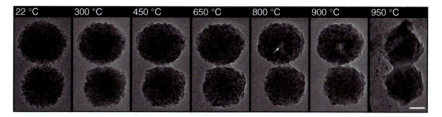

Figure 3.14 TEM images recorded at different temperatures during in-situ heating of two ThO$_2$ grains (50 nm in diameter) at low magnification (scale bar = 20 nm). The white arrow reported on the image recorded at $T = 800\,°C$ shows the formation of a pore in the grain. Source: Podor et al. [85]/from Royal Society of Chemistry.

Annealing metal hydroxide precipitates, obtained using wet chemical methods, is one way to synthesize oxide ceramic materials. Heat treatment of Mg(OH)$_2$ flakes to form MgO nanoparticles that sinter to form good-quality dense ceramic products was first observed using hot stage in a HVEM to understand the sintering process [83]. Majidi et al. have reported a quantitative analysis scheme for in-situ sintering using 3% yttria-stabilized zirconia during heating in TEM [84]. They monitored the size of both the pores and the particles to reveal different stages of sintering. They developed a MATLAB-based image processing tool to calculate the projected area of the agglomerate with and without internal pores and addressed the projection problem by obtaining low energy-loss analysis combined with STEM imaging. The shrinkage curves obtained through in-situ TEM analysis and modeling can be used to discern the fundamental mechanisms of sintering [84].

In principle, sintering of precipitate precursors can be considered a two-step process: first the nanocrystals form within individual precipitate or grain, coalesce to form a single crystal, and then the two crystals thus formed coalesce to form a larger crystal or attach to each other via a grain boundary. Ponder et al. applied in-situ heating to ThO$_2$ precipitates, synthesized by wet chemical methods, dried at 60 °C, and transferred to TEM heating chip. The diffraction contrast in the image recorded at 22 °C indicates that multiple small crystallites are present in an individual precipitate (Figure 3.14). These crystallites rearranged upon heating and coalescence to form a single crystal (Figure 3.14). Crystallites oriented along the same crystallographic directions attached easily to each other, and above 900 °C most of the voids and defects were eliminated. The rearrangement within the grain also impacted the contact between the grains and progressive evolution to form a neck. They also report that misorientation between two grains results in the formation of grain boundary [85]. Such advanced in-situ TEM observations can overcome the challenge of monitoring the self-assembly process and sintering during synthesis of ceramic powders.

Atomic-scale images of necking as first step of sintering have been commonly observed and reported for other systems, such as FePt magnetic nanoparticles [37], and for sub-nanometer-sized Ag rods [86]. Other examples include, but are not limited to, sintering of nano-sized ZrO$_2$ particles [87], silver and copper nanoparticles on (001) copper observed by in-situ ultrahigh vacuum transmission electron microscopy [88], low-temperature sintering behavior of UO_2 NPs formed by radiolysis [89].

3.4.4 Thermal Stability of Materials

Environmental conditions, such as temperature, of functioning devices may change during operation, resulting in device failure or reduced performance. Therefore, in-situ TEM observations have been widely employed to evaluate the materials stability as function of temperature at nanoscale. For example, stability of perovskite solar cells upon heating defines their operating temperature [90].

3.4.4.1 Alloys

Application of alloys ranges from kitchen utensils to automobile parts where temperature is routinely cycled between RT and hundreds of degrees. The performance and stability of alloys, such as steel or Al-based alloys, is directly controlled by the composition and morphology of their micro/nano structures. In-situ TEM observations as a function of temperature can be used to correlate the synthesis process with structural changes. The degradation process in alloys can proceed by solute segregation at grain boundaries [91], leaching out of elements in form of precipitates at grain boundaries, dislocation formation and motion, etc. TEM has been extensively used to characterize the composition and structure of individual grains and grain boundaries at nanoscale, which are in turn controlled by the synthesis conditions such as environment, subsequent heat treatment, cooling rate, and operating temperature.

The effect of Al content in formation of secondary phase in Al_xCoCrFeNi high entropy alloy, the phase stability, and deformation mechanisms of both single-phase and multi-phase alloys at elevated temperatures has been investigated [92]. Other examples are understanding the microstructural changes in a rapidly solidified Fe–Si–C alloy [93] and understanding the stress-relief mechanisms during melting of sub-micrometer Al–Si alloy particles [94]. In-situ heating has been also employed to understand the phase transition behavior during different processing treatments that can assist us in developing new alloys with higher stability and desired properties.

3.4.4.2 Core–Shell Structures

Thermal stability of core–shell structured nanoparticles is of vital importance for their practical applications at elevated temperature. Understanding the evolution of chemical distribution and the crystal structure of core–shell nanostructures with temperature at the nanoscale will open the route for finding applications and property enhancement routes through controlled design of new nanomaterials.

For example, core–shell non-stoichiometric Cu_5FeS_4 icosahedral nanoparticles have been investigated by in-situ heating in TEM [95]. EDS analysis of nanoparticles confirmed an Fe-rich core surrounded by Cu shell (Figure 3.15a) that was structurally and compositionally stable up to 250 °C (Figure 3.15b,c). However, the compositional difference between core and shell disappeared upon heating to 300 °C due to interdiffusion of Cu and Fe (Figure 3.15d,e) that persisted upon

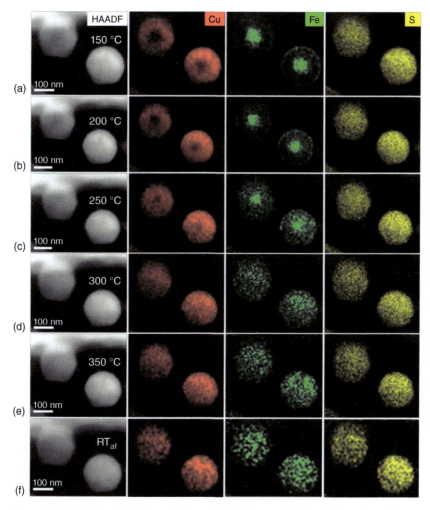

Figure 3.15 In-situ chemical evolution of Cu_5FeS_4 nanoparticles during thermal annealing. The groups of HAADF images and the corresponding EDS mappings for Cu_5FeS_4 nanoparticles annealed at (a) 150 °C, (b) 200 °C, (c) 250 °C, (d) 300 °C, (e) 350 °C, and (f) cooled to RT. Source: Zhang et al. [95]/from MDPI, CC BY 4.0.

cooling back to RT (Figure 3.15f). In contrast, structural difference between the core and shell, as determined by high-resolution images, remained intact even after heating to 350 °C, indicating a high structural stability. The inconsistency between chemical composition and crystal structure could be due to the intrinsic strain existing in icosahedral configuration and existence of various structures in this material system.

A combination of HAADF images and EDS maps was also used to evaluate the phase segregation, decomposition, and alloying behaviors of Au–Cu_2O core–shell nanoparticles as a process to generate extra fine structures in hybrid nano-objects [96]. Another example is the homogenization of nanoparticle with Pd-rich core and

Au-rich shell after heating above 550 °C as revealed by HAADF-STEM images [13]. On the other hand, thermal stability of Pd-Pt core–shell nanoparticles was found to depend on the shape (cubic versus octahedron) and composition [97]. The temperature for shape change and alloying temperatures were observed to be opposite to each other, i.e. while the octahedron shape was preserved until 900 °C, the core–shell structure was lost upon heating to 600 °C due to interdiffusion of Pt and Pd. In contrast, the faceted shape of cubes was lost at 500 °C, but the alloying did not start until 800 °C. The DFT calculations were used to rationalize the shape stability and alloying behavior [97]. These findings open new opportunities for revealing the thermal stability of core–shell nanostructures for various applications and are helpful for a controlled design of new core–shell nanostructures.

3.4.4.3 2-D Materials

Recently, 2-D materials have become popular due to their unique electronic, sensing, and catalytic properties. However, their applications often require high-temperature conditions making it important to understand and determine their thermal stability and degradation process. Although electron-beam-induced degradation is possible, careful measurements have been used to reveal the structural modifications with temperature. For example, Wang et al. have reported the thermal behavior of vertically aligned MoS_2 layers during in-situ heating up to 1000 °C [98]. The degradation was observed to start around 875 °C with formation of voids (Figure 3.16a) between domains of vertically aligned MoS_2 layers that continued to grow (Figure 3.16b–h). The onset of voids at multiple locations and their growth with increasing temperature are marked by the shaded areas in five different colors in Figure 3.16. The void formation could be due to (i) compression of vertically aligned layers in c direction, (ii) release of stress caused by lattice expansion due to heating, or (iii) electron-beam-induced loss of S. In-situ measurements of interplanar distances confirm an expansion of around 5% upon heating, ruling out the compression as possible factor for void formation. However, such a large vdW gap expansion will accumulate stress in tightly packed domains of vertically aligned 2D MoS_2. Therefore, the deformation started when the strain generation peaked at 875 °C and the domains had no way to release it. Loss of S was also detected during observation period that resulted in formation of Mo particles above 950 °C [98]. In conclusion, a combination of lattice expansion and loss of S resulted in the deformation of tightly packed domains of vertically aligned 2D MoS_2. This example provides technologically important method for materials design and optimization in their potential high-temperature applications.

3.4.5 Phase Transformation

Phase transformation with temperature is commonly observed phenomenon in a large set of materials and affects their mechanical, electronic, magnetic, and conducting properties, to name a few. Moreover, not only the phase transformation but also existence of transient and metastable phases may be different for nanomaterials than their bulk counterparts [99]. Therefore, understanding the local, even

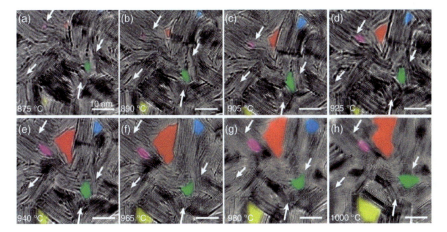

Figure 3.16 (a–h) With increasing temperature, the voids denoted by the shaded areas of red, blue, green, yellow, and magenta start to form at ≈ 875 °C and continuously expand up to 1000 °C. The white arrows mark the evolution of amorphous regions due to loss of S from the lattice. Source: Wang et al. [98]/from American Chemical Society.

atomistic, phase transformation mechanisms and understanding of the kinetics are invaluable in identifying nanoscale features of transformations that may play a key role in the future design of nanomaterials with desired properties. In-situ heating has been successfully employed to decipher atomic-scale structural transformation mechanisms controlling properties such as superconductivity and ferromagnetism [99]. For example, high-temperature tetragonal phase of HfO_2 has high dielectric constant and large bandgap, but it is unstable at lower temperatures. However, in-situ STEM images by Hudak et al. show that the transformation temperature for monoclinic to tetragonal phase can be lowered to 600 °C for polycrystalline nanorods with nm scale crystallites, instead of over 1000 °C for bulk [100].

Bourgeois et al. have followed the evolution of diffuse scattering and satellite diffraction spots in electron diffraction patterns upon heating to elucidate the structural modulations in polycrystalline $LuFe_2O_4$ due to oxygen charge order [101]. This study is good example of exploiting electron diffraction to follow complicated structural transformations in a single crystal upon heating to understand the coexistence and competition among the multitype ordered states in a system to establish structure–property relationships.

Yu et al. have employed low magnification BF TEM and HAADF imaging to follow the formation and growth of face-centered cubic (FCC)-Ti platelets from RT hexagonal closed packed (HCP) Ti matrix, their stability and mechanical properties [102], upon heating. It is possible that formation of FCC structure is responsible for the high level of plasticity observed in Ti films. They prepared electron-transparent regions from 100 μm thick Ti film using a twin-jet polisher for the TEM heating experiments. BF TEM images, recorded at 600 °C, show the presence of dark lines that could be due to dislocations (Figure 3.17a). Dark spots start to appear and their density increased during next few seconds of observation (Figure 3.17b,c). Their size grew to form platelets, but the number decreased after

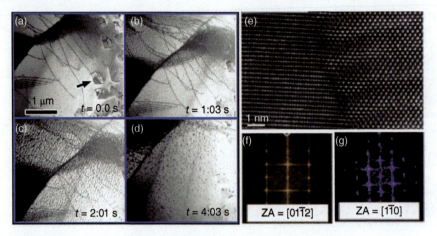

Figure 3.17 (a–d) Time-resolved BF TEM images recorded at 600 °C during the in-situ TEM heating experiment showing the precipitation and growth of FCC phase (dark spots) in the HCP matrix. (e) HAADF-STEM image of HCP and FCC Ti interface. FFT patterns of (f) HCP and (g) FCC phases. Source: Yu et al. [102]/from Elsevier.

4.03 seconds of heating, and the dark lines also disappeared (Figure 3.17d). The growth could be classified as Ostwald ripening, but detailed analysis was not performed to unequivocally determine the process. These FCC platelets remained unchanged upon cooling. The high-resolution HAADF image (Figure 3.17e) at the boundary between the platelet and matrix and their FFTs (Figure 3.17f,g) confirmed the structures to belong to Ti FCC and Ti HCP phases, respectively. The FCC platelets thus formed within the HCP matrix phase were also found to be stable under mechanical deformation, and in-situ nanomechanical testing revealed similar mechanical deformation behavior as the parent HCP phase.

Some of other representative examples include, but are not limited to, the phase transition from $H_2Ti_3O_7$ to TiO_2 (B) and orientation relationship between the two phases in a 1D single nanocrystal [103], phase transformation dynamics and temperature for the tetragonal to the cubic structures in $K_{0.5}Bi_{0.5}TiO_3$ perovskite ceramics [104], phase transformation nucleation/crystallization of Ta thin film upon heating using electron diffraction; BCC to FCC phase transformation in thin Ta films [105], crystalline to liquid phase transformation of Al nanoparticles upon melting using EELS [106], multiple-step martensitic transformations in Ni-rich NiTi shape memory alloys [107], mechanism of thermal reduction of Co_3O_4 nanoparticles to CoO [108], effect of heating rate on phase transformation, and disintegration of atomically thin MoS_2 layers [109].

3.4.6 Materials Synthesis

Synthesis is the fundamental step to obtain materials with novel properties. Although TEM is not ideally suited for this purpose, it has potential to reveal atomic-scale mechanisms and metastable phases that have potential for novel applications. Several materials are synthesized via solid-state reactions at high temperatures via diffusion of one element into other or bilateral diffusion.

Moreover, high-temperature diffusion in epitaxial layers can limit their operational temperature. In-situ TEM experiments deliver an atomic-scale identification of the diffusing specie, diffusion rate, and/or mechanism that can provide a better control on the synthesis and/or operating conditions. Fabrication of nanowires or functional materials is one such example as their size (diameter and length) and composition controls their electronic and sensing properties.

Boston et al. have used in-situ heating in a TEM to follow the growth conditions and mechanism of quaternary, Y_2BaCuO_5 (Y211) nanowires, which follows a micro-crucible growth mechanism [110]. The starting material for TEM observations was prepared by mixing stoichiometric yttrium, barium and copper nitrates with sodium alginate in deionized water and subjected to rigorous heat treatment. In-situ TEM images show that $BaCO_3$ nanoparticles melted and nanowires started to grow out of the surface within one minute and continue to grow with time. Detailed phase analysis of the quenched sample using lattice resolution images and EDS analysis show a Y211 nanowire (red) extruding from amorphous mixture of Y, Cu, and Ba (blue) as well as presence of CuO (yellow) and $BaCO_3$ (green) particles. Video recording revealed that $BaCO_3$ nanoparticles diffused through a porous matrix containing copper and yttrium oxides to the surface that act as a catalyst for the Y211 nanowire growth. Their video recording suggested that sites on the rough surface of the porous matrix act as microcrucible to drive metal oxide nanowire growth at 500 °C. It is proposed that $BaCO_3$ particles melt at 450 °C and dissolve Y and Cu cations while diffusing through the porous matrix. The faceted end and uniform diameter of the nanowires grown by microcrucible mechanism confirm them to be similar to nanowires grown by other methods and should have the same physical properties [110].

Solid-state diffusion as way to tune the stoichiometry of platinum silicide (Pt_xSi) thin films has also been reported [111]. In-situ TEM measurement during heating shows that the precise control on the thickness of the precursor film, temperature, and the growth period produced metal silicide thin films with selective stoichiometry (Pt_3Si, Pt_2Si, and $PtSi$). Therefore, stochiometric films of other metal silicides (Ti, Mo, Cr, Ni, W, Co) can be synthesized by changing the precursor film thickness (limiting the solid-state diffusion) and the growth kinetics (temperature and time). Other examples include formation of onion-like structures during annealing of WS_2 and MoS_2 films [112], structure evolution during the decomposition of $Cu(OH)_2$ to form Cu nanowires ($Cu(OH)_2 \rightarrow CuO \rightarrow Cu_2O \rightarrow Cu$) [113], mechanism of heat-induced fusion of silver nanowires [114], thermal stability of 4H-gold nanowires [115].

3.5 Limitations and Possibilities

Despite the successful applications provided in this chapter, there are limitations above and beyond the experimental design considerations described in Section 3.3 as described below:

> Forget Icarus, Fly as Close To The Sun As You Want!
> – Kyle Hill

Temperature limit: 1500 °C is currently the maximum achievable temperature for special MEMS-based heating chips and heating wire holder [9, 15] that is below the temperature needed to observe reactions in ceramic materials.

EDS mapping: In-situ EDS heating experiments are limited due to detector count saturation caused by infrared photon emission at high temperature. Mele et al. have reported that the design of NanoEx™-i/v allows EDS elemental maps to be acquired below 700 °C due to low infrared photon emission from the small heater volume [15].

Possible solutions: Solutions to existing problems and limitations are continuously being explored. For example, beam-induced changes in FePt nanoparticles during high-temperature annealing were reduced by periodic electron beam blanking and comparing the in-situ results with those from ex-situ experiments to verify the reliability of the in-situ data [37]. Kooi et al. evaluated the effect of electron beam on the crystallization of amorphous $Ge_2Sb_2Te_5$ films by varying the electron density and operating voltage of the TEM to find suitable conditions to mitigate the electron beam effects [36].

New MEMS-based heating chips are also being explored, for example, the temperature limit has increased from 1000 to 1500 °C since their inception [13, 15]. Heating chips have also been modified to make them compatible with EDS mapping [15, 116].

3.6 Chapter Summary

In-situ heating experiments using TEM-based techniques provide an indispensable opportunity to understand the reaction mechanisms, reveal the existence of metastable phases, measure the kinetics that can lead to design target materials with specific properties.

Both furnace and MEMS-based holders are commercially available and could be further modified if needed.

However, electron beam effects, sample preparation, stability of the sample, reactivity of the sample with the TEM grid and/or support, achievable temperature, and thermal drift should be evaluated for experimental setup.

References

1 Butler, E.P. and Hale, K.F. (1981). Dynamic experiments in the electron microscope. In: *Practical Methods in Microscopy*, vol. 9 (ed. A.M. Glauert), 109–182. Amsterdam/New York/Oxford: North Holland Publishing Company.

2 Whelan, M.J. (1958). A high temperature stage for the Elmiskop 1. *4th International Congress of Electron Microscopy*, Berlin, Germany (10–17 September 1958). Berlin, Heidelberg: Springer.

3 Fisher, R.M., Swann, P.R., and Nutting, J. (1960). A new objective pole-piece and specimen heating stage for the Elmiskop. *2nd European Regional Conference Electron Microscopy*, The Dutch Society for Electron Microscopy, Delft, Netherlands (6–9 July 1960).

4 Hedley, J.A. and McGeagh, J. (1963). Specimen heating with temperature measurement from −150 °C to 2200 °C inside the EM6 electron microscope. *Journal of Scientific Instruments* 40: 484.

5 Swann, P.R. (1972). Side-entry single tilt speciment holders for heating and stress corrosion cracking of electron microscope specimens. *Proceedings of the fifth European Regional Conference of Electron Microscopy*, Manchester, England (5–12 September 1972). Academic Press.

6 Valdre, U. (1965). A double-tilting heating stage for an electron microscope. *Journal of Scientific Instruments* 42: 853.

7 Swann, P.R. (1978). Specimen devices for in situ experiments. *Ninth International Congress of Electron Microscopy*, Toronto, Canada (1–9 August 1978). Toronto: Microscopical Society of Canada.

8 Butler, E.P. and Hale, K.F. (1981). High temperature microscopy. In: *Dynamic Experiments in the Electron Microscope* (ed. A.M. Glauert), 1–457. Amsterdam/New York/Oxford: North Holland Publishing Company.

9 Kamino, T. and Saka, H. (1993). Newly developed high resolution hot stage and its applications to materials science. *Microscopy Microanalysis Microstructures* 4: 127–135.

10 Kamino, T. and Saka, H. (1997). HREM in situ experiment at very high temperature. In: *In-Situ Mircroscopy in Materials Research* (ed. P.L. Gai), 173–200. Boston/Dordrecht/London: Kluwer Academic Publishers.

11 Tanigaki, T., Ito, K., Nagakubo, Y. et al. (2009). An in situ heating TEM analysis method for an interface reaction. *Journal of Electron Microscopy* 58 (5): 281–287.

12 Zhang, M., Olson, E.A., Twesten, R.D. et al. (2005). In situ transmission electron microscopy studies enabled by microelectromechanical system technology. *Journal of Materials Research* 20 (7): 1802–1807.

13 Allard, L.F., Bigelow, W.C., Jose-Yacaman, M. et al. (2009). A new MEMS-based system for ultra-high-resolution imaging at elevated temperatures. *Microscopy Research and Technique* 72 (3): 208–215.

14 Creemer, J.F., Helveg, S., Kooyman, P.J. et al. (2010). A MEMS reactor for atomic-scale microscopy of nanomaterials under industrially relevant conditions. *Journal of Microelectromechanical Systems* 19 (2): 254–264.

15 Mele, L., Konings, S., Dona, P. et al. (2016). A MEMS-based heating holder for the direct imaging of simultaneous in-situ heating and biasing experiments in scanning/transmission electron microscopes. *Microscopy Research and Technique* 79 (4): 239–250.

16 van Omme, J.T., Zakhozheva, M., Spruit, R.G. et al. (2018). Advanced microheater for in situ transmission electron microscopy; enabling unexplored analytical studies and extreme spatial stability. *Ultramicroscopy* 192: 14–20.

17 Spruit, R.G., Omme, J.T.V., Ghatkesar, M.K. et al. (2017). A review on development and optimization of microheaters for high-temperature in situ studies. *Journal of Microelectromechanical Systems* 26 (6): 1165–1182.

18 Harris, C.T., Martinez, J.A., Shaner, E.A. et al. (2011). Fabrication of a nanostructure thermal property measurement platform. *Nanotechnology* 22 (27): 275308.

19 Hwang, W.-J., Shin, K.-S., Roh, J.-H. et al. (2011). Development of micro-heaters with optimized temperature compensation design for gas sensors. *Sensors* 11 (3): 2580–2591.

20 Verheijen, M.A., JJTM, D., JFP, T. et al. (2004). Transmission electron microscopy specimen holder for simultaneous in situ heating and electrical resistance measurements. *Review of Scientific Instruments* 75 (2): 426–429.

21 Picher, M., Mazzucco, S., Blankenship, S., and Sharma, R. (2015). Vibrational and optical spectroscopies integrated with environmental transmission electron microscopy. *Ultramicroscopy* 150: 10–15.

22 Grapes, M.D., LaGrange, T., Friedman, L.H. et al. (2014). Combining nanocalorimetry and dynamic transmission electron microscopy for in situ characterization of materials processes under rapid heating and cooling. *Review of Scientific Instruments* 85 (8): 084902.

23 Zhang, Z.L. and Su, D.S. (2009). Behaviour of TEM metal grids during in-situ heating experiments. *Ultramicroscopy* 109 (6): 766–774.

24 Sharma, R., Chee, S.W., Herzing, A. et al. (2011). Evaluation of the role of au in improving catalytic activity of Ni nanoparticles for the formation of one-dimensional carbon nanostructures. *Nano Letters* 11 (6): 2464–2471.

25 Egerton, R.F. (2013). Control of radiation damage in the TEM. *Ultramicroscopy* 127: 100–108.

26 Russo, C.J. and Egerton, R.F. (2019). Damage in electron cryomicroscopy: lessons from biology for materials science. *MRS Bulletin* 44 (12): 935–941.

27 Lian, J., Wang, L.M., Sun, K., and Ewing, R.C. (2009). In situ TEM of radiation effects in complex ceramics. *Microscopy Research and Technique* 72 (3): 165–181.

28 Watanabe, M., Someya, T., and Nagahama, Y. (1970). Temperature rise of specimen due to electron irradiation. *Journal of Physics D: Applied Physics* 3 (10): 1461–1468.

29 Robison, W., Sharma, R., and Eyring, L. (1991). Observation of gold silicon alloy formation in thin-films by high-resolution electron-microscopy. *Acta Metallurgica et Materialia* 39 (2): 179–186.

30 Hinode, H., Sharma, R., and Eyring, L. (1990). A study of the decomposition of neodymium hydroxy carbonate and neodymium carbonate hydrate. *Journal of Solid State Chemistry* 84 (1): 102–117.

31 Kang, Z.C. and Eyring, L. (1987). Surface processes observed by high-resolution electron microscopy: beam-induced transformation and reduction in a modified β-PbO_2 crystal. *Ultramicroscopy* 23 (3): 275–281.

32 Ko, D.-H. and Sinclair, R. (1994). In-situ dynamic high-resolution transmission electron microscopy: application to Pt/GaAs interfacial reactions. *Ultramicroscopy* 54 (2): 166–178.

33 Takeda, S. and Yoshida, H. (2013). Atomic-resolution environmental TEM for quantitative in-situ microscopy in materials science. *Microscopy* 62 (1): 193–203.

34 Sinclair, R., Ponce, F.A., Yamashita, T. et al. (1982). Dynamic observation of defect annealing in CdTe at lattice resolution. *Nature* 298 (5870): 127–131.

35 Sinclair, R. and Parker, M.A. (1986). High-resolution transmission electron microscopy of silicon re-growth at controlled elevated temperatures. *Nature* 322 (6079): 531–533.

36 Kooi, B.J., Groot, W.M.G., and Hosson, J.T.M.D. (2004). In situ transmission electron microscopy study of the crystallization of $Ge_2Sb_2Te_5$. *Journal of Applied Physics* 95 (3): 924–932.

37 Wittig, J.E., Bentley, J., and Allard, L.F. (2017). In situ investigation of ordering phase transformations in FePt magnetic nanoparticles. *Ultramicroscopy* 176: 218–232.

38 Vijayan, S., Wang, R., Kong, Z., and Jinschek, J.R. (2022). Quantification of extreme thermal gradients during in situ transmission electron microscope heating experiments. *Microscopy Research and Technique* 85 (4): 1527–1537.

39 Thornburg, D.D. and Wayman, C.M. (1973). Specimen temperature increases during transmission electron microscopy. *Physica Status Solidi (A)* 15 (2): 449–453.

40 Gaulandris, F., Simonsen, S.B., Wagner, J.B. et al. (2020). Methods for calibration of specimen temperature during in situ transmission electron microscopy experiments. *Microscopy and Microanalysis* 26 (1): 3–17.

41 Kim, T.-H., Bae, J.-H., Lee, J.-W. et al. (2015). Temperature calibration of a specimen-heating holder for transmission electron microscopy. *Applied Microscopy* 45 (2): 95–100.

42 Vijayan, S. and Aindow, M. (2019). Temperature calibration of TEM specimen heating holders by isothermal sublimation of silver nanocubes. *Ultramicroscopy* 196: 142–153.

43 Winterstein, J.P., Lin, P.A., and Sharma, R. (2015). Temperature calibration for in situ environmental transmission electron microscopy experiments. *Microscopy and Microanalysis* 21 (6): 1622–1628.

44 Niekiel, F., Kraschewski, S.M., Müller, J. et al. (2017). Local temperature measurement in TEM by parallel beam electron diffraction. *Ultramicroscopy* 176: 161–169.

45 Wehmeyer, G., Bustillo, K.C., Minor, A.M., and Dames, C. (2018). Measuring temperature-dependent thermal diffuse scattering using scanning transmission electron microscopy. *Applied Physics Letters* 113 (25): 253101.

46 Mecklenburg, M., Hubbard, W.A., White, E. et al. (2015). Nanoscale temperature mapping in operating microelectronic devices. *Science* 347 (6222): 629–632.

47 Miller, B., Pérez Garza, H.H., and Mecklenburg, M. (2018). Local in situ temperature measurements from aluminum nanoparticles. *Microscopy and Microanalysis* 24: 1924–1925.

48 Wang, C., W, CD, Y., Bruma, A. et al. (2020). Endothermic reaction at room temperature enabled by deep-ultraviolet plasmons. *Nature Materials* 20: 346–352.

49 Hu, X., Yasaei, P., Jokisaari, J. et al. (2018). Mapping thermal expansion coefficients in freestanding 2D materials at the nanometer scale. *Physical Review Letters* 120 (5): 055902.

50 Hage, F.S., Nicholls, R.J., Yates, J.R. et al. (2018). Nanoscale momentum-resolved vibrational spectroscopy. *Science Advances* 4 (6): eaar7495.

51 Yan, X., Liu, C., Gadre, C.A. et al. (2019). Unexpected strong thermally induced phonon energy shift for mapping local temperature. *Nano Letters* 19 (10): 7494–7502.

52 Yu, S., Wu, Y., Wang, S. et al. (2019). Thermodynamic analysis of a MEMS based differential scanning calorimeter model. *Sensors and Actuators A: Physical* 291: 150–155.

53 Pérez Garza, H.H., Morsink, D., Xu, J. et al. (2017). MEMS-based nanoreactor for in situ analysis of solid–gas interactions inside the transmission electron microscope. *Micro & Nano Letters* 12 (2): 69–75.

54 Lee, H., OFN, O., Kim, G.Y. et al. (2021). TEM sample preparation using micro-manipulator for in-situ MEMS experiment. *Applied Microscopy* 51 (1): 8.

55 Vijayan, S., Jinschek, J.R., Kujawa, S. et al. (2017). Focused ion beam preparation of specimens for micro-electro-mechanical system-based transmission electron microscopy heating experiments. *Microscopy and Microanalysis* 23 (4): 708–716.

56 Duchamp, M., Xu, Q., and Dunin-Borkowski, R.E. (2014). Convenient preparation of high-quality specimens for annealing experiments in the transmission electron microscope. *Microscopy and Microanalysis* 20 (6): 1638–1645.

57 Smith, D.J., Petford-Long, A.K., Wallenberg, L.R., and Bovin, J.-O. (1986). Dynamic atomic-level rearrangements in small gold particles. *Science* 233 (4766): 872–875.

58 Balk, T.J., Dehm, G., and Arzt, E. (2003). Parallel glide: unexpected dislocation motion parallel to the substrate in ultrathin copper films. *Acta Materialia* 51 (15): 4471–4485.

59 Shrestha, S.L., Xie, K.Y., Ringer, S.P. et al. (2013). The effect of clustering on the mobility of dislocations during aging in Nb-microalloyed strip cast steels: in situ heating TEM observations. *Scripta Materialia* 69 (6): 481–484.

60 Gong, C., He, K., Chen, Q. et al. (2016). In situ high temperature atomic level studies of large closed grain boundary loops in graphene. *ACS Nano* 10 (10): 9165–9173.

61 Ma, K., Hu, T., Yang, H. et al. (2016). Coupling of dislocations and precipitates: impact on the mechanical behavior of ultrafine grained Al–Zn–Mg alloys. *Acta Materialia* 103: 153–164.

62 Arakawa, K., Amino, T., and Mori, H. (2011). Direct observation of the coalescence process between nanoscale dislocation loops with different Burgers vectors. *Acta Materialia* 59 (1): 141–145.

63 Inkson, B.J., Dehm, G., and Wagner, T. (2002). In situ TEM observation of dislocation motion in thermally strained Al nanowires. *Acta Materialia* 50 (20): 5033–5047.

64 Zhou, T., Babu, R.P., Hou, Z., and Hedström, P. (2022). On the role of transmission electron microscopy for precipitation analysis in metallic materials. *Critical Reviews in Solid State and Materials Sciences* 47 (3): 388–414.

65 Bagmut, A.G. and Bagmut, I.A. (2020). Kinetics of electron beam crystallization of amorphous films of Yb_2O_2S. *Journal of Non-Crystalline Solids* 547: 120286.

66 Lee, H.-J., Ni, H., Wu, D.T., and Ramirez, A.G. (2005). Experimental determination of kinetic parameters for crystallizing amorphous NiTi thin films. *Applied Physics Letters* 87 (11): 114102.

67 Ramirez, A.G., Ni, H., and Lee, H.-J. (2006). Crystallization of amorphous sputtered NiTi thin films. *Materials Science and Engineering: A* 438–440: 703–709.

68 Liu, C., Malladi, S.K., Xu, Q. et al. (2017). In-situ STEM imaging of growth and phase change of individual $CuAl_x$ precipitates in Al alloy. *Scientific Reports* 7 (1): 2184.

69 Leonard, H.R., Rommel, S., Li, M.X. et al. (2021). Precipitation phenomena in a powder-processed quasicrystal-reinforced Al–Cr–Mn–Co–Zr alloy. *Materials Characterization* 178: 111239.

70 Liu, X., Man, T.H., Yin, J. et al. (2018). In situ heating TEM observations on carbide formation and α-Fe recrystallization in twinned martensite. *Scientific Reports* 8 (1): 14454.

71 Kim, H., Marshall, A., PC, M.I., and Saraswat, K.C. (2004). Crystallization kinetics and microstructure-dependent leakage current behavior of ultrathin HfO_2 dielectrics: in situ annealing studies. *Applied Physics Letters* 84 (12): 2064–2066.

72 Sunde, J.K., Wenner, S., and Holmestad, R. (2020). In situ heating TEM observations of evolving nanoscale Al–Mg–Si–Cu precipitates. *Journal of Microscopy* 279 (3): 143–147.

73 Ivanov, Y.P., Meylan, C.M., Panagiotopoulos, N.T. et al. (2020). In-situ TEM study of the crystallization sequence in a gold-based metallic glass. *Acta Materialia* 196: 52–60.

74 Albu, M., Meylan, C.M., Panagiotopoulos, N.T. et al. (2020). Microstructure evolution during in-situ heating of AlSi10Mg alloy powders and additive manufactured parts. *Additive Manufacturing* 36: 101605.

75 Nakajima, H., Tsukasaki, H., Ding, J. et al. (2021). Crystallization behaviors in superionic conductor Na_3PS_4. *Journal of Power Sources* 511: 230444.

76 Falqui, A., Loche, D., and Casu, A. (2020). In situ TEM crystallization of amorphous iron particles. *Crystals* 10 (1): 41.

77 Singh, M., Ghosh, C., Miller, B., and Carter, C.B. (2021). Direct visualization of the earliest stages of crystallization. *Microscopy and Microanalysis* 27 (4): 659–665.

78 Sohn, S., Jung, Y., Xie, Y. et al. (2015). Nanoscale size effects in crystallization of metallic glass nanorods. *Nature Communications* 6 (1): 8157.

79 Yu, H., Yao, Z., Long, F. et al. (2017). In situ transmission electron microscopy study of the thermally induced formation of δ'-ZrO in pure Zr and Zr-based alloy. *Journal of Applied Crystallography* 50 (4): 1028–1035.

80 Costa, D., Rodrigues, M.S., Roiban, L. et al. (2021). In-situ annealing transmission electron microscopy of plasmonic thin films composed of bimetallic Au–Ag nanoparticles dispersed in a TiO_2 matrix. *Vacuum* 193: 110511.

81 Ross, F.M., Tersoff, J., and Tromp, R.M. (1998). Ostwald ripening of self-assembled germanium islands on silicon(100). *Microscopy and Microanalysis* 4 (3): 254–263.

82 Walsh, M.J., Yoshida, K., Gai, P.L., and Boyes, E.D. (2010). In-situ heating studies of gold nanoparticles in an aberration corrected transmission electron microscope. *Journal of Physics: Conference Series* 241: 012058.

83 Thangaraj, N., Westmacott, K.H., and Dahmen, U. (1991). HVEM studies of the sintering of MgO nanocrystals prepared by $Mg(OH)_2$ decomposition. *Ultramicroscopy* 37 (1): 362–374.

84 Majidi, H., Holland, T., and van Benthem, K. (2015). Quantitative analysis for in situ sintering of 3% yttria-stablized zirconia in the transmission electron microscope. *Ultramicroscopy* 152: 35–43.

85 Podor, R., Trillaud, V., Nkou Bouala, G.I. et al. (2021). A multiscale in situ high temperature high resolution transmission electron microscopy study of ThO_2 sintering. *Nanoscale* 13 (15): 7362–7374.

86 Dong, M., Wang, W., Wei, W. et al. (2019). Understanding the ensemble of growth behaviors of sub-10-nm silver nanorods using in situ liquid cell transmission electron microscopy. *The Journal of Physical Chemistry C* 123 (34): 21257–21264.

87 Rankin, J. and Sheldon, B.W. (1995). In situ TEM sintering of nano-sized ZrO_2 particles. *Materials Science and Engineering: A* 204 (1): 48–53.

88 Yeadon, M., Yang, J.C., Averback, R.S. et al. (1998). Sintering of silver and copper nanoparticles on (001) copper observed by in-situ ultrahigh vacuum transmission electron microscopy. *Nanostructured Materials* 10 (5): 731–739.

89 Nenoff, T.M., Jacobs, B.W., Robinson, D.B. et al. (2011). Synthesis and low temperature in situ sintering of uranium oxide nanoparticles. *Chemistry of Materials* 23 (23): 5185–5190.

90 Divitini, G., Cacovich, S., Matteocci, F. et al. (2016). In situ observation of heat-induced degradation of perovskite solar cells. *Nature Energy* 1 (2): 15012.

91 Rokni, M.R., Widener, C.A., and Champagne, V.R. (2014). Microstructural stability of ultrafine grained cold sprayed 6061 aluminum alloy. *Applied Surface Science* 290: 482–489.

92 Rao, J.C., Diao, H.Y., Ocelík, V. et al. (2017). Secondary phases in Al_xCoCrFeNi high-entropy alloys: an in-situ TEM heating study and thermodynamic appraisal. *Acta Materialia* 131: 206–220.

93 Chen, Y.C., Chen, C.M., Su, K.C., and Yang, K. (1991). Microstructural investigation of a rapidly solidified Fe–Si–C alloy. *Materials Science and Engineering: A* 133: 596–600.

94 Storaska, G.A. and Howe, J.M. (2004). In-situ transmission electron microscopy investigation of surface-oxide, stress-relief mechanisms during melting of sub-micrometer Al–Si alloy particles. *Materials Science and Engineering: A* 368 (1): 183–190.

95 Zhang, B., Zhao, X., Dong, T. et al. (2020). Structural core–shell beyond chemical homogeneity in non-stoichiometric Cu_5FeS_4 nano-icosahedrons: an in situ heating TEM study. *Nanomaterials* 10 (1): 4.

96 Li, D., Wang, Z.L., and Wang, Z. (2017). Phase separation prior to alloying observed in vacuum heating of hybrid Au/Cu_2O core–shell nanoparticles. *The Journal of Physical Chemistry C* 121 (2): 1387–1392.

97 Vara, M., Roling, L.T., Wang, X. et al. (2017). Understanding the thermal stability of palladium–platinum core–shell nanocrystals by in situ transmission electron microscopy and density functional theory. *ACS Nano* 11 (5): 4571–4581.

98 Wang, M., Kim, J.H., Han, S.S. et al. (2019). Structural evolutions of vertically aligned two-dimensional MoS_2 layers revealed by in situ heating transmission electron microscopy. *The Journal of Physical Chemistry C* 123 (45): 27843–27853.

99 Yu, L., Hudak, B.M., Ullah, A. et al. (2020). Unveiling the microscopic origins of phase transformations: an in situ TEM perspective. *Chemistry of Materials* 32 (2): 639–650.

100 Hudak, B.M., Depner, S.W., Waetzig, G.R. et al. (2017). Real-time atomistic observation of structural phase transformations in individual hafnia nanorods. *Nature Communications* 8 (1): 15316.

101 Bourgeois, J., Hervieu, M., Poienar, M. et al. (2012). Evidence of oxygen-dependent modulation in $LuFe_2O_4$. *Physical Review B* 85 (6): 064102.

102 Yu, Q., Kacher, J., Gammer, C. et al. (2017). In situ TEM observation of FCC Ti formation at elevated temperatures. *Scripta Materialia* 140: 9–12.

103 Lei, Y., Sun, J., Liu, H. et al. (2014). Atomic mechanism of predictable phase transition in dual-phase $H_2Ti_3O_7/TiO_2$ (B) nanofiber: an in situ heating TEM investigation. *Chemistry – A European Journal* 20 (36): 11313–11317.

104 Otoničar, M., Škapin, S.D., Jančar, B. et al. (2010). Analysis of the phase transition and the domain structure in $K_{0.5}Bi_{0.5}TiO_3$ perovskite ceramics by in situ XRD and TEM. *Journal of the American Ceramic Society* 93 (12): 4168–4173.

105 Janish, M.T., Mook, W.M., and Carter, C.B. (2015). Nucleation of fcc Ta when heating thin films. *Scripta Materialia* 96: 21–24.

106 Palanisamy, P., Md, J., Asta, M., and Howe, J.M. (2015). Examination of the electronic structure of crystalline and liquid Al versus temperature by in situ electron energy-loss spectroscopy (EELS). *Micron* 76: 14–18.

107 Dlouhy, A., Khalil-Allafi, J., and Eggeler, G. (2003). Multiple-step martensitic transformations in Ni-rich NiTi alloys – an in-situ transmission electron microscopy investigation. *Philosophical Magazine* 83 (3): 339–363.

108 Chen, X., van Gog, H., and van Huis, M.A. (2021). Transformation of Co_3O_4 nanoparticles to CoO monitored by in situ TEM and predicted ferromagnetism at the Co_3O_4/CoO interface from first principles. *Journal of Materials Chemistry C* 9 (17): 5662–5675.

109 Kumar, P., Horwath, J.P., Foucher, A.C. et al. (2020). Direct visualization of out-of-equilibrium structural transformations in atomically thin chalcogenides. *npj 2D Materials and Applications* 4 (1): 16.

110 Boston, R., Schnepp, Z., Nemoto, Y. et al. (2014). In situ TEM observation of a microcrucible mechanism of nanowire growth. *Science (AAAS)* 344 (6184): 623–626.

111 Streller, F., Agarwal, R., Mangolini, F., and Carpick, R.W. (2015). Novel metal silicide thin films by design via controlled solid-state diffusion. *Chemistry of Materials* 27 (12): 4247–4253.

112 Zink, N., Therese, H.A., Pansiot, J. et al. (2008). In situ heating TEM study of onion-like WS_2 and MoS_2 nanostructures obtained via MOCVD. *Chemistry of Materials* 20 (1): 65–71.

113 Wang, Z.L., Kong, X.Y., Wen, X., and Yang, S. (2003). In situ structure evolution from $Cu(OH)_2$ nanobelts to copper nanowires. *The Journal of Physical Chemistry B* 107 (33): 8275–8280.

114 Kim, C.-L., Lee, J.-Y., Shin, D.-G. et al. (2020). Mechanism of heat-induced fusion of silver nanowires. *Scientific Reports* 10 (1): 9271.

115 Wang, Q., Zhao, Z.L., Cai, C. et al. (2019). Ultra-stable 4H-gold nanowires up to 800 °C in a vacuum. *Journal of Materials Chemistry A* 7 (41): 23812–23817.

116 Zaluzec, N.J., Burke, M.G., Haigh, S.J., and Kulzick, M.A. (2014). X-ray energy-dispersive spectrometry during in situ liquid cell studies using an analytical electron microscope. *Microscopy and Microanalysis* 20 (2): 323–329.

4

In-Situ Cryo-TEM

> *To appreciate the beauty of a snowflake it is necessary to stand out in the cold.*
> – Aristotle

Visualization of materials at low or cryogenic temperatures has evolved from being an indispensable technique for probing the electron beam–sensitive materials to several other material systems. There are two ways to observe samples at cryogenic temperatures: using a dedicated cryo-TEM with a cryo holder or by using a cryo sample holder in a regular TEM. Former has been routinely employed by the biologist as a dedicated cryo-TEM is equipped with a thick copper shield around the sample that is maintained at cryogenic temperature using an autofill system from a liquid N_2 reservoir, thus the internal environment surrounding sample is kept at cryogenic temperature. Also, it is compatible with a cartridge to load multiple samples at one time, which helps to maintain low temperature for long periods. Cryo techniques have provided a way to image biological samples in their native environment without staining. The samples are prepared by plunge freezing to create a film of vitreous ice that keeps the samples hydrated and blocks the electron radiation damage. The profound success of employing cryo-TEM for imaging biological structures and fragile biomolecules lead to the 2017 Nobel Prize to Dubochet, Frank, and Henderson in Chemistry. On the other hand, a TEM specimen holder with a liquid N_2 or helium reservoir attached on the end, which is outside the TEM column, is often employed by physical scientists. Although low-temperature TEM has also been used in the field of physical science for long time, it has gained new enthusiasm since 2017, and its applications for in-situ TEM is a growing field [1, 2]. Here we provide the history, current status, and performing in-situ experiments using cryo-TEM holders in chemical, physical, and materials sciences field with relevant examples. The value of cryo-TEM for materials research has been realized by funding agencies, and a recent DOE-BES workshop report is now available (https://science.osti.gov/-/media/bes/pdf/reports/2021/CryoEM_RT_Report.pdf).

In-Situ Transmission Electron Microscopy Experiments: Design and Practice, First Edition. Renu Sharma.
© 2023 WILEY-VCH GmbH. Published 2023 by WILEY-VCH GmbH.

4.1 Historical Perspective

Historically, the motivation to observe samples at low temperature has been to (i) preserve the biological samples in their native form by freezing them and (ii) to mitigate the electron beam damage [3]. Honjo et al. reported the design of a TEM holder, with sufficient degrees of freedom to vary the temperature from −80 to 200 °C, to study low-temperature crystalline structures and phase transitions in 1956 [4]. The principles described for the construction and functioning of their side-entry cryo holder have remained about the same over the years. In the 1960s, a number of other research groups also reported design of specimen holders using cryogens (Liquid N_2 or He) that were specific for a particular TEM [5–10], including a top-entry liquid He-stage with the sample surrounded by transverse magnetic field [10]. These holders could be categorized based on the (i) nature of the cryogen, (ii) cooling method, and (iii) holder type (side or top entry). In most cases, liquid N_2 and liquid He were the cryogens used and were either contained in a reservoir attached to the holder or by circulating the cryogen around the sample using a flow system. In early days, up to the late 1990s, top-entry specimen holders were preferred for high-resolution work as they were more stable and less prone to be affected by external environment. During this period, both types of holders were modified for low-temperature work. Cold gas instead of liquid [11] and cryogenic refrigeration technique (Joule–Thompson cooling) were also tried [6]. Later, the technological developments in the construction of the TEM column and environmental control of the TEM room made it possible to use side-entry holders without losing spatial resolution.

Most of these specimen holder or stages, as they are sometimes referred, were used to make static observations such as structure of ice. However, a few examples of in-situ observations can also be found, such as nucleation and growth of condensed gases (Ar, Xe, and Kr) [12], martensitic phase transformation in β Cu–Zn alloy [13], annealing of ordered and disordered Cu_3Au alloy at low temperatures [14]. It is important to note that these stages were designed and built by research groups for their specific need and instrument, i.e. the make and model of the TEM in their lab. However, in recent years, side-entry specimen holders with a reservoir for cryogenic cooling using liquid N_2 or He have become commercially available, e.g. https://www.gatan.com/products/tem-specimen-holders/cooling-situ-holders, https://hummingbirdscientific.com/products/cryo-biasing, https://www.melbuild.com/cooling-1st.php, http://hennyz.com, a holder with Peltier cooling device is also available (https://www.kitano-seiki.co.jp/en/product/tem/TEM.html).

4.2 Specimen Holder Design and Function

As mentioned above, cryo-TEM holders are commercially available, and there is not a specific need to design and build a new holder except to include other stimuli for correlative microscopy that are not readily available. However, understanding the design and functioning of a cryo holder will help us to strategize (i) the purchase of a new holder and (ii) design our experiments to fully exploit its capabilities.

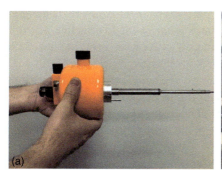
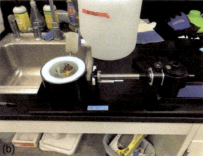

Figure 4.1 (a) Liquid N$_2$ holder with reservoir (orange) and (b) cryo transfer station and holder.

First and foremost, we should prepare a list of our needs that could be for a single researcher, his/her team, or a group of researchers and their team. Latter situation is quite common as most of TEMs reside in a central facility that caters to the needs of a larger community. One of the essential questions is the lowest temperature required to choose between the two cryogens, liquid N$_2$ or He, needed to cool the samples. Note that the achievable temperatures are $\approx 100\,K$ for liquid N$_2$ and $\approx 15\,K$ for liquid He, depending on the construction of holders, heat shielding, condition of the reservoir, etc. Since the holder material and construction of the reservoir are different for these two cryogens, the holder must be chosen according to the temperature requirements. Apart from cryo holders for liquid N$_2$ and He, cryo-transfer holders are also available and should be used to transfer a frozen sample to the TEM.

A typical cooling holder has the same general features as shown in Figure 5.1, except it has a reservoir that is used to contain cryogen (Figure 4.1a). Basic requirements are similar to the other TEM holders, discussed in detail in Chapter 5, i.e. the body should be made of nonmagnetic material, the tip should fit in the space between objective pole pieces, the sample well should be compatible with geometry of the sample support to be used (3 mm grid or MEMS based chip), etc. Additionally, the thermal conductivity and the heat transfer from reservoir to the sample tip are important considerations. Copper is the best thermal conductor and machinable material, but it gets tarnished by the environment easily. This problem can be solved by using a copper alloy or enclosing it with other material such as nonmagnetic stainless steel.

The reservoir is essentially a metallic vacuum thermos with special construction that includes a zeolite or activated carbon as moisture trap and heating element for temperature control. It also has an access port to regenerate the zeolite and evacuate the space between outer and inner walls to restore its thermal insulating capability. The zeolite trap keeps it moisture-free and retards any ice build-up around the reservoir. In fact, the ice build-up is a visual indicator that the reservoir is not performing well, i.e. cryogen will boil-off sooner than specified period by the manufacturer, generally around four hours. Moreover, it will cause a continuous boiling off the liquid that will result in (i) vibrations in the sample, (ii) sample drift, and (iii) temperature fluctuation and (iv) reduce the accessible experimental period.

As mentioned above, the construction of the reservoir for liquid He is different than for liquid N_2, mainly due to its much lower boiling point. In principle, a continuous flow approach, instead of reservoir, is more effective when using liquid helium. However, the construction of such system is much more complicated with additional requirements. As the continuous boiling of the liquid generates sample vibrations, it is imperative to keep the cryogen temperature below its boiling point. Therefore, liquid helium reservoir is often made of double containers, where outer container is filled with liquid N_2 to keep the He in the inner container below its boiling point.

The cryo holder is also equipped with a heating element, thermocouple and connected to a temperature controller box. Whereas thermocouple is used to measure the temperature, the heater serves dual purpose: it expedites the evaporation of the cryogen after finishing experiments and brings the reservoir to RT; second is to provide a way to achieve variable sample temperature. The second function is important for performing temperature-resolved experiments below RT. Both the thermocouple and heating capabilities are controlled by an external electronic control box. A cryo transfer holder with a cryo station (Figure 4.1b) is also commercially available to keep the samples in frozen state during sample transfer to TEM column.

Recently, a MEMS-based cryo holder (Figure 4.2), with heating capability incorporated within the chip providing continuous tunable temperature, has been reported [16, 17]. Moreover, this unique design, with a large cryogen container and a bunch of copper wires to transport the low temperature to the sample for extended period of time and low drift rate, improves our capability to acquire atomic-scale STEM images. Although this holder is not commercially available currently, but

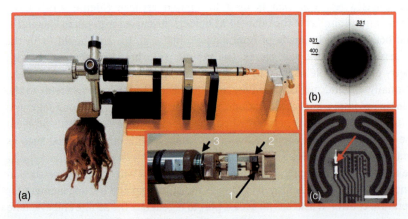

Figure 4.2 (a) Photograph of the HennyZ liquid-nitrogen double-tilt specimen holder for low-temperature investigations. A high-magnification image of the holder tip is shown as an inset, with numbers 1, 2, and 3 indicating the location of the MEMS chip, the six needles for electrical biasing, and the active temperature regulator, respectively. (b) Fast Fourier transform of a single-frame high-resolution TEM image where the 400 and 331 Au reflections are indicated by the respective arrows. (c) Low-magnification micrograph of a FIB lamella deposited onto a MEMS chip for low-temperature observations. The scale bar is 50 μm. Source: Tyukalova et al. [15]/from American Chemical Society.

its potential applications for time and temperature-resolved in-situ cryo-TEM observations below and above RT have been reported (e.g. Figure 4.12) [15].

Another upcoming modification is to retrofit specimen chamber with an extra cryo clip or a box sleeve that will reduce the ice build-up and extend the observation period at cryo temperatures for materials.

4.3 Specimen Design and Preparation

There are two ways to perform cryo-TEM experiments: first is to start at the room temperature and cool it down to the desired temperature for observations; the second is to start with cryo-samples and examine them while heating to RT or above. Note that most of the commercial holders are capable of heating the samples only up to $\approx 200\,°C$. When starting from RT, any of the specimen preparation techniques described in Section 2.4 can be used. Again, the choice will depend on the nature of the sample (powder, bulk, thin film, epitaxial film, etc.) and the receiving geometry of the specimen well of the holder (3 mm grid or disc or microchip).

Plunge freezing of supported suspensions and/or suspended thin films [18], cryo-sectioning or freeze fracturing of plunge frozen "bulk," and vitreous sectioning of high pressure frozen film, are some of the sample preparation techniques employed if the experimental observations are to be started with frozen sample [19]. The samples are then loaded in cryo transfer holder using cryo station (Figure 4.1b) and transferred to the microscope. Note that this procedure allows the sample to be kept at cryogenic temperature, starting from preparation to observation time. The support films on TEM grids can be made hydrophilic, a requirement to have very thin liquid layer for plunge freezing, either by plasma cleaning or heating under ultraviolet lamp.

However, a special fast freezing technique has been reported for extracting time-resolved samples during a reaction to decipher the reaction mechanisms and identify intermediate or metastable phases. Lu et al. have reported the construction of an advanced device to mix the reaction components and spray the mixture at fixed time intervals on a hydrophilic TEM grid, as it is being plunged into cryogen (Figure 4.3a,d) and its application for biological molecules [20]. This arrangement reduces the mixture extraction and freezing time to ms as all steps are accomplished using a microfabricated monolithic device that encompasses a mixer for liquids (Figure 4.3b), reaction channel, and a pneumatic sprayer (Figure 4.3c) on the same chip. This system employs humidified N_2 to generate atomization that produces sufficiently thin microdroplets that are collected on a hydrophilic carbon support film (Figure 4.3e). The size of the microfluidic channel controls the reaction period and can be changed to achieve desired time resolution [20].

Another similar approach for preparing vitrified time-resolved samples has been reported by Dandey et al. [21]. They have developed a spray-mixing approach on Spotiton robot [22] that takes advantage of piezo-controlled liquid dispenser that dispenses a stream of 50 pl droplets on to a nanowire grid on its way to vitrification. Although these techniques are designed for biological materials, they can be used to

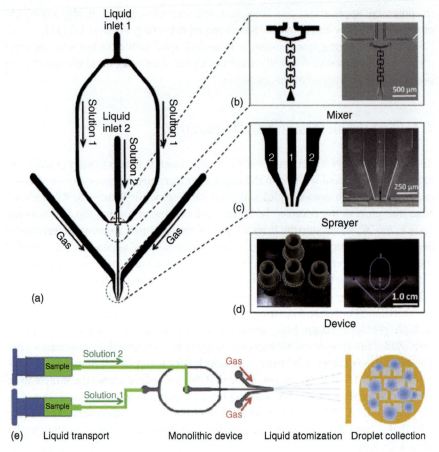

Figure 4.3 Monolithic micro mixing–spraying device. (a) Device layout. (b) Details of the micro-mixer configuration (left) and corresponding SEM image (right). (c) Details of the fabricated nozzle (left) and SEM image (right). "1" indicates the liquid-filled reaction channel, and "2" denotes a channel carrying pressurized gas. (d) Photograph showing the top view of a finished device, with four nano ports attached for tubing connections (left) and a bottom view of the device, showing the transparent glass wafer cover through which the device's micro-mixer and the micro sprayer can be seen (right). (e) Experimental setup for droplet collection. Source: Lu et al. [20]/from Elsevier.

freeze fluidic reactions at select time interval for cryo-TEM observation to determine reaction steps and uncover metastable phases.

> Low temperature enables adsorption and e-beam deposition of precursor material all over the sample instead of targeted area. The fraction of adsorbed organometallic precursor can be reduced by ion beam exposure that hardens and increases the electrical conductivity of the sample.
> – Michael Zachman, Oak Ridge National Laboratory

As mentioned in Chapter 2, cryo-FIB can also be used to make site-specific lamella samples for cryo-TEM observations [23–25]. The setup for cryo-FIB is different than a regular FIB lamella lift-out process. First, the sample is frozen in bulk form, using plunge freezing technique (see Chapter 2), a transfer system is needed to bring the frozen sample in to the FIB column where the sample stage is kept cold to avoid contamination or condensation, the ion source must be introduced at a controlled rate to avoid ions depositing on the sample, and last but not the least, a cryo manipulator with cryo transfer is needed to attach the lamella to the TEM grid and subsequently transfer to the TEM [23, 26–29]. Cryo-FIB sample preparation is especially advantageous to explore solid–liquid interphases of Li ion batteries (see Figure 4.7 in applications section) [30].

4.4 Practical Aspects of Performing Cryogenic Cooling

Cryo-TEM observations and experiments are need-based and as mentioned in the sample preparation (Section 4.2), we either start with RT and then examine the sample at set temperatures during cooling down or we start with a frozen sample. Another distinction to keep in mind is whether we are using a dedicated cryo-TEM or cryo-holder. Following sections are based on the assumption that physical scientists are invariably taking the second approach due to the restrictions posed by specimen loading cartridge for the former. Starting from vitrified frozen samples has become more common in recent years. The current side-entry cryo-TEM holders use a bulky reservoir, filled with cryogen that sits outside the TEM column. First and foremost, follow the manufacturer's directions. Following are some general directions for getting the best performance from your cryo-holder:

1. Make sure that the reservoir is evacuated during the night before you plan to use it.
2. The zeolite or active carbon is regenerated.
3. Allow time for cryogen to boil-off before refilling and wait until it stops boiling and steady temperature is reached. Cryogen will continue to evaporate, refill it as often as possible.
4. Make sure that all the connections are well isolated and not source of sample vibrations. Best test is to image the sample and measure the resolution using Young's fringes before and after connecting the controller to the holder.

Sample preparation is the most crucial part if you plan to start your observations from a vitrified frozen sample. For vitrified samples, there is delicate balance between the amount of liquid covering the sample and plunge freezing speed. Too much liquid will result in having too thick layer of liquid on top of your solid sample and reduce signal-to-noise ratio and too thin layer will evaporate quickly reducing the observation time [18]. Generally, a blotting paper is used to remove the excess of liquid after dipping the TEM grid in the liquid and quickly plunge frozen. Further details can be found in Refs. [18, 31–34]. Cryo-FIB lift-out offers an attractive alternative to make site-specific samples for cryo-TEM [23, 24, 26].

Another important step is to transfer the vitrified sample to the TEM without (i) warming it up and/or (ii) crystallizing the liquid. The cryo station to load the sample and cryo transfer holder are required but still the time of transfer plays an important role. Also, the time the sample sits in the pre-evacuation chamber of the TEM is also crucial for keeping the samples in vitrified state. A dedicated ETEM can be advantageous as the pre-evacuation time can be reduced to a few seconds as its column is being pumped using turbo pumps instead of ion pump.

> While cryo TEM provides unique results that can't be obtained by other methods, they are currently challenging to perform and relatively low throughput. Keep this in mind while designing experiments.
> – Michael Zachman, Oak Ridge National Laboratory

4.5 Some Noteworthy Applications

As mentioned earlier, cryo-TEM has been predominantly used for static imaging. In-situ observations of samples below room temperature have been limited. There are some interesting phenomena, such as phase transformation in superconducting oxides, magnetic behavior, glass transition temperature, etc. that have been explored [15, 35]. Structural and morphological changes with temperature can be obtained by comparing images recorded at set temperature intervals. Also, controlling the electron beam–induced degradation can provide extended observation period required to perform in-situ experiments for beam-sensitive materials [36].

Areas of research taking advantage of cryo-TEM can be divided into three main categories: biology, materials science, and physical science. As mentioned earlier, biologists have been using the technique in a dedicated cryo-TEM for a long time that has earned a Nobel Prize in 2017 (Dubochet, Frank, and Henderson). Most of the cryo-TEM techniques applied in materials and physical science category, reported so far, can be divided into five main categories: (i) static imaging at set temperatures; (ii) pseudo in-situ; and (iii) in-situ temperature resolved, (iv) in-situ time and temperature-resolved observations; and (v) correlative in-situ Cryo-TEM. Traditionally, the motivation for using cryo-TEM as characterization technique has been to decrease radiation damage and explore low-temperature phase transformation. However, the application field has expended due to availability of aberration-corrected high-resolution TEM and direct electron detectors for fast acquisition times [37].

Figure 4.4 shows various fields of applications. Reduced radiation damage in cryo samples has opened the door to investigate soft materials [38], polymers [39], zeolites, and metal–organic frameworks (MOFs) [40]. Other application areas include, but are not limited to, energy materials [41], stability of 2D materials [42], reactions in liquids [43], quantum materials [44, 45], low-temperature phase transformations in superconducting, ferroelectric and ferromagnetic materials. The challenges and application of in-situ and operando cryo-TEM can also be found in Refs. [15, 35, 37, 46]. It is important to note that the availability of aberration-corrected microscopes has improved our ability to perform cryo-TEM/STEM experiments. In addition, atomic-resolution information from different phases, forming as a function

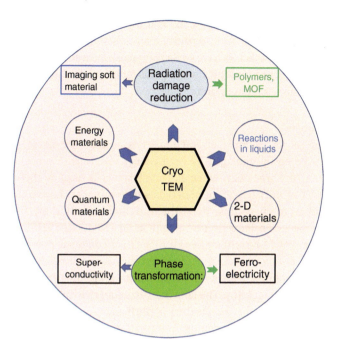

Figure 4.4 Block diagram showing the various fields of application of cryo-TEM in the general area of physical and materials sciences.

of temperature, can be obtained by acquiring images with the fast electron detectors and adding multiple images after drift correction using post-processing algorithm.

In the following sections, we will provide representative examples from these application areas, even if some of them may not fall into in-situ category, the characterization techniques developed have potential to be used for in-situ time and temperature-resolved observations in future.

4.5.1 Mitigating Radiation Damage

We know that interaction of high-energy electron often results in damaging or altering the structure of materials and is especially crucial for soft materials. There are many ways to control the damage [3], cooling the sample is one of them that has been successfully exploited by the biologist. An estimated 5- to 10-fold reduction by beam damage in the samples at cryogenic temperature has been reported [3, 47]. For example, Tyukalova and Duchamp report that the electron dose needed for spinal to rock salt phase transition in $LiNi_{0.5}Mn_{1.5}O_4$ and $ZnCo_{1.8}Ni_{0.2}O_4$ is much lower at cryogenic temperature than at RT [47]. They were able to determine the native phase for both compounds to be spinel from atomic-resolution cryo-STEM images while electron beam induced the formation of other phases at RT. Moreover, Z-contrast in cryo-STEM images also provides possibility to obtain compositional information from images for beam-sensitive materials [48]. 4-D STEM observations have also been enabled by using liquid N_2 holder [49].

Combination of aberration-corrected microscope and high-speed detectors has opened the door to characterize several beam-sensitive materials as described below. We should keep in mind that fast detectors have two advantages; one is to record low dose images, the second is to record multiple images at ms temporal resolution and add them after drift correction to obtain optimum SNR.

4.5.1.1 Structure of Polymers

Polymers are considered as an example of soft materials as they are made of organic chains that can be easily damaged by high-energy electrons. Moreover, they are non-conductive and charge easily when electron beam is focused on them. Therefore, these are a part of the category of materials where scientists have used bio-inspired cryo techniques to image their nanostructures and formation mechanisms [50]. For example, organic solar cells are being investigated due to their low cost and high performance. Poly(3-hexylthiophene) or P3HT is a conjugate polymer that is one of the most used components in organic electronics as photovoltaic inks [51]. A lamellar structure is reported to form due to the π–π stacking, and this ordering is known to increase the hole mobility. The lamellae can adopt two different orientations during the drying process that affects their performance. Wirix et al. have used cryo-TEM images to investigate the conjugate formation process of samples in two different solutions, in toluene as a non-halogen and in 1,2-dichlorobenzene (oDCB), as a halogenated solvent, respectively, aged for >7 days [52]. No noticeable difference was observed between images recorded from vitrified solution of P3HT in toluene at cryogenic temperature or from dried sample at RT. On the other hand, a higher concentration of P3HT wires was observed in the dried sample (Figure 4.5a) compared with vitrified sample (Figure 4.5b in oDCB), indicating that aggregate formation occurred during drying period. They also used tomography to elucidate the 3D structure of this photovoltaic ink in its native state using cryo-TEM [52].

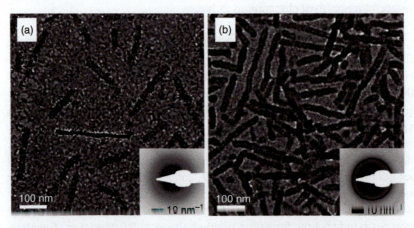

Figure 4.5 (a) Cryo-TEM image of vitrified solutions of P3HT 1 wt%, aged for >7 days in oDCB showing initial dispersion (b) TEM image of the same solutions drop-cast and blotted on a carbon support film of a TEM grid. Insets are low-dose ED patterns from two samples. Source: Wirix et al. [52]/from American Chemical Society.

3D tomography was also used to reveal the structure of hydrated Nafion, an ion-containing random copolymer that is used as a solid electrolyte in many electrochemical applications. Allen et al. reported that the dry Nafion membrane contained ≈ 3.5 nm spherical clusters corresponding to the hydrophilic sulfonic-acid-containing phase. On the other hand, an interconnected channel-type network, with a domain size of ≈ 5 nm, was present in the hydrated Nafion sample in which the proton transport occurs [53].

Morphological evolution during the self-assembly process of block copolymers and other soft-materials has also been reported [54]. Core–shell structures, with inorganic core and polymeric shell, have also been investigated using cryo-TEM. For example, the volume change in thermosensitive polymer particles with core–shell structures as function of temperature was found to be confined to the change in volume of the shell [55]. Cryo-TEM imaging and spectroscopy were also used to determine the structure, morphology, and chemistry of hairy hybrid Janus-type catalyst, made of silica core covered by two distinct polymeric shells [56].

4.5.1.2 Structure of MOF and Zeolites

MOFs form a family of materials with potential applications for gas storage, separation, and catalysis due to their porous structure with large channels. Their structure can be described as framework channels made of metal ions linked together by organic molecules making it possible to tune their chemistry and crystallinity for specific applications [57, 58]. Their chemistry and porous structure made them unsuitable for TEM observation as they were damaged within seconds by electron beam. Recently, this impediment has been overcome by using cryo-TEM and the arrangement of large channels has been imaged [59]. Furthermore, low dose combined with direct electron detectors has also made it possible to obtain atomic-resolution images for some of the MOFs [60, 61].

Li et al. have taken a step further to infer their growth mechanisms and successfully imaged the MOF channels filled with CO_2 molecules [40]. Figure 4.6a shows representative images from their report showing stability of the structure at $-170\,°C$ for long enough time to record low-dose images that were suitable to extract atomic-resolution information by using image processing techniques such as denoising [40]. They report that the vacuum-dried zeolitic imidazolate framework with cubic structure, which became amorphous under normal imaging condition at RT, was stable at $-170\,°C$ under similar imaging conditions (Figure 4.6a). The step edges (Figure 4.6b) present on the surface of as-synthesized crystals provided insight into their growth mechanism. More importantly, they show that plunge freezing can immobilize the gas molecules trapped in the channels. The images recorded at $-170\,°C$ of empty (Figure 4.6c,d) and CO_2 filled (Figure 4.6e,f) ZIF-8 crystal oriented in $\langle 111 \rangle$ direction clearly show the presence of CO_2 in the channels. This observation was further confirmed by a 3% increase lattice spacing for CO_2-filled crystals [40]. Furthermore, DFT calculations have reported two distinct binding sites for CO_2 within the ZIF-8 channels that are predicted to be energetically favorable [62].

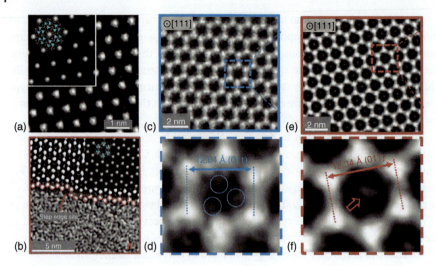

Figure 4.6 (a) HRTEM cryo-TEM image of ZIF-8 particles taken along ⟨111⟩ direction at −170 °C with electron dose of ≈ 50 e$^-$ Å^{-2} with ZIF-8 atomic structure overlaid onto simulated image calculated using 225 nm underfocus and 100 nm sample thickness. The simulated HRTEM image matches well with the experimental cryo-TEM image. (b) Denoised and magnified image of ZIF-8 after synthesis (not vacuum dried) recorded at 250 nm overfocus with electron dose rate of ≈ 4.5 e$^-$ Å^{-1} s^{-1} for 1.5 seconds. The ZIF-8 atomic structure is overlaid onto simulated image calculated using experimental parameters. The Zn clusters on the surface are double-coordinated and are outlined in red circles. A step-edge site is indicated by red arrow. (c) CTF-corrected cryo-TEM image of dried, empty ZIF-8, bright regions correspond to the mass density. (d) Magnified image of the blue box in (c), corresponding to single unit cell. Density (circled in blue) near the interior edge of the unit cell may correspond to the organic imidazolate linkers. (e) CTF-corrected denoised Cryo-EM image of CO_2-filled ZIF-8 particle taken with same imaging parameters as used for image shown in (c). Contrast in the center of the 6-ring window is clearly observed for multiple unit cells. (f) Magnified image of a single unit cell from the red boxed region in (e). Density at the center of the unit cell (indicated by red arrow) likely corresponds to CO_2 adsorbed within ZIF-8. Source: Li et al. [40]/from Elsevier.

Formation mechanism of mesoporous SBA-15 (zeolite) as a function of colloidal stability was also investigated using cryo-TEM [63]. Time-resolved images were recorded by extracting samples from the reaction mixture at different synthesis time. The particles were observed to grow via formation of silica–Pluronic–water "flocs," followed by their coalescence in an arbitrary manner. However, the final particles had well-defined hexagonal shape that could be due to the interface between flocs and surrounding media are covered by Pluronic molecules providing steric stabilization. As the flocs grow, the coverage of polymers at the interface increased until a stable size was reached [63]. These reports provide a proof of principle to observe the nucleation and growth mechanisms of similar electron beam–sensitive materials.

4.5.1.3 Cryo-TEM for Energy Materials

Most of the materials with energy applications, such as Li ion batteries, hybrid perovskite solar cells, MOFs, fall into the category of materials that are easily damaged by electron beam radiation. Recent progress in cryo-TEM for physical sciences offers new tools and methods to address many crucial yet unanswered questions for energy materials research [41]. High-resolution imaging and spectroscopy have provided a much-needed breakthrough in following the structural evolution during battery cycling, at the solid–electrolyte interphase (SEI), nucleation sites, lattice ordering of Li, etc. [64]. Here we describe a few examples where cryo-TEM has been successfully used.

New technologies are being continuously developed to get the optimum performance from the Li ion–based solid-state batteries. TEM imaging and spectroscopy have been employed to understand and measure the SEI of both charged and discharged states. The ease of lithiation and de-lithiation process, starting from outside domains to inside domains in hybrid MoS_2/C anode material, imaged at −20 °C, has been proposed as a reason for their improved cycling stability [65]. Apart from in-situ observations of lithiation process (also see Chapter 6 – solid–liquid interactions), it is desirable to determine the structure of a charged and discharged commercial battery. One of the failure mechanisms for Li batteries is formation of metal dendrites structure at SEI. TEM imaging is challenging as both the Li and its halides are unstable in electron beam. Therefore, mitigation of electron beam damage under cryo conditions has been utilized to obtain the structural and chemical information from the Li dendritic structures in a commercial battery.

Kourkoutis's group has used cryo FIB cross-sectioning of a coin battery to obtain 3D morphology and report existence of two types of dendrites (Figure 4.7a,b) [30]. These two lamellae were cryo transferred to the TEM column, and cryo-STEM imaging and EELS were used to elucidate the structure, morphology, and chemistry of Li dendrites [30]. An extended SEI layer, ≈ 300 to 500 nm thick, was observed on type I dendrite but was absent in case of type II dendrite (Figure 4.7c,d). The authors propose that part of the SEI may have been lost during sample prep and actual loss of Li to SEI may be larger than observed in the images. Increased concentration of the Li and oxygen compared with the electrolyte is observed in EELS maps, whereas F is not detected in the type I dendrite (Figure 4.7e). While extended SEI is absent in the type-II dendrite, a thin carbon-free layer containing Li and oxygen is present (Figure 4.7f). On the other hand, nearly spherical large structures containing carbon, oxygen, lithium, and fluorine are present near both types of dendrites (Figure 4.7d) [30]. Thus, a combination of cryo-FIB and analytical cryo-STEM techniques mitigated the loss of Li and revealed the structure and composition of intact SEI in Li-metal batteries at nanoscale.

Mitigation of electron beam–induced damage at cryo temperatures (−173 °C) has also been demonstrated for electrochemically deposited Li where deposited material was found to be amorphous with uneven crystalline LiF crystalline surface [66]. Similarly, Tyukalova and Duchamp reported that $LiNi_{0.5}Mn_{1.5}O_4$, used as a cathode material for lithium-ion batteries, transforms from the spinel into the rocksalt phase under electron beam irradiation at room temperature [47]. However, they found a threefold tolerance to electron dose at cryogenic temperatures, making it possible to record atomic-resolution images of the structural transformation.

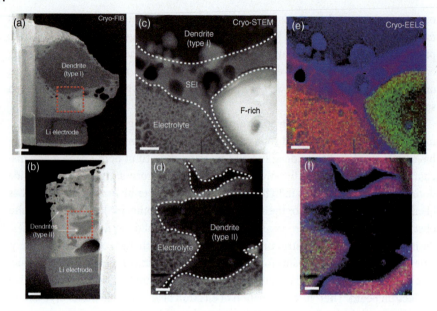

Figure 4.7 Structure and elemental composition of dendrites and their interphase layers in electron-transparent lamellae. (a, b) Electron transparent cryo-FIB lift-out lamellae of type I (a) and type II (b) dendrites. (c, d) HAADF cryo-STEM imaging reveals an extended SEI layer on the type I dendrite (c), but not on the type II dendrite (d). (e, f) EELS elemental mapping shows that both SEIs are oxygen-rich, but that the type II SEI contains no carbon (contrast has been adjusted for clarity. (e) The type I dendrite has an appreciable oxygen content, (f) whereas the type II dendrite does not. Fluorine-rich structures were often observed near both dendrite types. Scale bars, 1 μm (a, b), 300 nm (c–f). Source: Zachman et al. [30]/from Springer Nature.

4.5.1.4 Reactions in Liquids

Liquid-cell holders are now routinely used to investigate the structure of liquid–solid interfaces and to follow the liquid–solid reactions (see Chapter 6). However, electron beam has been reported to ionize the liquids and change the reaction paths [67]. Therefore, a pseudo in-situ method for time-resolved observation has been adapted by terminating the reaction at various stages by plunge freezing TEM sample grids and examining the sample using cryo-TEM holder [43]. This method was first proposed by Talmon et al., where they used a controlled environment vitrification system for cryofixation of solutions on a TEM grid at set time interval to examine them in TEM under cryo conditions [68]. The reaction parameters such as time, reaction temperature, pH of the solution and/or solvents were predetermined using differential scanning calorimetry, X-ray diffraction, etc. Then reaction is performed ex-situ and arrested to take snapshots of reaction in progress using cryo-TEM. It was successfully employed to capture the intermediate structures formed during a biological phase transformation associated with phospholipid membrane fusion as a function of pH [68].

Physical scientists have also employed this technique to observe time-resolved reaction steps in solution by freezing samples at set time intervals. For example,

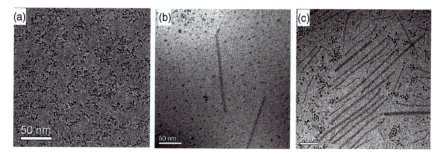

Figure 4.8 Cryo-TEM images of (a) freshly prepared suspension of ferrihydrite after dialysis (pH 4), after (b) 5 days, and (c) 24 days of aging at 80 °C. Note the formation of oriented aggregates by the self-assembly of nanoparticles to form goethite nanorods. Source: Yumono et al. [69].

Yuwono et al. have reported the mechanism of oriented aggregation of ferrihydrite nanoparticles, which resulted in formation of meso-crystals, by vitrifying and examining liquid samples that were aged ex-situ [69]. Freshly prepared suspension ferrihydrite after dialysis for days in solution with pH 4 contained randomly dispersed nanoparticles (Figure 4.8a). Time-resolved evolution revealed that the formation of goethite nanorod-like structure (Figure 4.8b) is an intermediate step for growing large, oriented aggregates of nanocrystals (Figure 4.8c) [69].

Another example reported is the evolution of mesoporous SBA-15 with time, revealed by extracting and vitrifying a droplet from the reaction mixtures of varying chemical compositions at fixed time intervals for cryo-TEM observation. The images revealed that at early stages, floc formation resulted in nucleation of mesoporous material, and their growth rate and colloidal stabilization depended on composition of the starting solution [70]. Their aggregation leads to the nucleation of amorphous nanoparticles in solution. In another example, stearic acid monolayer was used as a template to follow the template-directed mineralization process in solution [71]. Time-resolved extraction of vitrified solutions of $Ca(HCO_3)_2$ revealed the formation of prenucleation clusters of $CaCO_3$, which gradually transformed to polycrystalline particles before forming single crystal particles [71]. On the other hand, Smeets et al. reported a slightly different nucleation mechanism of $CaCO_3$ that proceeded with formation of intermediate phases in controlled dilute solutions that follow the fundamental concepts of nucleation theory [72]. Similar time-resolved experiments were conducted to observe the formation of monodispersed silica nanoparticle and their growth via agglomeration [73], progression of the biomimetic formation of amorphous calcium carbonate under Langmuir monolayers of a self-organized surfactant [74].

4.5.1.5 Quantum and 2-D Materials

Electronic properties of 2-D materials such metal dichalcogenides depend on their stacking order that results in generating charge density waves (CDW) and periodic lattice distortion (PLD). The modulated and broken symmetry states couple strongly to their electronic properties such as superconductivity or metal to insulator transitions [75]. Atomic-resolution STEM imaging at cryogenic temperatures has made it possible to image and understand the nature of CDW and PLDs in these materials [76]. For example, temperature-dependent electronic and structural reshaping, identified as CDW and PLD, respectively, is directly related to their electronic properties [77]. Therefore, it is important to understand the atomic origins of these charge-ordered phases to establish a structure–property relationship.

Hovden et al. have investigated the structural transformation of exfoliated 1T-TaS$_2$ flakes to understand the metal-to-insulator phase transition as the PLD becomes commensurate with the crystal lattice. They used atomic-resolution HAADF-STEM images to follow non-commensurate to commensurate phase transition upon in-situ cooling from 20 to −178 °C [42]. A non-commensurate PLD structure observed in a 34-layer-thick flake (Figure 4.9a) transitioned to a commensurate structure (Figure 4.9b) upon cooling to −178 °C. The composite image obtained by superimposing the FFTs from the two HAADF-STEM images further confirms this transition (Figure 4.9c,d). In Figure 4.9c, the six bright peaks (circled white) mark the Bragg spots of the parent TaS$_2$ lattice, and the PLD wave vectors can be identified by the additional sets of peaks surrounding the central beam. The second-order harmonic peaks are at the corners of large triangles marked, i.e. three of the large triangles for each phase are drawn over the FFT (Figure 4.9c). It is important to note that the red triangle from the room temperature image (red) is rotated and smaller than the blue triangles from the low-temperature image, due to the transition from non-commensurate to commensurate phase, which is more noticeable at low frequencies (Figure 4.9d). In addition, stacking transitions in the atomic positions, occurring due to bond length shift, can also be inferred from these

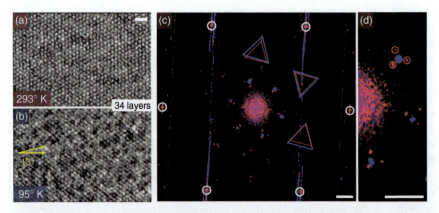

Figure 4.9 Atomic-resolution HAADF-STEM imaging of non-commensurate-to-commensurate phase transition of thin 1T-TaS$_2$. Upon cooling to −178 °C, the room-temperature phase transitions from the NC (a) to the C phase (b) with C-PLD ordering visible in the HAADF image of 34-layer TaS$_2$ (c and d) FFT of NC (red) and C (blue) images show hallmark PLD peaks within the hexagonal Bragg peaks (marked white). The second-order PLD peaks in the NC phase appear as singular spots that form large triangles (red triangles) that rotate and expand upon commensuration (blue triangles) at lower temperature. (Scale bar: 10 μm for (a) and (b); 0.05 μm^{-1} for (c) and (d). Source: Hovden et al. [42]/from National Academy of Science.

observations. The superstructures observed at low temperatures suggest both in- and out-of-plane ordering of CDWs [42].

While out-of-plane ordering of CDW clusters may be responsible for its insulating ground state and its exotic metastable phase for 1T-TaS$_2$, interlayer interactions were found to be prominent for 1T'-TaTe$_2$, a prototype 2D-layered structure. Here, electron diffraction patterns and atomic-resolution STEM images recorded at RT and −180 °C reveal strong contrast modification at Ta sites at low temperature [78]. After careful analysis of HAADF images and computing phonon frequencies using supercell frozen phonon approach, it was concluded that the trimerization along the *b* axis, along with subtle distortion of Te positions in the structure, generates interlayer interactions and CDW [78].

Intercalation of other material between the loosely bound transition metal dichalcogenides (TMD) follows the same pattern as graphite intercalated compounds used for battery applications. While some of the metals strongly bind with the sulfur in the lattice and form compounds with altered charge distribution in the layers, Hg intercalates in TaS$_2$ without bonding and moves freely between the layers. Therefore, in order to observe the deintercalation mechanism of Hg from fully interacted Hg$_{1.25}$TaS$_2$, the sample was cryo transferred to the TEM (Figure 4.10a) and slowly heated from −170 °C to RT [79]. It was interesting to note that no specific on-set point among various flakes was observed. Deintercalation often started from the outer most layers as well in the middle of the flake and proceeded in random order, instead of layer-by-layer mechanism (Figure 4.10b), due to charge density redistribution between the un-intercalated and intercalated layer.

> I happen to have discovered a direct relation between magnetism and light, also electricity and light, and the field it opens is so large and I think rich
> – Michael Faraday

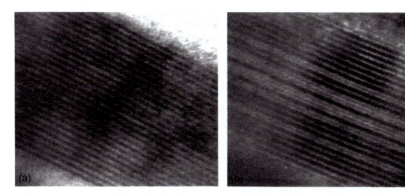

Figure 4.10 (a) TEM image of Hg$_{1.25}$TaS$_2$ crystal at −170 °C prior to deintercalation (b) the same crystal region after the onset of deintercalation in internal layers and after deintercalation has begun to progress away from the onset layers; Note the irregular termination and bending of the internal layers near the crystal edge. Source: McKelvy et al. [79]/from American Chemical Society.

4.5.2 Phase Transformations Below RT

Modifications in materials properties, such as mechanical, electrical, and magnetic, are often connected to phase transformations. Low-temperature phases may have desirable properties such as superconductivity and mechanical strength or undesirable properties that restrict their applications to low temperature. As mentioned before, most of the cryo observations are performed at a few set temperatures to decipher the phase transformation that can explain the change in materials property.

High T_c superconductors with T_c above −233 °C have been for interest for some time. It has been proposed that atomic-scale structural changes upon cooling should be responsible for metal to superconducting properties. However, a clear picture of related phase transformations in these materials could not be achieved as (i) the temperatures lower than −173 °C were hard to achieve (ii) and the vibrations and/or sample drift transferred to the sample from the boiling of cryo liquid in the reservoir.

Magnetic and ferroelectric phase transitions are also known to occur at low temperatures; therefore, cryo-TEM is ideally suited for comprehensive study to determine the mechanism and atomic domain structure with temperature to establish structure–property relationships. In 1967, a He-cooled TEM holder was first used to observe the presence and disappearance of magnetic domain structures in CrI_3 ionic ferromagnet below and above Curie temperature, respectively [9]. Understanding the changes in the microstructure of magnetic shape memory alloys is a way to explain their magnetic and physical behavior. Co–Ni–Ga form magnetic shape memory alloys with body-centered cubic structure that undergoes martensitic phase transformation at low temperature. Zak et al. report a diffusionless martensitic phase transformation in Co–Ni–Ga shape memory alloys to start around 140 K (−133 °C) using a modified side-entry cooling holder [80].

Recently, Mun et al. have used cryo-STEM imaging to directly visualize the intriguing domain system in (111) barium titanate (BTO; $BaTiO_3$) thin films, grown on strontium titanate (STO; $SrTiO_3$), as a function of temperature [81]. A polymorphic nanodomain state in BTO is reported to form during growth due to the threefold rotational symmetry enforced by the (111)-STO substrate. Domains ranging from 1 to 10 nm in size, with tetragonal, orthorhombic, and rhombohedral structures, formed due to displacement of Ti in the octahedra, coexist at RT and exhibit exceptional dielectric and piezoelectric properties. Therefore, the atomic displacement vectors with respect to Ba were measured as function of temperature to determine the domain structure distribution. HAADF image recorded at 300 K (RT) clearly reveals the presence of nanodomains with different symmetries (Figure 4.11 left), which transformed to pure rhombohedral phase upon cooling to 140 K (−133 °C) (Figure 4.11 center). An enhancement in ferroelectricity, evidenced by an increase in domain size and polarization magnitude, is attributed to be due to the rhombohedral phase. However, the single rhombohedral domain collapsed into complex nanodomains upon further cooling to 95 K (−178 °C) (Figure 4.11 right). The authors attributed this unexpected domain phase transition to the mechanical constraint arising from the STO substrate [81]. This example illustrates

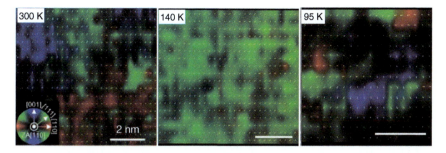

Figure 4.11 2D displacement vector maps of the (111)-BTO film acquired from an atomic-resolution HAADF-STEM image at 300 K (27 °C), 140 K (−133 °C) and 95 K (−178 °C). Each vector was measured according to the relative displacement of Ti compared with the center of the Ba cage with picometer (pm)-precision. The field was drawn based on the color wheel (inset). The mean values of the two clusters are plotted as yellow dots. Scale bar = 2 nm. Source: Mun et al. [81]/from American Chemical Society.

the application of cryo-TEM to elucidate the structure property relationship at low temperature for other functional materials.

As mentioned in the holder design section (Section 4.2; Figure 4.2), a new holder designed by Henny Zandbergen [16] has been tested to observe in-situ temperature-resolved domain structures in $BaTiO_3$ (BTO), a ferroelectric prototype [15]. We know that piezoelectric properties are controlled by the ferroelectric domain structure and that their redistribution across the different phases strongly depends on temperature. By taking advantage of the versatility of the HennyZ holder, Tyukalova et al. have reported formation of ferroelectric domains in real space and their movement upon cooling from 400 K (127 °C) to 200 K (−73 °C) and heating back to 400 K (127 °C). Figure 4.12 shows a set of images, extracted from a video, recorded during temperature cycling. The typical bend contours, seen as dark lines, arising due to slightly varying diffraction conditions, are only visible at 400 K (127 °C) when the crystal is in cubic form. The characteristic dark lines from the domain walls can be seen to move as the crystal undergoes phase transformations below the curie temperature ($\approx$ 393 K; 120 °C) to tetragonal (at $\approx$ 278 K; 5 °C) and finally to orthorhombic phase around 183 K (−90 °C) [15]. All of the three phases exhibit spontaneous polarization in different directions. The possibility to record images continuously with decreasing temperature will lead to more in-situ work at cryogenic temperatures possible and reveal phase transformations at atomic scale.

It is important to mention that electron spectroscopy, especially EELS, has also been used to follow the phase transformations below RT. Zhao et al. have recently used a combination of atomic-resolution HAADF-STEM and EELS mapping to image direct electron transfer related to superconducting pairing at $FeSe/SrTiO_3$ (STO) interface [82]. They deposited 1-, 8-, and 14-unit cell layers of FeSe on STO. These layers were capped by depositing a 10 nm of Te (Figure 4.13a) to keep the surface layer from contaminating during sample transport to TEM column and EELS mapping. Simulations based on Green's function for core-loss spectra at 10 K (−263 °C) for Fe-L_3 edge show a blue shift for all thicknesses (Figure 4.13b). Moreover, EELS maps, recorded at 10 K (−263 °C), show this shift to extend all

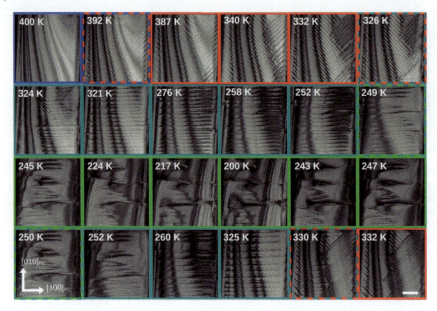

Figure 4.12 BF-TEM micrographs of a temperature cycle of BaTiO$_3$ (heating to high temperature, cooling to the lowest point [200 K], and heating back to high temperature sequentially). The scale bar is 400 nm. Source: Tyukalova et al. [15]/from American Chemical Society.

the way into FeSe layers for all thickness (Figure 4.13c–e). These measurements indicate a transfer of electrons from STO to FeSe layer at 10 K (−263 °C) that is accumulated within the first two layers of FeSe film near STO. These results were further confirmed using electrical transport measurements and indicated that back gate applied from STO is particularly effective in enhancing T_c of the films without appreciably changing the carrier density [82].

As we know, EELS in principle represents the density of states and can be used to observe the electronic states of individual atoms in a compound. Klie et al. have used EELS to explore the spin states of Co atoms in LaCoO$_3$ that exhibits an anomaly in its magnetic susceptibility around 80 K (−193 °C). They acquired EELS data from LaCO$_3$ single crystal at RT (300 K), slightly above the transition temperature (85 K; −188 °C) and below the room temperature (10 K; −263 °C) (Figure 4.14). It is obvious that neither the near-edge structure nor the intensities of Co L$_2$/L$_3$ lines changed with temperature (Figure 4.14, inset). However, the pre-peak of O-K edge (marked "a" in Figure 4.14) increased as temperature was decreased. This pre-edge feature is supposed to be due to the filling of the hybridized O 2p and Co 3d states and the second peak (b) is generally attributed La 5d band. The fine features of O K-edge, representing the bonding between O 2p – La 5d, and O 2p – Co 4sp, remain unchanged during in-situ cooling experiment. On the other hand, Co 3d – O 2p hybridized band changes with the onset of the spin state transition of the Co^{3+}, resulting in different filling of the Co 3d state with temperature. They show that O K edge pre-peak intensity is a direct measure of Co^{3+} spin state as confirmed by DFT calculations included in the report [83].

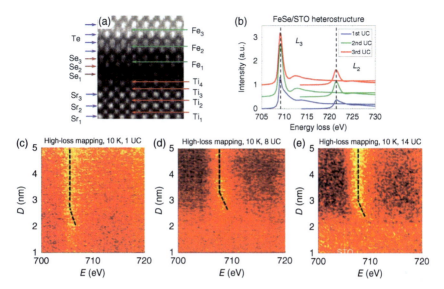

Figure 4.13 (a) The HAADF-STEM image of the 1-unit cell FeSe film on $SrTiO_3$ with FeTe capping layers. (b) FEFF simulation of the core-loss EELS spectra at 10 K (−263 °C) using the super cell from image in (a). Core-loss EELS mapping at 10 K (−263 °C) with an energy range between 680 and 740 eV for layer thickness of (c) 1 unit cell (d) 8 unit cells, (e) 14 unit cell. The dash lines are shown as guides for eyes to follow the energy shift as expected from simulated spectra shown in (b). Source: Zhao et al. [82]/from American Association for the Advancement of Science, CC BY-NC 4.0.

Moreover, Rui and Klie have successfully combined atomic-scale imaging and EELS acquired at RT and 95 K (−178 °C) to reveal the oxygen ordering associated with the antiferrodistortive phase transition at low temperature in $La_{0.5}Sr_{0.5}CoO_{3-\delta}$ thin films grown on STO [84]. They used Co white-line ratio to determine the Co valence state and found that the vacancy ordering to be present only in the films grown on $SrTiO_3$, but films grown on $LaAlO_3$ did not change with temperature indicating the influence of the substrate on the thin film properties.

4.5.3 Correlative In-Situ Experiments at Low Temperature

4.5.3.1 Mechanical Testing

Measurement of mechanical properties and their relationship to the defect and dislocation density at low temperature is of technological interest. Gruber et al. have performed in-situ tensile tests to measure the yield, tensile strength, and the resulting elongation at low temperature for Al alloys [85]. Such experiments can be performed using a commercially available TEM holder that combines cooling with mechanical deformation (Gatan, Inc.). A comparison between TEM images, recorded at RT and 77 K (−196 °C) under varying strain conditions for various Al alloys, was used to explain their mechanical behavior [85]. Their in-situ TEM observations exhibited a reduction in slip lines and uniform dislocation arrangement when deformed at low temperatures compared with RT that could be related to the measured increase in yield strength.

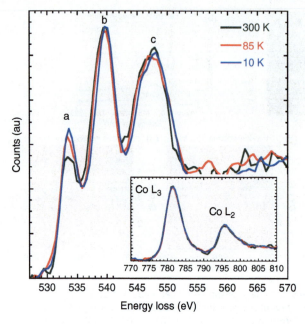

Figure 4.14 EELS spectrum of the O K edge at 300 K (27 °C), 85 K (−188 °C), and 10 K (−263 °C). The Co L edge at these temperatures is shown in the inset. Source: Klie et al. [83]/American Physical Society.

Plasticity of CoCrFeMnNi (Cantor) alloy has also been tested using similar methods to understand their exceptional performance at cryogenic temperatures [86]. Oliveros et al. reported time-resolved data collected after applying short strain pulsed at 100 K (−173 °C) and RT (300 K; 27 °C). An example of two pile-up of perfect dislocation, located on (−1–11) and (−111) planes, can be seen in the images extracted from in-situ cryo-TEM straining experiments at 104 K (−169 °C) (Figure 4.15a). The plastic deformation was observed to intensify on the (−1–11) plane after 2 μm displacement increment was applied on the holder and resulted in the propagation of partial dislocations that leave the stacking fault (Figure 4.15b). On the other hand, the perfect dislocations moving on (−111) were found to be more active. After a detailed quantitative analysis of overall 60 different grain orientations, they found that although twin formation and perfect dislocation glide could be triggered at the onset of plasticity at both temperatures, more frequent occurrence of twinning at low temperature may explain the improved performance of the alloy at low temperatures [86].

4.5.3.2 Magnetic Field

Similar to some of the other examples presented here, no in-situ time or magnetic field–resolved studies have been reported so far. However, cryo-TEM is used to trap the nanoparticles in solution such that movement of particles is due to the magnetic field only. One such example is magnetic field–induced assembly of superparamagnetic magnetite (Fe_3O_4) in colloidal suspension. It is reported that whereas long

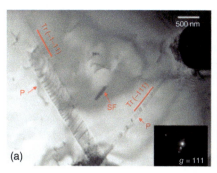

Figure 4.15 Mix of perfect and partial dislocations in the development of plastic deformation during an in-situ TEM experiment performed at 104 K (−169 °C) in a part of the foil where the thickness is 165 nm. (a) Low magnification image of a zone deforming plastically, containing perfect dislocations pile-up (P) and one widely split dislocation separated by a stacking fault (SF). (b) Glide of partial dislocations on (−111) leaves longer stacking faults upon straining. Source: Oliveros et al. [86]/from Elsevier, CC BY-NC-ND 4.0.

dipole chains, up to millimeter long, form in dried sample inside the magnetic field, only short chains with random orientation were present in vitrified samples at cryo temperatures [87].

A cryo biasing holder is also commercially available from Hummingbird Scientific, but its practical applications have not been reported so far.

4.6 Benefits and Limitations

Original applications of cryo-TEM or low temperature observations were directed toward unveiling the structure transformation mechanisms, dislocation formation, or magnetic and ferroelectric domain wall movements. Understanding these low-temperature phenomena helped in building structure–property relationships. For example, plasticity of alloys at low temperature was important to determine their performance under low temperature conditions. Currently, mitigating the radiation damage, a valuable lesson learnt from biologists, has expanded the cryo-TEM field. Structures of large number of electron beam–sensitive materials, such as polymers and MOFs, have now been imaged to reveal atomic-scale nuances of their structure and interactions with liquids or gases.

Quasi in-situ (time-resolved) experiments can be performed by extracting and freezing samples at set time intervals from bench-top reaction chamber for cryo-TEM observations. Current in-situ cryo-TEM experiments have been limited to observing the material at set temperatures without or with addition of another stimulus, such as magnetic field and/or electrical biasing.

The entire cryo-TEM operation, from sample preparation to imaging, is tedious and time-consuming. Furthermore, currently available commercial holders impose restrictions on high-resolution imaging due to the vibrations arising from continuous bubbling of the cryogenic liquid in the reservoir. Also, time resolution is restricted as it takes more than 20 minutes for sample to stop drifting after reaching the desired observation temperature.

However, time and temperature-resolved dynamic imaging may become possible using a new design of cryo holder (HennyZ) and that will open doors for more in-situ cryo-TEM experiments [16].

4.7 Chapter Summary

In-situ Cryo-TEM is a growing field of applications as it provides us the opportunities to observe synthesis and transformation in the electron beam–sensitive materials and reactions in liquids. An increased interest can be attributed to the increased availability of aberration-corrected TEM, fast detectors, and image processing algorithms.

Improvements in the design of liquid N_2 or liquid He, especially for liquid He holders, are still needed. HennyZ cryo holder is an improvement in the right direction as it provides an improved time resolution by decreasing the sample stabilization time and reduced drift. Moreover, the vibrations generated from the boiling of cryogen are not transferred to the sample as the container is not directly connected to the holder.

A redesign of sample cartridge for dedicated cryo-TEM is a desirable improvement to combine imaging with spectroscopy and a cooling box for the sample stage of the regular TEM will also help.

References

1 Milne, J.L., Borgnia, M.J., Bartesaghi, A. et al. (2013). Cryo-electron microscopy – a primer for the non-microscopist. *The FEBS Journal* 280 (1): 28–45.
2 Frank, J. (2002). Single-particle imaging of macromolecules by cryo-electron microscopy. *Annual Review of Biophysics and Biomolecular Structure* 31: 303–319.
3 Egerton, R.F. (2013). Control of radiation damage in the TEM. *Ultramicroscopy* 127: 100–108.
4 Honjo, G., Kitamura, N., Shimaoka, K., and Mihama, K. (1956). Low temperature specimen method for electron diffraction and electron microscopy. *Journal of the Physical Society of Japan* 11 (5): 527–536.
5 Venables, J.A., Ball, D.J., and Thomas, G.J. (1968). An electron microscope liquid helium stage for use with accessories. *Journal of Physics E: Scientific Instruments* 1 (2): 121–126.
6 Rudman, M.R., Critchell, J.W., and Flewitt, P.E.J. (1971). A simple design for a low temperature electron microscope stage. *Micron* 3 (3): 396–405.
7 Watanabe, H. and Ishikawa, I. (1967). A liquid helium cooled stage for an electron microscope. *Japanese Journal of Applied Physics* 6: 83.
8 Butler, E.P. and Hale, K.F. (1981). High temperature microscopy. In: *Dynamic Experiments in the Electron Microscope* (ed. A.M. Glauert), 1–457. Amsterdam/New York/Oxford: North Holland Publishing Company.

9 Matricardi, V.R., Lehmann, W.G., Kitamura, N., and Silcox, J. (1967). Electron microscope observations of ferromagnetic domains in chromium tribromide. *Journal of Applied Physics* 38 (3): 1297–1298.

10 Hörl, E.M. (1968). Liquid helium cooled stage with a transverse magnetic field. *Review of Scientific Instruments* 39 (7): 1027–1028.

11 Heide, H.G. and Urban, K. (1972). A novel specimen stage permitting high resolution electron microscopy at low temperatures. *Journal of Physics E: Scientific Instruments* 5: 803.

12 Venables, J.A. and Ball, D.J. (1971). Nucleation and growth of rare-gas crystals. *Proceeding of the Royal Society A* 322: 331.

13 Rapacioli, R. and Ahlers, M. (1973). An electron microscopy study of martensite formation in B-Cu-Zn. *Scripta Materialia* 7: 977.

14 Fukushima, H., Shimomura, Y., Guinan, M.W. et al. (1991). Annealing experiments of low temperature 14 MeV neutron-irradiated ordered and disordered Cu_3Au by TEM. *Journal of Nuclear Materials* 179–181: 943–946.

15 Tyukalova, E., Vimal Vas, J., Ignatans, R. et al. (2021). Challenges and applications to operando and in-situ TEM imaging and spectroscopic capabilities in a cryogenic temperature range. *Accounts of Chemical Research* 54 (16): 3125–3135.

16 Goodge, B.H., Goodge, B.H., Bianco, E. et al. (2020). Atomic-resolution cryo-STEM across continuously variable temperatures. *Microscopy and Microanalysis* 26 (3): 439–446.

17 Bell, D. and Zandbergen, H. A JEOL-based cooling holder with a low specimen drift allowing sub 1 Å STEM imaging. In: *European Microscopy Congress 2016: Proceedings*, 352–353. Wiley online library https://doi.org/10.1002/9783527808465.EMC2016.6871.

18 Cabra, V. and Samsó, M. (2015). Do's and don'ts of cryo-electron microscopy: a primer on sample preparation and high quality data collection for macromolecular 3D reconstruction. *Journal of Visualized Experiments* 95: e52311.

19 Gao, M., Kim, Y.-K., Zhang, C. et al. (2014). Direct observation of liquid crystals using cryo-TEM: specimen preparation and low-dose imaging. *Microscopy Research and Technique* 77 (10): 754–772.

20 Lu, Z., Shaikh, T.R., Barnard, D. et al. (2009). Monolithic microfluidic mixing–spraying devices for time-resolved cryo-electron microscopy. *Journal of Structural Biology* 168 (3): 388–395.

21 Dandey, V., Budell, W., Wei, H. et al. (2020). Time-resolved cryo-EM using Spotiton. *Nature Methods* 17 (9): 897–900.

22 Jain, T., Sheehan, P., Crum, J. et al. (2012). Spotiton: a prototype for an integrated inkjet dispense and vitrification system for cryo-TEM. *Journal of Structural Biology* 179 (1): 68–75.

23 Kuba, J., Mitchels, J., Hovorka, M. et al. (2021). Advanced cryo-tomography workflow developments – correlative microscopy, milling automation and cryo-lift-out. *Journal of Microscopy* 281 (2): 112–124.

24 Schreiber, D.K., Perea, D.E., Ryan, J.V. et al. (2018). A method for site-specific and cryogenic specimen fabrication of liquid/solid interfaces for atom probe tomography. *Ultramicroscopy* 194: 89–99.

25 Smeets, M., Bieber, A., Capitanio, C. et al. (2021). Integrated cryo-correlative microscopy for targeted structural investigation in situ. *Microscopy Today* 29 (6): 20–25.

26 Zachman, M.J., Asenath-Smith, E., Estroff, L.A. et al. (2016). Site-specific preparation of intact solid–liquid interfaces by label-free in situ localization and cryo-focused ion beam lift-out. *Microscopy and Microanalysis* 22 (6): 1338–1349.

27 Rubino, S., Akhtar, S., Melin, P. et al. (2012). A site-specific focused-ion-beam lift-out method for cryo transmission electron microscopy. *Journal of Structural Biology* 180 (3): 572–576.

28 Mahamid, J., Schampers, R., Persoon, H. et al. (2015). A focused ion beam milling and lift-out approach for site-specific preparation of frozen-hydrated lamellas from multicellular organisms. *Journal of Structural Biology* 192 (2): 262–269.

29 Parmenter, C.D. and Nizamudeen, Z.A. (2021). Cryo-FIB-lift-out: practically impossible to practical reality. *Journal of Microscopy* 281 (2): 157–174.

30 Zachman, M.J., Tu, Z., Choudhury, S. et al. (2018). Cryo-STEM mapping of solid–liquid interfaces and dendrites in lithium-metal batteries. *Nature* 560 (7718): 345–349.

31 Dobro, M.J., Melanson, L.A., Jensen, G.J. et al. (2010). Plunge freezing for electron cryomicroscopy, Chapter 3. In: *Methods in Enzymology* (ed. G.J. Jensen), 63–82. Academic Press.

32 Galway, M.E., Heckman, J.W., Hyde, G.J. et al. (1995). Advances in high-pressure and plunge-freeze fixation, Chapter 1. In: *Methods in Cell Biology* (ed. D.W. Galbraith, H.J. Bohnert and D.P. Bourque), 3–19. Academic Press.

33 Tivol, W.F., Briegel, A., and Jensen, G.J. (2008). An improved cryogen for plunge freezing. *Microscopy and Microanalysis* 14 (5): 375–379.

34 Dubochet, J. and McDowall, A.W. (1981). Vitrification of pure water for electron microscopy. *Journal of Microscopy* 124 (3): 3–4.

35 Lyumkis, D. (2019). Challenges and opportunities in cryo-EM single-particle analysis. *The Journal of Biological Chemistry* 294 (13): 5181–5197.

36 Russo, C.J. and Egerton, R.F. (2019). Damage in electron cryomicroscopy: lessons from biology for materials science. *MRS Bulletin* 44 (12): 935–941.

37 McComb, D., Lengyel, J., and Carter, C.B. (2019). Cryogenic transmission electron microscopy for materials research. *MRS Bulletin* 44 (12): 924–928.

38 Watt, J., Huber, D., and Stewart, P. (2019). Soft matter and nanomaterials characterization by cryogenic transmission electron microscopy. *MRS Bulletin* 44 (12): 942–948.

39 Oostergetel, G.T., Esselink, F.J., and Hadziioannou, G. (1995). Cryo-electron microscopy of block copolymers in an organic solvent. *Langmuir* 11 (10): 3721–3724.

40 Li, Y., Wang, K., Zhou, W. et al. (2019). Cryo-EM structures of atomic surfaces and host–guest chemistry in metal–organic frameworks. *Matter* 1 (2): 428–438.

41 Zhang, Z., Cui, Y., Vila, R. et al. (2021). Cryogenic electron microscopy for energy materials. *Accounts of Chemical Research* 54 (18): 3505–3517.

42 Hovden, R., Tsen, A.W., Liu, P. et al. (2016). Atomic lattice disorder in charge-density-wave phases of exfoliated dichalcogenides (1T-TaS$_2$). *Proceedings of the National Academy of Sciences* 113 (41): 11420–11424.

43 Matatyaho Ya'akobi, A. and Talmon, Y. (2021). Extending cryo-EM to nonaqueous liquid systems. *Accounts of Chemical Research* 54 (9): 2100–2109.

44 Zhu, Y. (2021). Cryogenic electron microscopy on strongly correlated quantum materials. *Accounts of Chemical Research* 54 (18): 3518–3528.

45 Minor, A.M., Denes, P., and Muller, D.A. (2019). Cryogenic electron microscopy for quantum science. *MRS Bulletin* 44 (12): 961–966.

46 Li, Y., Huang, W., Li, Y. et al. (2020). Opportunities for cryogenic electron microscopy in materials science and nanoscience. *ACS Nano* 14 (8): 9263–9276.

47 Tyukalova, E. and Duchamp, M. (2020). Atomic resolution enabled STEM imaging of nanocrystals at cryogenic temperature. *Journal of Physics: Materials* 3 (3): 034006.

48 Elbaum, M., Seifer, S., Houben, L. et al. (2021). Toward compositional contrast by cryo-STEM. *Accounts of Chemical Research* 54 (19): 3621–3631.

49 Bustillo, K.C., Zeltmann, S.E., Chen, M. et al. (2021). 4D-STEM of beam-sensitive materials. *Accounts of Chemical Research* 54 (11): 2543–2551.

50 Franken, L.E., Boekema, E.J., and Stuart, M.C.A. (2017). Transmission electron microscopy as a tool for the characterization of soft materials: application and interpretation. *Advanced Science* 4 (5): 1600476.

51 Sirringhaus, H., Brown, P.J., Friend, R.H. et al. (1999). Two-dimensional charge transport in self-organized, high-mobility conjugated polymers. *Nature* 401: 685–688.

52 Wirix, M.J.M., PHH, B., Friedrich, H. et al. (2014). Three-dimensional structure of P3HT assemblies in organic solvents revealed by cryo-TEM. *Nano Letters* 14 (4): 2033–2038.

53 Allen, F.I., Comolli, L.R., Kusoglu, A. et al. (2015). Morphology of hydrated as-cast nafion revealed through cryo electron tomography. *ACS Macro Letters* 4 (1): 1–5.

54 Patterson, J.P., Xu, Y., Moradi, M.-A. et al. (2017). CryoTEM as an advanced analytical tool for materials chemists. *Accounts of Chemical Research* 50 (7): 1495–1501.

55 Crassous, J.J., Ballauff, M., Drechsler, M. et al. (2006). Imaging the volume transition in thermosensitive core–shell particles by cryo-transmission electron microscopy. *Langmuir* 22 (6): 2403–2406.

56 Kirillova, A., Schliebe, C., Stoychev, G. et al. (2015). Hybrid hairy Janus particles decorated with metallic nanoparticles for catalytic applications. *ACS Applied Materials & Interfaces* 7 (38): 21218–21225.

57 Furukawa, H., Cordova, K.E., O'Keeffe, M., and Yaghi, O.M. (2013). The chemistry and applications of metal–organic frameworks. *Science* 341 (6149): 1230444.

58 Li, H., Eddaoudi, M., O'Keeffe, M., and Yaghi, O.M. (1999). Design and synthesis of an exceptionally stable and highly porous metal–organic framework. *Nature* 402 (6759): 276–279.

59 Lebedev, O.I., Millange, F., Serre, C. et al. (2005). First direct imaging of giant pores of the metal–organic framework MIL-101. *Chemistry of Materials* 17 (26): 6525–6527.

60 Zhu, Y., Ciston, J., Zheng, B. et al. (2017). Unravelling surface and interfacial structures of a metal–organic framework by transmission electron microscopy. *Nature Materials* 16 (5): 532–536.

61 Zhang, D., Zhu, Y., Liu, L. et al. (2018). Atomic-resolution transmission electron microscopy of electron beam–sensitive crystalline materials. *Science* 359 (6376): 675–679.

62 Fischer, M. and Bell, R.G. (2014). Interaction of hydrogen and carbon dioxide with *sod*-type zeolitic imidazolate frameworks: a periodic DFT-D study. *CrystEngComm* 16 (10): 1934–1949.

63 Ruan, J., Kjellman, T., Sakamoto, Y., and Alfredsson, V. (2012). Transient colloidal stability controls the particle formation of SBA-15. *Langmuir* 28 (31): 11567–11574.

64 Liu, Y., Ju, Z., Zhang, B. et al. (2021). Visualizing the sensitive lithium with atomic precision: cryogenic electron microscopy for batteries. *Accounts of Chemical Research* 54 (9): 2088–2099.

65 Liu, X., Wang, Y., Yang, Y. et al. (2020). A MoS_2/carbon hybrid anode for high-performance Li-ion batteries at low temperature. *Nano Energy* 70: 104550.

66 Wang, X., Zhang, M., Alvarado, J. et al. (2017). New insights on the structure of electrochemically deposited lithium metal and its solid electrolyte interphases via cryogenic TEM. *Nano Letters* 17 (12): 7606–7612.

67 Schneider, N.M., Norton, M.M., Mendel, B.J. et al. (2014). Electron–water interactions and implications for liquid cell electron microscopy. *The Journal of Physical Chemistry C* 118 (38): 22373–22382.

68 Talmon, Y., Burns, J.L., Chestnut, M.H., and Siegel, D.P. (1990). Time-resolved cryotransmission electron microscopy. *Journal of Electron Microscopy Technique* 14 (1): 6–12.

69 Yuwono, V.M., Burrows, N.D., Soltis, J.A., and Penn, R.L. (2010). Oriented aggregation: formation and transformation of mesocrystal intermediates revealed. *Journal of the American Chemical Society* 132 (7): 2163–2165.

70 Ruthstein, S., Schmidt, J., Kesselman, E. et al. (2006). Resolving intermediate solution structures during the formation of mesoporous SBA-15. *Journal of the American Chemical Society* 128 (10): 3366–3374.

71 Pouget, E.M., PHH, B., JACM, G. et al. (2009). The initial stages of template-controlled $CaCO_3$ formation revealed by cryo-TEM. *Science* 323 (5920): 1455–1458.

72 Smeets, P.J.M., Finney, A.R., WJEM, H. et al. (2017). A classical view on nonclassical nucleation. *Proceedings of the National Academy of Sciences* 114 (38): E7882–E7890.

73 Carcouët, C.C.M.C., de Put MWP, v., Mezari, B. et al. (2014). Nucleation and growth of monodisperse silica nanoparticles. *Nano Letters* 14 (3): 1433–1438.

74 Pichon, B.P., PHH, B., Frederik, P.M., and NAJM, S. (2008). A quasi-time-resolved cryoTEM study of the nucleation of $CaCO_3$ under langmuir monolayers. *Journal of the American Chemical Society* 130 (12): 4034–4040.
75 Morosan, E., Zandbergen, H.W., Dennis, B.S. et al. (2006). Superconductivity in Cu_xTiSe_2. *Nature Physics* 2 (8): 544–550.
76 Bianco, E. and Kourkoutis, L.F. (2021). Atomic-resolution cryogenic scanning transmission electron microscopy for quantum materials. *Accounts of Chemical Research* 54 (17): 3277–3287.
77 Wilson, J.A., Di Salvo, F.J., and Mahajan, S. (1975). Charge-density waves and superlattices in the metallic layered transition metal dichalcogenides. *Advances in Physics* 24 (2): 117–201.
78 El Baggari, I., Sivadas, N., Stiehl, G.M. et al. (2020). Direct visualization of trimerized states in 1T′-$TaTe_2$. *Physical Review Letters* 125 (16): 165302.
79 McKelvy, M., Sidorov, M., Marie, A. et al. (1994). Dynamic atomic-level investigation of deintercalation processes of mercury titanium disulfide intercalates. *Chemistry of Materials* 6 (12): 2233–2245.
80 Zak, A., Danczak, A., and Dudzinski, W. (2020). Low-temperature martensite relaxation in Co–Ni–Ga shape memory alloy monocrystal revealed using in situ cooling, transmission electron microscopy and low rate calorimetry. *Acta Crystallographica Section B* 76 (4): 563–571.
81 Mun, J., Peng, W., Roh, C.J. et al. (2021). In situ cryogenic HAADF-STEM observation of spontaneous transition of ferroelectric polarization domain structures at low temperatures. *Nano Letters* 21 (20): 8679–8686.
82 Zhao, W., Li, M., Chang, C.Z. et al. (2018). Direct imaging of electron transfer and its influence on superconducting pairing at $FeSe/SrTiO_3$ interface. *Science Advances* 4 (3): eaao2682.
83 Klie, R.F., Zheng, J.C., Zhu, Y. et al. (2007). Direct measurement of the low temperature spin transition in $LaCoO_3$. *Physical Review Letters* 99 (4).
84 Rui, X. and Klie, R.F. (2019). Atomic-resolution in-situ cooling study of oxygen vacancy ordering in $La_{0.5}Sr_{0.5}CoO_{3-\delta}$ thin films. *Applied Physics Letters* 114 (23): 233101.
85 Gruber, B., Weißensteiner, I., Kremmer, T. et al. (2020). Mechanism of low temperature deformation in aluminium alloys. *Materials Science and Engineering A* 795: 139935.
86 Oliveros, D., Fraczkiewicz, A., Dlouhy, A. et al. (2021). Orientation-related twinning and dislocation glide in a cantor high entropy alloy at room and cryogenic temperature studied by in situ TEM straining. *Materials Chemistry and Physics* 272: 124955.
87 Wu, J., Aslam, M., and Dravid, V.P. (2008). Imaging of magnetic colloids under the influence of magnetic field by cryogenic transmission electron microscopy. *Applied Physics Letters* 93 (8): 082505.

5

Designing Liquid and Gas Cell Holders

With the shrinking of the materials world, understanding of the synthesis and functioning of nanomaterials has become from "desirable" to "essential" to meet the growing need of relating structure, chemistry, and properties at nanoscale. TEM-related techniques are most suited to decipher the nanoscale structure and chemistry, synthesis processes, and/or functioning. Therefore, researchers have been earnestly working to convert the TEM sample chamber into a nano-laboratory over the last few decades. Two approaches: (i) to modify the TEM column and (ii) to design and build dedicated TEM holders, are being pursued for this purpose. Designing and building sample holders that can fit within the pole-piece gap of a TEM to provide controlled environment or stimuli that replicate real-life synthesis or working conditions have proven to be simpler and cost-effective. MEMS technology has made it possible to fabricate microchips with multiple stimuli including windows to contain liquids and/or gas environment. These microchips are often called "lab on a chip" that can be loaded into appropriately modified TEM holders. Such holders include, but are not limited to, heating, cryo, liquid and/or gas cell, indentation/straining, biasing, etc. Although most of the specialty TEM holders mentioned above are now commercially available, understanding the design philosophy and nuances of building them is not only helpful for their effective use but also will assist in modifying an existing holder for a specific function that is not in the market yet. Moreover, most of the vendors of TEM holders are open to collaborate with researchers in improving or adding another functionality to their existing holder. Keeping with the theme of this book, we will concentrate on the liquid and gas environment control and combining certain other stimuli, such as heating, cooling, or biasing, etc. Some of the "one of a kind" TEM holders with multifunctionality are described in Chapter 8. Here we will use the term "window cell" for environment control that could be liquid or gas. An overall review with relevant references for other holders with specific stimuli is also described in Section 1.7.

Briefly, in this chapter we will discuss the history of designing or modifying holders, provide general description and design principle for commercial holders.

5.1 Historical Perspective

As mentioned in Chapters 3 and 4 for heating and cryo, history of modified TEM holders for specific applications is almost as old as the TEM itself. In early days, most of the TEMs were either dedicated for HREM or STEM imaging or chemical analysis, also abbreviated as HRTEM, STEM, and ATEM. Top-entry specimen holders were used for HREM and STEM imaging as they were entirely immersed in the TEM column to reduce the effect of environmental instabilities, such as sound and air drafts on effective image resolution. Also, the sample could be fitted within a smaller objective pole-piece gap with reduced C_s. Although incorporating external stimuli to top-entry holder was challenging as they required external control, thereby reducing the image resolution, both heating and cooling (cryo) capabilities were attempted as described in Chapters 3 and 4. Containment of liquids between two electron transparent windows (amorphous carbon or SiO_2) was also successfully achieved [1, 2].

In 1990s, there was revolutionary shift to develop technology for objective pole-piece design that reduced the C_s value for medium-voltage microscopes and lifted some of the environmental constraints for housing a TEM. Development and incorporation of aberration correction over last two decades have been an additional step forward in improving image resolution (see Chapter 1 for detail). As a result, a modern microscope equipped with analytical capabilities can be operated in either HRTEM or STEM mode without compromising its performance. Another advantage, most relevant for in-situ microscopy, is that top-entry holders were replaced by side-entry holders, making it easier to add external stimuli with external control. As mentioned in Chapter 1, specimen holders to incorporate different stimuli have been developed, and many of them are commercially available. Side-entry specimen holders have been modified to incorporate various stimuli such as heating (Chapter 3), cooling (Chapter 4), nanoindentation, biasing, optical illumination (Section 1.7, 8.1.2 and 8.1.4), liquid (Chapter 6) or gas (Chapter 7) environment, etc.

It is important to remember that most of the current technologies require microfabrication (often mentioned as MEMS-based) and require access to cleanroom technology (see Section 3.2.2). Here we will try to understand the design and fabrication of window holders that are used to contain gas and liquid environment around the sample but are compatible to include other stimuli. Specific design and applications of liquid and gas environments will be further described in Chapters 6 and 7, respectively.

5.2 Design Philosophy

There are two schools of thought among the researchers who design and incorporate novel capabilities to the TEM platform. The first group works with a vision that if they can design and develop a new functionality to the TEM holder, others will find an application for it and use it to decipher novel phenomenon. The second group identifies a specific research problem with specific questions to be answered and make the modifications accordingly. These two design philosophies are not completely apart, and both have positive and negative outcomes. For example, we get a TEM holder that can be used for multiple materials systems in the first case while in the second case it may be restricted for one specific application only. On the other hand, we may end up with a holder that may not find its niche application in first case, while the other at least has one.

It is difficult to describe the path leading to the first scenario as it is up to your imagination, but it is relatively easy to describe the steps leading to the second, where we identify a problem that can only be answered using in-situ TEM. In most cases, our goal is to decipher structural and chemical changes under the influence of external stimuli. Therefore, we can start as follows:

1. Identify a specific problem that cannot be answered using existing characterization techniques, i.e. requires dynamic atomic-scale imaging and/or spectroscopy.
2. **Identify existing holder sources:** purchase one that is available in the market, contact a research group that has developed a holder suitable to your needs but not commercially available. If you can't find anything, go to step 3.
3. Contact a TEM holder vendor with your problem and design idea to find out if they will collaborate or modify their existing holder to suit your needs. If it does not work, go to step 4.
4. Design and fabricate a new holder using your institution's facilities such as clean room and machine shop.

If step 4 is our only option, let us start by understanding the design and functioning of a regular side-entry TEM holder. Major components of a single-tilt side-entry holder (Figure 5.1) are the tip, pin, rod assembly, O-ring seal, and handle. The tip is a thin plate-like section where the sample loading well is located. The thickness, width, and length of this section depend on the TEM column design characteristics; in other words, it depends on the make and model of the microscope. For example, specimen tip is wider and thinner for JEOL microscopes than for Thermo-Fischer (FEI), and slight variations are possible based on the model. The tip section could be an integral part of the rod or welded (attached) to a round rod, whose length and diameter are also microscope-dependent. The rod assembly connects the specimen tip to a handle that is used for holding and inserting the specimen into the TEM column. Often the wiring to connect a stimulus to external controller or motion controllers (piezo drive) is incorporated in the handle. The rod has two parts, first is right next to the specimen tip and its length and diameter again are TEM-dependent. Second part is in principle a continuation of the first part but with larger diameter with an angled step between the two parts. This part has groves where one or two O-rings (again depends on the TEM make and/or model) are placed to seal the specimen chamber from the liner tube walls. The pin opens the sample loading chamber to a vacuum pump to evacuate the sample transfer chamber. TEM manufacturer most probably will not share the design of their holder with you so you will need to do reverse engineering, i.e. carefully examine and measure each part of the regular sample holder to design your own. It is important to remember that the location and size of each part are specific to a particular TEM and may or may not have large tolerances. However, it is possible for TEM manufacturers engineering

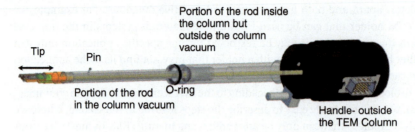

Figure 5.1 Schematic image of a side-entry TEM holder for a Titan microscope. The size of various sections depends on the make and model of the TEM. Note that the hollow rod section can be used to connect the tip section, where sample or microchip is loaded, to the external controllers for gas/liquid flow, heating, cooling, biasing, etc. Courtesy: Glenn Holland, NIST.

team to agree to look at your CAD design and make suggestions and often willing to sign an non-disclosure agreement (NDA) such that your idea is safe from possible encroachments. All of the internal parts must be made of non-magnetic material.

> Consult with the technical staff of the TEM manufacturer to determine the material and functioning of each section before building your own holder.

As mentioned above, the overall design, diameter, and length of various sections of the holder are defined by the specific TEM model and cannot be changed. On the other hand, the tip section of TEM holders can be modified within the dimensions available for a particular holder. Here we will consider the MEMS-based microchips with windows where we want to incorporate other special functionalities. We have two possibilities: (i) we design and fabricate a microchip to load our samples on that will fit in the existing holder, or (ii) build a new holder with redesigned sample loading area and sample-well to accommodate our modified microchip. In either case, the first step is to design and fabricate a microchip with windows.

In general, window cells are fabricated for multiple-purpose applications, such as to confine liquid or gas along with other stimuli. Here in we will use the term "microchip" or "chip" in short that contains a window cell for loading liquid or gas and may or may not be equipped with other stimuli. In the following Sections (5.3 and 5.4), we describe the basic design and working of such microchips and fundamental steps for building a new holder. Other methods to confine liquid or gas around the sample, such as open or graphene cell, are discussed in Chapters 6 and 7, respectively.

> The weight of the handle of the TEM Holder should be considered when packing the external connections to various stimuli in it. It is important to remember that the end of the holder rod inside the column vacuum is not clamped. The weight of the handle can tip the holder such that it may be impossible to reach eucentric height using Z controls
> – Penghan Lu, Ernst Ruska Center, Jülich

5.3 Windows

Windows are invariably a crucial part of the liquid and gas cell but can also be important for heating and biasing holders. The ability for electrons to travel through the windows, sample, and liquid or gas with minimum multiple scattering events is the key factor when designing a microchip. In other words, the same image formations rules apply here as for the TEM imaging and spectroscopy in vacuum. Although the contributions of electron scattering from windows and liquid or gas cannot be completely abated, it should be minimized as much as possible. Following are some of the important considerations:

5.3.1 Image Resolution: Thickness and Material Properties of the Windows

The image resolution achieved when using a microchip depends on the film's (window's) (i) electron transparency and (ii) electron path length. Electron transparency or electron path length affects signal-to-noise ratio (SNR), which in turn affects both the effective image resolution and intensity of the EELS peaks. Whereas electron transparency is a materials property of the window materials, electron path length depends on the combined thickness of the windows and gas or liquid within the two windows. In principle, the resolution includes both elastic and inelastic electron scattering through the sample for the phase-contrast TEM imaging. There are two ways of electron scattering from window material and liquid can affect the resolution; one is the reduction in SNR due to multiple scattering, and other is an increase in the value to chromatic aberrations (C_c) due to inelastic scattering.

TEM resolution in liquids:

$$d_{TEM} = A_L \cdot \alpha \cdot C_c \cdot t/E^2 \quad (5.1)$$

where A_L is liquid dependent constant, α is objective semi-angle, t is thickness, C_c is the chromatic aberration constant, and E is energy (Figure 5.2a), and for STEM [11], d_{STEM} is proportional to $t^{1/2}$ (Figure 5.2b).

More precisely, the resolution in STEM mode is dependent on the location of the probe within the sample with respect to the windows. The sample or nanoparticles located near the top window are scanned by an unperturbed electron probe, while the beam broadening by the liquid affects the probe size and thereby the resolution for the sample near the bottom window.

As expected, aberration-corrected microscopes improve the resolution not only by compensating for C_c (improved dose efficiency), the C_s correction improves the interpretability of TEM images [10]. As the SNR is strongly dependent on the dose rate, the spatial resolution is also found to be proportional to (dose rate)$^{-1/4}$ [3]. De Jonge et al. have reported achievable resolution based on the dose-limited bright-field contrast and phase-contrast resolution. They predict the phase-contrast TEM resolution for carbon nano-objects in thin water layers to be 2 nm for $D = 10^2$ e$^-$ Å^{-2} or 6 nm for $D = 1$ e$^-$ Å^{-2}, where D is the dose that matched well with experimental results [12, 13]. Moreover, the intensity of the EELS peaks is reduced with increased thickness, both in low-loss and core-loss region, as the

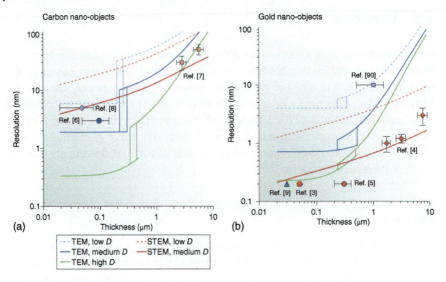

Figure 5.2 Theoretical maximum image resolution versus thickness of water. Calculated TEM and STEM resolutions for spherical nano-objects composed of (a) carbon and (b) gold in water. The resolutions are calculated for different electron densities, D. For TEM, phase-contrast and bright-field contrast equations apply to thin and thick water layers, respectively. The exact transition thickness is unknown, as indicated by the vertical lines. The resolution was calculated without taking limitation by diffraction or lens aberrations in account, A beam energy of 200 keV, 5 mrad objective aperture for bright-field contrast, TEM camera detector quantum efficiency of 20%, an optimized STEM annular dark-field detector opening angle, the objects located at the electron entrance (for STEM) or exit (for TEM) surfaces, and a negligible window thickness. High $D = 10^5$ e$^-$ Å^{-2}, low $D = 1$ e$^-$ Å^{-2}, and medium $D = 10^2$ and 10^3 e$^-$ Å^{-2} for carbon and gold, respectively, reflecting typical experimental values. Experimental data points acquired at approximately corresponding D are included for comparison. Source: Jonge et al. [10]/Springer Nature.

contribution from the multiple scattering increases. The calculated resolution for window cell in TEM is found to be limited to about 0.5 nm for 50-nm-thick SiN at 200 kV. The recommended liquid thickness for low loss is below 130 nm and lower for core-loss spectroscopy [4]. However, atomic-resolution images, using graphene liquid cell, have been reported [14, 15].

5.3.2 Strength and Flexibility

Several window materials such as graphene, graphene oxide, hexagonal boron nitride (hBN), stoichiometric silicon nitride, silicon dioxide, and polyimide have been considered suitable for fabricating liquid/gas cell windows. Whereas graphene and hBN can be made extremely thin and electron transparent, they are not very compatible with mainstream cleanroom techniques for large scale production or incorporation of other stimuli. Silicon dioxide (SiO$_2$) has low Young's modulus, is electron transparent, and stress-free thin films can be easily grown but has low etch rate in potassium hydroxide (KOH) used for wafer-scale fabrication. However, SiN$_x$ membranes have become a material of choice due to their clean room compatibility, high strength, low Young's modulus, and high level of fracture toughness. Non-stoichiometric SiN_x is better suited for fabrication as its expansion coefficient can be matched to Si wafer to avoid stress in the film upon heating. Commonly used standard processes in nanofabrication facilities can produce films with tensile stress as low as 200 MPa.

5.3.3 Tolerance for the Pressure Difference

Microchip windows not only need to withstand the pressure generated by the liquid or gas enclosed by the windows but also the pressure difference as the TEM column is kept in vacuum. This difference creates a deflection in the window that results in a nonuniform electron path length across the cell and adversely effects the spectroscopy measurements (Section 6.5.3) [16]. The pressure difference in a square membrane is given by:

$$P = (3:93 t\delta\sigma_0/a^2) + 1:834 Y t \delta^3/(1-v_f)a^4) \tag{5.2}$$

where σ_0 is the initial stress in the membrane, Y the Young's modulus, v the Poisson's ratio, t the film thickness, a the halfwidth of the membrane, and δ the deflection (note that for a given value of a, the deflection is approximately a factor of 2 larger for a long rectangular membrane) [17]. Equation (5.2) is only valid for the situation when the thickness t is much smaller than the half-width, a; otherwise, finite element analysis (FEA) is necessary to estimate the deflection. The options for controlling the deflection are limited: the size of membrane, a, can be reduced, or the initial stress can be increased. However, reducing width of a window is limited as the wafer thickness variations cause membrane widths to vary by as much as a factor of 2 below 20 µm. Also, the initial stress cannot be increased above 1 GPa without compromising the strength of the membrane [18].

This deflection results in what is known as bulging of the windows and for thin films, it can be characterized by making load-deflection measurements. The change in the focal plane of a thin film, after applying pressure on one side, is measured using an optical microscope that has been calibrated to measure the deflection. The load-deflection behavior is reported to follow the following relationship for a square membrane [19]:

$$(Yt/a^4)\delta^3 + (1:66 t\sigma_0/a^2)\delta = 0:547p \tag{5.3}$$

where δ is the maximum membrane deflection, t is the film thickness, a is the side of the square membrane, Y is Young's modulus, σ_0 is the residual stress, and p is the external differential pressure [20]. Dwyer and Harb have used this equation to model deflection of a rectangular membrane, with $2a$ representing the smaller lateral dimension of the rectangle [5]. Figure 5.3 shows the calculated results for the maximum membrane deflection, δ, as a function of differential pressure, p, assuming a residual tensile stress of 200 MPa, and a Young's modulus of 300 GPa for a SiN_x membrane. It is also assumed that the thin membrane will be able to withstand the pressure up to the maximum amount displayed in the plots. However, it is possible for membrane to fracture at some point as the pressure is increased.

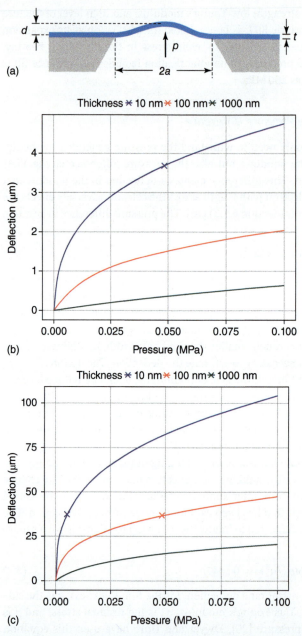

Figure 5.3 (a) Cross-sectional view depicting bulging of a square-shaped silicon nitride membrane of side $2a$ and thickness t, due to net differential pressure p. The parameter of interest is the maximum deflection d at the center of the membrane. Maximum deflection of a 100 mm square membrane (b) and a 1 mm square membrane (c) are plotted as a function of differential pressure up to 1 atm and for different membrane thicknesses (10, 100, and 1000 nm). The cross marks represent calculated fracture points. The fracture model assumes that the membrane has a preexisting crack line that is 10 mm in length. Absence of a fracture point indicates that the membrane is able to withstand the pressure at least up to the top end of the calculated range.

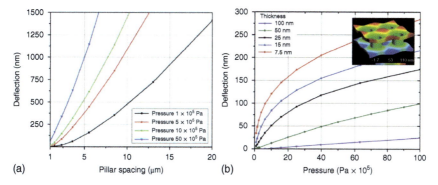

Figure 5.4 (a) Membrane deflection as a function of pillar spacing for different pressures. (b) Membrane deflection as a function of pressure for membranes of different thicknesses and a constant pillar edge-to-edge spacing of 1 μm. In all cases the initial membrane stress is 180 MPa. The inset shows a finite element simulation of the deflections of a structure with a membrane thickness of 50 nm and a support pillar pitch and edge-to-edge spacing of 2 and 1 μm, respectively. The vertical displacements in the image are exaggerated by ≈ 5 Å for clarity. Source: Tanase et al. [18]/Cambridge University Press.

Therefore, the fracture toughness of the membrane is an important materials property that is directly related to the lateral stress at which a membrane containing crack will fail [5].

Based on these calculations, Tanase et al. designed and built a monolithic liquid cell with small viewing windows, separated by pillars, to prevent the windows from bulging [18]. The small windows between the membrane support pillars keep deflections at the few-nanometer level (<10 nm) when the cell is in operation and prevents capillary forces from collapsing and sticking the membranes together when the cell is dried after etching, or when fluid is introduced during sample loading (Figure 5.4a). An attractive feature of such a cell is its ability to support very high pressures – up to 5×10^6 Pa (≈ 50 atm) before there is any danger of the membrane breaking (Figure 5.4b). Mueller et al. have also evaluated the controlled flow, and liquid path length for practical applications [21]. Similar monolithic cells have been microfabricated that enable membrane geometry on the scale of a few micrometers [22]. However, such microchips have not become readily available, partially due to the difficultly to incorporate a nanofluidic system to flow liquids through the microchip (see Chapter 6 for more information).

> Round windows are shown to be more robust than square or rectangular windows.

5.3.4 Inert or Corrosion Resistant

Reaction of the liquid or gas with the window material may generate holes and destroy the windows; therefore, it is obvious that windows should be made of a material that is inert to the commonly used liquids, such as water, salt solutions, and organic solvents, and gasses such as H_2, O_2, N_2, CO, CO_2, etc. Again, SiN_x is currently an accepted window material and considered revolutionary for the development of in-situ TEM technology. Still, its compatibility with the liquid or gas or other stimuli to be used should always be checked as explained more in Chapters 6 and 7.

5.4 Microfabricated Window Cell (Microchips)

As mentioned in Chapters 3 and 4, the development of MEMS technology has played a crucial role in the development of dedicated in-situ holders. Broadly, microchips for window cells are made by fabricating two microchips that consist of bottom and top pieces with windows, with or without flow control, which can be bonded, glued, or clamped together using spacers [5, 23–27]. The two-microchip configuration allows for us to vary the height of the cell, thereby controlling the mean free pathlength (λ) of the gas or liquid enclosed between the two windows by changing the thickness of the spacer. The exact alignment of the two windows can be challenging but possible using appropriate markers. For example, microsphere ball lenses can be placed in alignment holes etched during fabrication of the microchips [28]. Also, design for self-aligning cell, composed of out- and in-frame microchips with SiN_x window, which employs surface tension mechanism generated using bio-sample droplet and does not need positioning tools, has been reported [24]. Fabrication of windows is the fundamental process needed to make microchips; therefore, we first describe the general process for making a window cell using two microchips for different types of cells with specific applications.

It is a wafer-scale process where multiple microchips with windows are fabricated on single wafer, with snap lines etched to separate them later (Figure 5.5a,c). The advantage is that several microchips can be fabricated simultaneously that have undergone the same fabrication process. For a simple two-microchip configuration, a spacer layer of SiO_2, Au, or In is deposited on the bottom microchip (Figure 5.5b) [25, 26]. Spacers could also be polystyrene microspheres that are placed at the four corners of the bottom chip, and their size controls the thickness of the liquid/gas layer [25]. Also, other stimuli, such as heating and/or biasing, liquid flow channels, are fabricated using etching or nanolithography on the bottom microchip. In the final step, individual microchips are diced along the etch lines using a dicing saw and are ready to be put together after loading the sample (Figure 5.5d). A brief description of microchip fabrication process is given below.

As mentioned earlier, microfabrication is a clean room process, and there is considerable "know-how" needed for designing and making wafer-size samples as explained in various publications referred in this section. The entire process can be more than 30 steps, depending on the configuration of the final product. Detailed process for fabricating specific type of the windowed cell can be found in Refs. [4, 23, 26, 27, 29–31]. A general toolbox for nanolithography, developed at NIST, can provide a guideline for sculpted patterning for nanoscale devices [6]. Here we will consider general layout and cleanroom procedures for fabricating static or a flow-controlled microchip with windows only.

Simple process steps of fabricating thin SiN_x windows are shown in Figure 5.6 and explained in more detail in Refs. [5, 32]. After etching the snap lines, we start

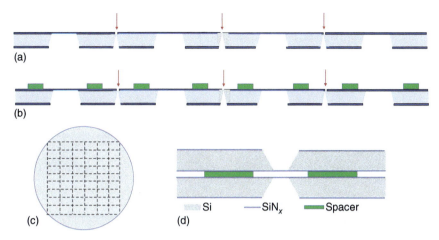

Figure 5.5 (a) Top microchip with SiN$_x$ windows and (b) bottom microchip with SiO$_2$ spacers, (c) top view of the Si wafer with etched lines defining individual microchips as marked by red arrows in (a) and (b). (d) A sandwiched cell showing the space for liquid.

with depositing thin SiN$_x$ films on both sides of a clean and polished Si wafer by low vapor chemical vapor deposition (LPCVD) (Figure 5.6a). Dichlorosilane and NH$_3$ are used as precursors with LPCVD at temperatures above 800 °C (above decomposition temperature of NH$_3$), and their ratio is used to control the value of x in SiN$_x$ thin films to be deposited. It is important to remember that the non-stoichiometry of SiN$_x$ is important to match its expansion coefficient with Si. The thickness of the deposited film is generally kept around 50 nm by varying the deposition time, as it is thick enough to sustain the pressure difference and thin enough to be electron transparent for high-resolution imaging [3, 33]. Fabrication of thinner windows, up to 20 nm, is possible but requires special procedure and handling [29].

Polymethyl methacrylate (PMMA, commonly used photoresist) is spin coated on both sides over the SiN$_x$ film in the next step to protect the film from damage during the patterning process (Figure 5.6b). A window feature is then created by standard photolithography process (Figure 5.6c), followed by reactive ion etching (RIE, CHF$_3$, and CF$_4$ are commonly used etching agents) to remove one side of the SiN$_x$ layer (Figure 5.6d). The size of the window feature is usually kept $20 < x < 300$ μm and is limited by the thickness variation of Si wafer. The exposed Si area is then etched away using KOH and the SiN$_x$ on the other surface serves as an etch stop (Figure 5.6e). The cleanliness of the KOH is very important to avoid deposition of contaminants on the microchips. Similarly, uniform temperature of the etching solution is important for uniform etching of the Si [26]. A protective layer is often deposited as the last step to provide added protection for the windows during transport and handling that should be removed before loading the sample. The wedge-shaped etch profile arises from preferential etching of the different crystallographic plane of Si (Figure 5.6f). Though KOH etching of Si(100) with a SiN$_x$ mask is a most popular approach as it leads to formation of 57° wall, that is helpful for EDS analysis (see Section 5.5.1 for

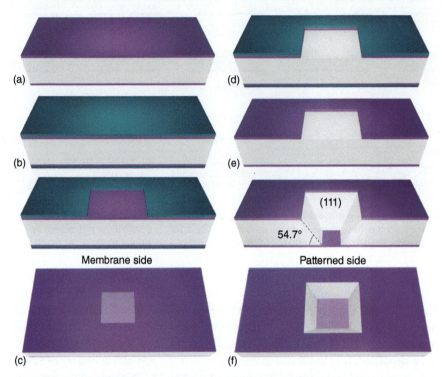

Figure 5.6 (top) Fabrication steps for free-standing SiN_x windows: (a) SiN_x is deposited on both sides of a double-side polished (100)-oriented Si wafer by LPCVD; (b) photoresist is sequentially coated onto both sides of the wafer. While photoresist is only needed on one side for photolithography, the photoresist on the back (membrane) side serves as a protective layer against etching of the SiN_x in step (d). (c) The window feature is created in the photoresist layer via standard photolithography (UV exposure and development). (d) Reactive ion etching (RIE) is used to transfer the window feature into the SiN_x layer. Note that RIE is isotropic; the vertical profile of the exposed area as shown in the figure does not accurately depict the actual isotropic etch profile, but because the depth of the etched region is much smaller than the lateral dimension of the exposed area, this effect can be ignored. (e) Photoresist is stripped from both sides using a suitable chemical bath. (f) Potassium hydroxide (KOH) is used to etch through the wafer and create the free-standing SiN_x structure. Potassium hydroxide stops at the SiN_x layer. Potassium hydroxide is anisotropic, with etch rates of the (100) and (110) planes much larger than the (111) planes. Thus, the etch front progresses along the (111) planes of the crystal, resulting in the narrowing of the window dimension relative to the dimension of the mask feature by a factor of the thickness of the wafer multiplied by ˇ2. To speed up the onset of the KOH process, it is recommended to dip the patterned wafer into a buffered oxide etch solution (BOE) prior to the KOH etch step. The buffered oxide etch solution removes the native oxide layer that forms naturally on the exposed Si. (bottom) Top down views of the fabricated membrane and supporting frame. The view from the membrane side shows only the window. The view from the patterned side shows the etch pit whose angles are determined by the silicon crystal planes. Source: Adapted from Dwyer and Harb [5].

details). Other unconventional etch profiles are possible to achieve by using alternative crystallographic orientations and mis-cut surface of Si wafers combined with different etch stops and etching solutions [34].

As mentioned above, this fundamental process is the starting point to make either static or flow cells with or without other stimuli as explained in Sections 5.4.1–5.4.4. Some variations in the clean room processing steps to fabricate windows and/or pattern contacts for various types of window cells have been reported and should be checked if you plan to fabricate your own microchips [7, 22, 26–30].

As described in Section 5.3.3, and discussed later in Section 6.5.3, bulging of windows due to pressure difference needs to be kept to a minimum value. One of the proposed solutions is to make multiple small circular windows separated from each other such as to distribute the thermal and mechanical stress. Pérez Garza et al. have used FEA to estimate the heat distribution across the window, size of the window, and inter-window distance to minimize window deflection in designing a gas nanoreactor [35].

A summary of various parameters and fabrication strategies for windowed microchips is given in Table 5.1.

5.4.1 Static Cells

This type of cell utilizes the microfabricated microchips with SiN_x windows to confine the liquid within the windows and is not usually used to observe gas–solid interactions. For static liquid experiments, a simple process is to use the top and bottom microchips made following the procedures shown in Figures 5.5 and 5.6. Then a fixed amount of liquid is drop casted on a microchip window (bottom) and sealed with the top microchip before the liquid dries out (Figure 5.7a,b). The amount of liquid is determined by the thickness and location of the spacers between the two microchips. Another way to add liquid in the cell by using reservoirs that are fabricated on the top microchip and are sealed using a top plate with a hole slightly larger than the viewing window (Figure 5.7c,d). Liquid, filled in the reservoir, enters the viewing window area via capillary action, and both reservoirs need to be filled to avoid any empty space within the cell. This method is available only if the microchips to be used are fabricated with reservoirs. Again, thickness of the liquid is determined by the height of the spacer layer between the two windows.

There are a number of ways to sandwich the two microchips such that the two windows are aligned and vacuum sealed [37]. Gluing them together using a vacuum-compatible glue is a simple method. High-temperature and high-pressure wafer bonding can also be used to form a hermetic seal between two wafers and was first used for a windowed-flow cell [38]. Squeezing the two microchips together (clamping), using O-ring seal, is yet another method where groves for an O-ring groves are etched in top and bottom microchips and the diameter of the O-ring controls the gap (thereby liquid thickness) between the two microchips [7]. However, alignment of the two windows is crucial to maintain the originally designed viewing area available. It is often achieved by aligning the two microchips

Table 5.1 Some common considerations for the fabricating microchips with windows.

	Comments	Reference
Substrate	Si wafer (100 or 110)	Commonly used	
Substrate thickness	100 to 500 μm	Should fit in the holder tip	[24, 29–31]
Window material	Preferably SiN$_x$	Universally accepted due to its suitable properties (Section 5.3)	
Window thickness	20 to 100 nm	Min. is constrained by wafer topography and strength; max. is constrained by effective SNR. 30 to 50 nm is preferred.	[4, 24, 27–30]
Window size	Thickness: <50 nm Size: 20 μm × <300 μm Circular: 30 μm diameter or 6 μm diameter	• Large viewing area-low deflection due to pressure difference • Round windows are mechanically more stable	[7, 27] [28] [35]
Etcher of SiN$_x$	Reactive ion or plasma	REI with SF$_6$ or CF$_4$ or CHF$_3$/Ar-based plasma	[26–29]
Etcher for Si	KOH	Commonly used	
Photoresist	PMMA	Commonly used	
Spacer material	SiO$_2$; Indium, polystyrene microspheres; O-ring	Spacer material varies	[7, 23, 26, 28, 30]
Sandwiching two chips	Gluing; wafer bonding using In or Au; O-ring seal	There are different methods to sandwich two microchips	[24, 26, 28, 30, 31, 36]
Liquid thickness	100 nm to 3 μm	Controlled by spacer thickness, higher for gas cells	[7, 23, 28, 30, 31, 35, 36]

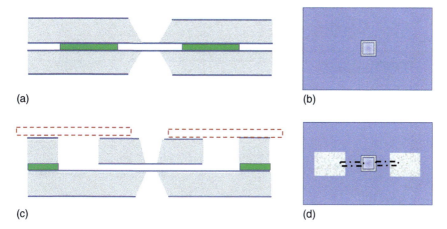

Figure 5.7 Static cells: (a, b) Side and top views of static cell where the liquid is drop casted on the bottom microchip and sealed by the top microchip. (c, d) Side and top views of a static cell with on microchip reservoirs to introduce liquid in the cell. The liquid thickness is controlled by the thickness of the spacer (solid green section) between top and bottom microchips.

using a high-power optical microscope. Precise location of groves for O-ring seal is also helpful. Li et al. also describe fabricating microchips with positioning marks that can be used for precise alignment of the windows [23].

These simple microfabricated chips containing static liquid can be used in a standard TEM holder depending on their size, shape, and thickness compatibility with regular TEM holder. Alternatively, the tip part of the standard holder can be modified to accommodate the microchips. On the other hand, entire holder needs to be rebuilt to accommodate microfluidic channels or other stimuli such as heating or biasing as explained in Sections 5.4.2–5.4.4.

5.4.2 Flow Cells

Although the basic steps for fabricating microchips with SiN_x windows are generally the same, the microchip fabrication for liquid flow requires additional steps. Here the liquid flow channels or vias are etched on one of the microchips following the same procedure as for making windows (Figures 5.5 and 5.6). The channels are coated with protective material and connected to gas or liquid flow system through flexible clean microfluidic tubes (e.g. PEEK tubes) that run through the specimen holder rod. Therefore, the specimen holder to accommodate a flow cell must be modified as discussed in Section 5.2.

There are several ways to flow the liquid through the cell (Figure 5.8). For example, the liquid could enter from one side of the window and is pumped out from the other (Figure 5.8a). Alternatively, both inlet and outlet could be from the same side (Figure 5.8b) or by using vias patterned on the top of the microchip (Figure 5.8c). Again, there are multiple ways to introduce liquid through the vias such as using a syringe and pump combination to control the liquid pathlength and flow [21]. On the

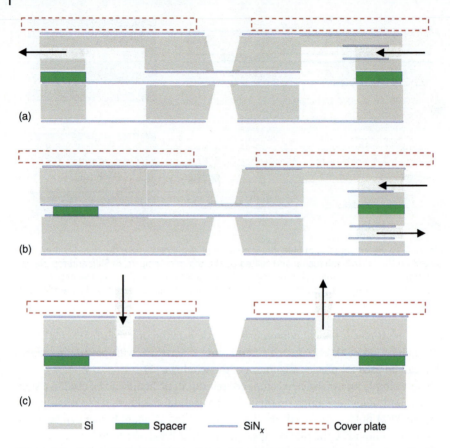

Figure 5.8 Schematic drawings showing various methods for incorporating gas or liquid flow in a window cell. Note that inlet and outlet channels can be located (a) on the opposite side, (b) same side or (c) on the top part of the microchip.

other hand, dual pumping system, one on the inlet and the other on the outlet side, is commonly used to avoid flow resistance due to capillary forces, and the flow rate is controlled by the diameter and length of the microfluidic tubes as well as by the pumping speed.

Menon et al. have reported an on-chip flow control system that does not require external flow channels through the TEM holder body [39]. They utilize the MEMS technology to fabricate on-chip reservoirs, and the flow is activated using an electromechanical pump. Most of their design follows the processes described above (Figures 5.5–5.7), except a hydrophobic barrier is patterned near the middle of viewing window (flow channel) dividing it into two parts. The electrical contacts are also patterned on the chip to connect a pair of electrodes. The liquid flow can be initiated in either direction by changing voltage across the pump electrodes (see Figure 6.5 for more details) [39].

5.4.3 Incorporation of Other Stimuli

Other stimuli, such as temperature and electric bias, are often incorporated in a microchip to follow their effect on structure and property under different environment. For example, heating of the sample is needed to follow the temperature effects on the synthesis or functioning of nanomaterials under controlled environments. It is critical for gas–solid interactions but is also advantageous for understanding nucleation and growth of nanoparticles in solution. Electrical biasing is important to follow the electrochemistry, especially for the functioning of batteries and solid-electrolyte reactions.

A combination of micro- and nano-lithography has been successfully employed to further modify the microchips to include heating and/or biasing elements in the microfabrication process [40]. Gas- and liquid-cells with heater and/or biasing are also commercially available (see the vendors list and link to their website in the reference section). On-chip joule heaters can be incorporated such that they are in the vicinity of the sample and provide effective heating for window gas cell experiments [35, 36, 41–43]. It is important to make sure that the Si chip remains at room temperature during sample heating, i.e. heat is confined to the window area, to avoid its thermal expansion that will cause sample drift. Finite element analysis using COMSOL, or similar platform, can be used to map the heat distribution in the window (see Section 3.3.3 and Figure 3.8). Laser heating has also been explored, where the laser is introduced through the holder rod and focused on the sample using a mirror located in the tip section of the holder [44]. Laser source has also been introduced using independent ports, if available in the sample area, to stimulate light-induced (photocatalytic) reactions [45–48] (see Chapter 8 for more information), which can also be used for laser heating. However, overall temperature, which can be estimated but not easily measured, is a major drawback laser heating.

> The biasing holders suffer from voltage drops with all the connections, which need to be calibrated or factored into quantitative biasing/electrochemistry experiments. We used a femtoampere-sensitivity potentiostat with a noise reduction preamp to mitigate the voltage differences between the supply and MEMS electrode, with reference electrodes.
> – Katherine Jungjohann, NREL, Denver

Liquid cell provides a natural environment for measuring and understanding electrochemical reactions that encompass variety of applications. Therefore, incorporation of electrodes for biasing has been successfully implemented in several commercial holders. Structural and chemical changes as a function of biasing voltage/current have made major impact in the area of battery research (see Section 6.6.7 for details). Incidentally, metal electrodes need to be lithographically patterned for both heating and biasing. Figure 5.9a shows a simple schematic for patterning four electrodes with connection pads on the same side of a chip. Another example is

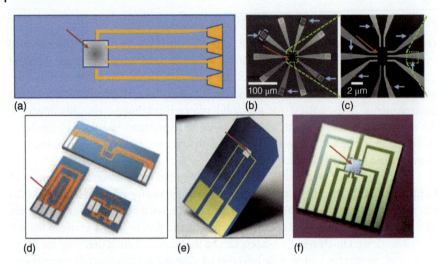

Figure 5.9 (a) Schematic for introducing heating and/or biasing connections to the gas or liquid cell. (b, c) Scanning electron images showing the 10 Ti electrodes, patterned at the center of the bottom chip (marked by blue arrows). Source: Leenheer et al. [49]/from American Chemical Society. (d–f) Images of commercially available chips for heating and biasing (see their websites for details). (d) DENSsolutions (https://qd-europe.com/at/en/product/wildfire-in-situ-tem-heating-series). (e) Norcada (https://www.norcada.com/products/in-situ-tem), and (f) Hummingbird (https://hummingbirdscientific.com/products/electrical-biasing-holder). Viewing windows are marked by red arrows.

a nanoaquarium, designed with flow channels and electrode in a more complicated way with connection pads on both side of the chip [31]. Leenheer et al. [49] have reported a star-shaped 10-electrode configuration for biasing the sample in different directions (Figure 5.9b,c). Figure 5.9d–f show designs of commercially available chips as shown on their websites. As mentioned earlier, a nanotechnology tool box, available from NIST, can be used to design any shape and configuration for nanolithography [6]. Like the gas/liquid flow lines, the electrical connections run through the sample holder rod and are connected to external controllers. Modern commercial systems come with computer programs for controlling the flow rate, temperature, and bias.

5.4.4 Monolithic Microchips

The lateral dimensions of the viewing windows in two chip systems described above are generally larger than 20 μm. This is because smaller membrane areas are difficult to make reproducibly, using etching schemes, and the accuracy required to align the two windows. Monolithic chips provide an alternative design strategy where microfluidic channels are created in a single chip [18, 22]. Here the first step is the same as described above, i.e. SiN_x layer is deposited on clean and polished Si wafer and etched from one side (Figures 5.6 and 5.10a). This membrane functions as a support for mechanical stability and substrate on which connection pads to the other stimuli, such as heating and/or biasing, can be patterned. Next a SiO_2 layer is deposited followed by the deposition of sacrificial polysilicon and then photoresist (Figure 5.10b).

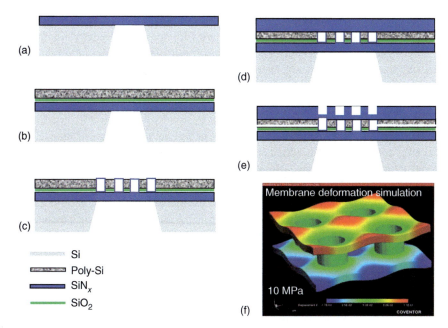

Figure 5.10 (a–e) Schematic of various fabrication steps for monolithic chips and (f) simulation showing minimum deflection in windows with 10 MPa pressure in the cell.

Multiple small windows, of the order of 2 μm, separated by pillars acting as channels, are then patterned using a UV lithography tool. In the next step, holes are created in the poly-Si/SiO$_2$ layer using the regular RIE as described in Section 5.4 and shown in Figure 5.10c. A thicker (200 nm) SiN$_x$ layer is then deposited on the top creating channels in the poly-Si/SiO$_2$ layer (Figure 5.10d). The thicker layer provides the mechanical strength and ease of handling and thinning down to create smaller viewing areas as shown in Figure 5.10e. A liquid inlet or outlet is also be created by etching on the side of the chip from where the liquid will enter due to capillary forces and can be sealed using UV curing system afterward. The smaller size of individual window reduces the window bulging due to pressure difference and multiple windows provide large viewing area. The simulation shows that minimum deflection of the windows is observed even with 10 MPa pressure inside the cell [18]. Such static cells are self-contained similar to the two chip systems shown in Figure 5.5.

Monolithic chips can be disposed to avoid cross-contamination and can also be designed to fit in the regular TEM holders as they are typically half the thickness of the sandwich systems. Although, reduced window deflection is a big advantage over the two chip sandwich systems, they are not very popular as they require more precision and cleanroom processing time, are only compatible with liquid or suspensions, and liquid flow through the micron size channels is not straightforward.

5.5 Examples of Modified Window Holders

We are in a time period when window-cell chips and modified holder with gas/liquid flow, heating, and biasing are commercially available. But as mentioned in Section 5.2 above, sometimes we find a problem that cannot be answered by existing technology. We have an option to work with one of the holder vendors and modify the existing chips, and/or holder to suit our need or design and fabricate new chips and the holder body. Here we provide some examples of both scenarios.

5.5.1 Redesigning the Microchips for Commercial Holder

As we know that spectroscopy is an important part of sample characterization. EELS and EDS are two commonly used spectroscopy methods to characterize and relate the structure and chemistry of the sample (see Chapters 6 and 7 for detailed applications for liquid and gas cell, respectively). While EELS have some issues due to the window bulging, which increases the total thickness that electron need to travel, there are simple solutions, such as collecting EELS data near the edges instead of the middle of the chip. Solutions for EDS data collection are not as straightforward. As the EDS detector arm is located slightly above but away from the sample in the sample chamber (Figure 5.11a). Normally the sample is tilted toward the detector by at least 20° to obtain good SNR. The window cells pose two problems, (i) it may be difficult to tilt a thicker specimen holder tip depending on the pole-piece gap of a TEM, and (ii) more importantly, the body of the window cell (penumbra) blocks the line of sight for X-rays coming from the sample to the detector. In summary, our efficiency to collect the EDS data is controlled by the tilt capability, the location and height of EDS detector with respect to specimen holder tip, as well as magnitude of the window cell side walls.

One of the solutions will be to use a TEM equipped with four in plane detectors, such as SuperX instead of one, to improve the SNR without tilting the sample [50]. Alternatively, Zaluzec et al. in collaboration with Protochips Inc. have redesigned the chip by shaving off the cell side walls (Figure 5.11b) to reduce the signal blockage by the cell penumbra [8]. The redesigned system is called "butterfly" configuration based on the shape (Figure 5.11b). As mentioned above, the line of sight depends on the location and height of the detector arm in the sample chamber. These values

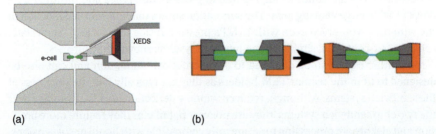

Figure 5.11 (a) Schematic showing the location of a window cell and EDS detector in objective lens pole-piece gap. Shadowed area illustrates the line of sight blocked by the penumbra of the cell to EDS detector, shown in red. Nominal dimensions of the objects sketched in this drawing are drawn approximately to scale. (b) Schematic diagram showing cross section of a window cell before (left) and after modifications (right), illustrating the removal of material that blocks the line of sight path to the X-ray detector. Source: Zaluzec et al. [8]/Cambridge University Press.

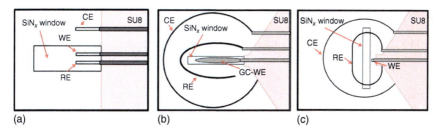

Figure 5.12 Schematic of the two electrode configurations commercially available from Protochips, Inc. (https://www.protochips.com) electrochemical chip: (a) linear configuration (b) glassy carbon (GC) electrode. (c) Schematic illustration of redesigned electrochemical chip and electrodes. Here the working electrode (WE) is $\approx 10\,\mu m$ outside the SiN_x window, whereas the reference (RE), counter (CE) electrodes circumscribe the SiN_x window. Source: Schilling et al. [54]/Cambridge University Press/CC BY 4.0.

can be used to calculate the angle of the cut needed to mitigate the blockage by the penumbra. Incorporation of tapered clamping plates can also reduce the signal blockage.

Another modification is made to improve the electrochemical measurements of corrosion. Corrosion of metals in aqueous medium is commonly observed phenomenon and liquid window cell provides an unprecedented opportunity to understand and measure the electrochemistry of the corrosion [9, 51–53]. Despite the electron beam effect that alters the chemistry of the water, commercially available chip designs for liquid biasing holders have been successfully employed to study battery materials and nanoparticle growth. However, this configuration was not found suitable for quantitative measurements of alloy materials, such as stainless steel under potentiostatic control.

Schilling et al. worked with Protochips (https://www.protochips.com) also to redesign the electrode configuration for quantitative electrochemical measurements of stainless steel [54]. In order to investigate the nanoscale changes occurring in electron transparent alloys under potentiostatic control to observe the potentiostatic polarization behavior, in-situ, they modified the electrode geometry from parallel (Figure 5.12a) to circular (Figure 5.12b). Also, the reference and counter electrodes (RE and CE) were modified to circumvent the working electrode (WE) such that near-uniform current lines can be achieved across the sample (Figure 5.12c). In addition, a WE is positioned outside the SiN_x such that the FIB lamella of alloy samples can be securely attached to the more robust Si chip, instead of SiN_x windows for corrosion testing. Glassy carbon, originally used to fabricate WE, had potential of introducing extra material with different potential that complicates the results and interpretation, is therefore, replace by Pt. They also used the "butterfly" design of the window cell (Figure 5.11b) to collect EDS data with reasonable SNR. The holder is connected to a current and potential control unit outside the microscope. They tested this microchip to monitor critical corrosion variables using open-circuit potential (OCP) and potentiodynamic polarization behavior and used it for in-situ measurements of OCP of an electron transparent stainless steel sample in H_2O with different pH values [54].

Karki et al. have reported cyclic voltammetry (CV) studies of some model compounds such as 0.01 M $CuSO_4$ and 20 mM $K_3Fe(CN)_6$/20 mM $K_4Fe(CN)_6$ in 0.1 M KCl solutions, using another setup for an electrochemical cell with specially optimized electrochemistry chips [55]. A custom-designed configuration for working electrode (WE), counter electrode (CE), and reference electrode (RE) was used to make quantitative measurements of electrochemical process that was similar to bulk measurements and is commercially available from Hummingbird Scientific (https://hummingbirdscientific.com/products/bulk-liquid-electrochemistry).

5.5.2 Modified Window Microchips and TEM Holder Combination

A number of window cells and holder designs have been reported where variations were made to improve the functionality of the window cell [25, 38, 56, 57]. Nanoaquarium is one such example where the liquid layer thickness was reduced to 100 nm, defined by SiO_2 spacer layer between two 50 nm SiN_x windows fabricated on Si wafers [31]. The two microchips are bonded using a plasma-activated wafer bonding process for hermetically seal the cell from the microscope vacuum. Both the inlet and the outlet are etched on the top microchip following the design similar to the one shown in Figure 5.7c. Embedded electrodes are also patterned for sensing and actuation during the wafer fabrication. Functionality of the nanoaquarium was demonstrated using aqueous suspension of Au and polystyrene microspheres [31].

Mehraeen et al. report a different design for fabricating of window gas cell and TEM holder that had in-built laser heating source developed in collaboration with E.A. Fischione Instruments, Inc. [44]. The tip of the holder is designed to accept 3 mm standard, dimpled or FIB prepared samples and the window spacing (liquid/gas thickness) can be varied from 0 to 250 µm. The built-in laser optics provides an added advantage for localized heating with a spot size between 30 and 300 µm and x- and y-translation capability. The laser source can also be used to follow photocatalytic reactions.

5.5.3 Non-window Cell Holder to Incorporate Other Stimuli

Here we include examples of TEM holders modified to incorporate stimuli other than heating, cooling, liquid, and gas, which are not discussed further in this book. For example, Dong et al. have modified a Nanofactory (this company is no longer in service) TEM holder with scanning-tunneling (STM-TEM) tip to incorporate a white LED to function as a photon source (Figure 5.13a,b). The system needed low power (0.01 V) and was used to measure the I–V curves under light illumination for ZnO nanowires and CdSe quantum dots along with atomic-scale imaging [58]. This combination provides a technique for simultaneous in-situ measurements of optical and electrical properties and relates them to the structural changes. Martis et al. have also reported the design and construction of a TEM holder with light input capabilities [59]. Other systems for in-situ illumination of TEM samples and measuring optical and optoelectronic properties combined with high-resolution imaging have been developed and reviewed by Fernando et al. [60] (see Chapter 8 for more examples).

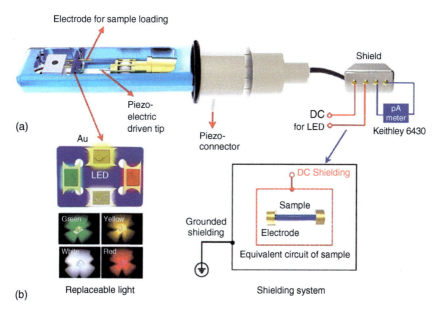

Figure 5.13 (a) The schematic for a modified Nanofactory scanning tunneling (STM)-TEM photoelectric holder for high-precision photocurrent measurement. An LED, driven by DC power, is placed on the sapphire substrate, which is inserted into the slot previously occupied by the electrical measurement system in the STM-TEM holder. A Keithley 6430 pA is used to obtain the photocurrent. (b) A schematic of the replaceable LED and design of the shielding system. Source: Dong et al. [58]/from Royal Society of Chemistry.

Micromachining has also been used to integrate two actuator and two beams on a microchip for double-tilt TEM holder to detect the effects of applied forces in two direction [61]; integrate nanojoule calorimeter for measuring heat capacity of nanoparticle samples [62]; include multiple electrical contacts that can be used for simultaneous heating, biasing, and mechanical actuators [63]; for multiprobe electronic measurements, such as three-probe field-effect, in-situ four-probe electronic measurements; and for in-situ nanomanipulation, and in-situ modification of nanodevices [64].

Other examples include, but are not limited to, optical excitation [58, 59] and in-plane magnetization and magnetic field measurements [65, 66]. Wire bonding is used to connect multiple probes on the microchip to the controllers outside the TEM column. These examples provide us with a know-how to design and build new in-situ holder for specific applications.

5.6 Take Home Message

Development and availability of MEMS technology have revolutionized the field of in-situ TEM applications. Fabricating a microchip with multiple stimuli is now limited by the need or imagination of the scientific community. Building a new holder to accommodate external connections to all multiprobes is also possible.

SiN$_x$ window on a Si wafer is the accepted platform for fabricating a microchip using a combination of cleanroom technologies, optical lithography, and FIB-based nanolithography.

Microchips for static or flow-controlled window cells are fabricated following similar process with subtle changes in design. Flow-controlled cells can also be used in static mode.

Thickness of the windows is controlled during the fabrication, but liquid/gas layer may be controlled during experimental setup by changing the O-ring diameter or size of polystyrene microspheres, depending on the microchip design.

Flow channels and/or reservoirs are etched on the Si chips and connected to external flow control system using clean flexible tubing.

Multiple probes are patterned on the chip using nanolithographic techniques and wire bonded to connect to the external controllers using wires running through the holder rod.

References

Commercial Vendors (in Alphabetic Order)

1. DENSsolutions, Inc.: https://denssolutions.com
2. Fischione: https://www.fischione.com
3. Hummingbird Scientific: https://hummingbirdscientific.com
4. Protochips, Inc.: https://www.protochips.com

Articles

1 Abrams, I.M. and McBain, J.W. (1944). A closed cell for electron microscopy. *Journal of Applied Physics* 15 (8): 607–609.
2 Double, D.D. (1973). Some studies of the hydration of Portland cement using high voltage (1 MV) electron microscopy. *Materials Science and Engineering* 12 (1): 29–34.
3 de Jonge, N. (2018). Theory of the spatial resolution of (scanning) transmission electron microscopy in liquid water or ice layers. *Ultramicroscopy* 187: 113–125.
4 Ring, E.A., Peckys, D.B., Dukes, M.J. et al. (2011). Silicon nitride windows for electron microscopy of whole cells. *Journal of Microscopy* 243 (3): 273–283.
5 Dwyer, J.R. and Harb, M. (2017). Through a window, brightly: a review of selected nanofabricated thin-film platforms for spectroscopy, imaging, and detection. *Applied Spectroscopy* 71 (9): 2051–2075.
6 Balram, K.C., Westly, D.A., Davanço, M. et al. (2016). The nanolithography toolbox. *Journal of Research of the National Institute of Standards and Technology* 121: 464–475.
7 Jensen, E. and Mølhave, K. (2017). Encapsulated liquid cells for transmission electron microscopy. In: *Liquid Cell Electron Microscopy* (ed. F.M. Ross), 35–55. Cambridge: Cambridge University Press.

8 Zaluzec, N.J., Burke, M.G., Haigh, S.J. et al. (2014). X-ray energy-dispersive spectrometry during in situ liquid cell studies using an analytical electron microscope. *Microscopy and Microanalysis* 20 (2): 323–329.

9 Hayden, S.C., Chisholm, C., Grudt, R.O. et al. (2019). Localized corrosion of low-carbon steel at the nanoscale. *npj Materials Degradation* 3 (1): 17.

10 de Jonge, N., Houben, L., Dunin-Borkowski, R.E., and Ross, F.M. (2019). Resolution and aberration correction in liquid cell transmission electron microscopy. *Nature Reviews Materials* 4 (1): 61–78.

11 Pu, S., Gong, C., and Robertson, A.W. (2020). Liquid cell transmission electron microscopy and its applications. *Royal Society Open Science* 7 (1): 191204.

12 Mirsaidov, U.M., Zheng, H., Casana, Y., and Matsudaira, P. (2012). Imaging protein structure in water at 2.7 nm resolution by transmission electron microscopy. *Biophysical Journal* 102 (4): L15–L17.

13 Park, J., Park, H., Ercius, P. et al. (2015). Direct observation of wet biological samples by graphene liquid cell transmission electron microscopy. *Nano Letters* 15 (7): 4737–4744.

14 Clark, N., Kelly, D.J., Zhou, M. et al. (2022). Tracking single adatoms in liquid in a transmission electron microscope. *Nature* 609 (7929): 942–947.

15 Zheng, H.M., Smith, R.K., Jun, Y.W. et al. (2009). Observation of single colloidal platinum nanocrystal growth trajectories. *Science* 324 (5932): 1309–1312.

16 Holtz, M.E., Yu, Y., Gao, J. et al. (2013). In situ electron energy-loss spectroscopy in liquids. *Microscopy and Microanalysis* 19 (4): 1027–1035.

17 Maier-Schneider, D., Maibach, J., and Obermeier, E. (1995). A new analytical solution for the load-deflection of square membranes. *Journal of Microelectromechanical Systems* 4 (4): 238–241.

18 Tanase, M., Winterstein, J., Sharma, R. et al. (2015). High-resolution imaging and spectroscopy at high pressure: a novel liquid cell for the transmission electron microscope. *Microscopy and Microanalysis* 21 (6): 1629–1638.

19 Allen, M.G., Mehregany, M., Howe, R.T., and Senturia, S.D. (1987). Microfabricated structures for the in situ measurement of residual stress, Young's modulus, and ultimate strain of thin films. *Applied Physics Letters* 51 (4): 241–243.

20 de Jonge, N. and Bonard, J.M. (2004). Carbon nanotube electron sources and applications. *Philosophical Transactions of the Royal Society of London, Series A: Mathematical, Physical and Engineering Sciences* 362 (1823): 2239–2266.

21 Mueller, C., Harb, M., Dwyer, J.R. et al. (2013). Nanofluidic cells with controlled pathlength and liquid flow for rapid, high-resolution in situ imaging with electrons. *The Journal of Physical Chemistry Letters* 4 (14): 2339–2347.

22 Jensen, E., Burrows, A., and Mølhave, K. (2014). Monolithic chip system with a microfluidic channel for in situ electron microscopy of liquids. *Microscopy and Microanalysis* 20 (2): 445–451.

23 Li, X., Mitsuishi, K., and Takeguchi, M. (2021). Fabrication of a liquid cell for in situ transmission electron microscopy. *Microscopy* 70 (4): 327–332.

24 Huang, T.-W., Liu, S.-Y., Chuang, Y.-J. et al. (2012). Self-aligned wet-cell for hydrated microbiology observation in TEM. *Lab on a Chip* 12 (2): 340–347.

25 Jonge, N.D., Peckys, D.B., Kremers, G.J., and Piston, D.W. (2009). Electron microscopy of whole cells in liquid with nanometer resolution. *Proceedings of the National Academy of Sciences* 106 (7): 2159.

26 Kim, B.H., Yang, J., Lee, D. et al. (2018). Liquid-phase transmission electron microscopy for studying colloidal inorganic nanoparticles. *Advanced Materials* 30 (4): 1703316.

27 Grant, A.W., Hu, Q.H., and Kasemo, B. (2004). Transmission electron microscopy 'windows' for nanofabricated structures. *Nanotechnology* 15 (9): 1175–1181.

28 Leenheer, A.J., Sullivan, J.P., Shaw, M.J., and Harris, C.T. (2015). A sealed liquid cell for in situ transmission electron microscopy of controlled electrochemical processes. *Journal of Microelectromechanical Systems* 24 (4): 1061–1068.

29 Neklyudova, M., Erdamar, A.K., Vicarelli, L. et al. (2017). Through-membrane electron-beam lithography for ultrathin membrane applications. *Applied Physics Letters* 111 (6): 063105.

30 Niu, K.-Y., Liao, H.-G., and Zheng, H. (2012). Revealing dynamic processes of materials in liquids using liquid cell transmission electron microscopy. *Journal of Visualized Experiments* 70: e50122.

31 Grogan, J.M. and Bau, H.H. (2010). The nanoaquarium: a platform for *in situ* transmission electron microscopy in liquid media. *Journal of Microelectromechanical Systems* 19 (4): 885–894.

32 Dwyer, J.R., Bandara, Y.M.N.D.Y., Whelan, J.C. et al. (2017). Silicon nitride thin films for nanofluidic device fabrication, Chapter 7. In: *Nanofluidics*, 2e (ed. J. Edel, A. Ivanov and M.J. Kim), 190–236. The Royal Society of Chemistry.

33 de Jonge, N., Poirier-Demers, N., Demers, H. et al. (2010). Nanometer-resolution electron microscopy through micrometers-thick water layers. *Ultramicroscopy* 110 (9): 1114–1119.

34 Seidel, H., Csepregi, L., Heuberger, A., and Baumgärtel, H. (1990). Anisotropic etching of crystalline silicon in alkaline solutions: I. Orientation dependence and behavior of passivation layers. *Journal of the Electrochemical Society* 137 (11): 3612–3626.

35 Pérez Garza, H.H., Morsink, D., and Xu, J. (2017). MEMS-based nanoreactor for in situ analysis of solid–gas interactions inside the transmission electron microscope. *Micro & Nano Letters* 12 (2): 69–75.

36 Creemer, J.F., Helveg, S., Kooyman, P.J. et al. (2010). A MEMS reactor for atomic-scale microscopy of nanomaterials under industrially relevant conditions. *Journal of Microelectromechanical Systems* 19 (2): 254–264.

37 de Jonge, N. and Ross, F.M. (2011). Electron microscopy of specimens in liquid. *Nature Nanotechnology* 6 (11): 695–704.

38 Creemer, J.F., Helveg, S., Hoveling, G.H. et al. (2008). Atomic-scale electron microscopy at ambient pressure. *Ultramicroscopy* 108 (9): 993–998.

39 Menon, V., Denoual, M., Toshiyoshi, H., and Fujita, H. (2019). Self-contained on-chip fluid actuation for flow initiation in liquid cell transmission electron microscopy. *Japanese Journal of Applied Physics* 58 (9): 090909.

40 van Omme, J.T., Wu, H., and Sun, H. (2020). Liquid phase transmission electron microscopy with flow and temperature control. *Journal of Materials Chemistry C* 8 (31): 10781–10790.

41 Alan, T., Yokosawa, T., Gaspar, J. et al. (2012). Micro-fabricated channel with ultra-thin yet ultra-strong windows enables electron microscopy under 4-bar pressure. *Applied Physics Letters* 100 (8): 081903.

42 Yokosawa, T., Alan, T., Pandraud, G. et al. (2012). In-situ TEM on (de)hydrogenation of Pd at 0.5–4.5 bar hydrogen pressure and 20–400 °C. *Ultramicroscopy* 112 (1): 47–52.

43 Allard, L.F., Overbury, S.H., Bigelow, W.C. et al. (2012). Novel MEMS-based gas-cell/heating specimen holder provides advanced imaging capabilities for in situ reaction studies. *Microscopy and Microanalysis* 18 (4): 656–666.

44 Mehraeen, S., McKeown, J.T., Deshmukh, P.V. et al. (2013). A (S)TEM gas cell holder with localized laser heating for in situ experiments. *Microscopy and Microanalysis* 19 (2): 470–478.

45 Yoshida, K., Nozaki, T., Hirayama, T., and Tanaka, N. (2007). In situ high-resolution transmission electron microscopy of photocatalytic reactions by excited electrons in ionic liquid. *Journal of Electron Microscopy* 56 (5): 177–180.

46 Cavalca, F., Laursen, A.B., Wagner, J.B. et al. (2013). Light-induced reduction of cuprous oxide in an environmental transmission electron microscope. *ChemCatChem* 5 (9): 2667–2672.

47 Liu, Q., Zhang, L., and Crozier, P.A. (2015). Structure–reactivity relationships of Ni–NiO core–shell co-catalysts on Ta_2O_5 for solar hydrogen production. *Applied Catalysis B: Environmental* 172–173: 58–64.

48 Picher, M., Mazzucco, S., Blankenship, S. et al. (2015). Vibrational and optical spectroscopies integrated with environmental transmission electron microscopy. *Ultramicroscopy* 150: 10–15.

49 Leenheer, A.J., Jungjohann, K.L., and Zavadil, K.R. (2015). Lithium electrodeposition dynamics in aprotic electrolyte observed in situ via transmission electron microscopy. *ACS Nano* 9 (4): 4379–4389.

50 Xu, W., Dycus, J.H., Sang, X., and LeBeau, J.M. (2016). A numerical model for multiple detector energy dispersive X-ray spectroscopy in the transmission electron microscope. *Ultramicroscopy* 164: 51–61.

51 Key, J.W., Zhu, S., Rouleau, C.M. et al. (2020). Investigating local oxidation processes in Fe thin films in a water vapor environment by in situ liquid cell TEM. *Ultramicroscopy* 209: 112842.

52 Song, Z. and Xie, Z.-H. (2018). A literature review of in situ transmission electron microscopy technique in corrosion studies. *Micron* 112: 69–83.

53 Chee, S.W., Pratt, S.H., Hattar, K. et al. (2015). Studying localized corrosion using liquid cell transmission electron microscopy. *Chemical Communications* 51 (1): 168–171.

54 Schilling, S., Janssen, A., Zaluzec, N.J., and Burke, M.G. (2017). Practical aspects of electrochemical corrosion measurements during in situ analytical transmission electron microscopy (TEM) of austenitic stainless steel in aqueous media. *Microscopy and Microanalysis* 23 (4): 741–750.

55 Karki, K., Serra-Maia, R., Stach, E. et al. (2020). Realistic bulk electrochemistry in liquid cell microscopy. *Microscopy and Microanalysis* 26 (S2): 1458–1459.

56 Williamson, M.J., Tromp, R.M., Vereecken, P.M. et al. (2003). Dynamic microscopy of nanoscale cluster growth at the solid–liquid interface. *Nature Materials* 2: 532–536.

57 Radisic, A., Vereecken, P.M., Hannon, J.B. et al. (2006). Dynamic nucleation and growth of Ni nanoparticle on high-surface area titania. *Surface Science* 600: 693–702.

58 Dong, H., Xu, T., Sun, Z. et al. (2018). Simultaneous atomic-level visualization and high precision photocurrent measurements on photoelectric devices by: in situ TEM. *RSC Advances* 8 (2): 948–953.

59 Martis, J., Zhang, Z., Li, H.-K. et al. (2021). Design and construction of an optical TEM specimen holder. *Microscopy Today* 29 (5): 40–44.

60 Fernando, J.F.S., Zhang, C., Firestein, K.L., and Golberg, D. (2017). Optical and optoelectronic property analysis of nanomaterials inside transmission electron microscope. *Small* 13 (45): 1701564. (12 pp).

61 Sato, T., Tochigi, E., and Mizoguchi, T. (2016). An experimental system combined with a micromachine and double-tilt TEM holder. *Microelectronic Engineering* 164: 43–47.

62 Zhang, M., Olson, E.A., Twesten, R.D. et al. (2005). In situ transmission electron microscopy studies enabled by microelectromechanical system technology. *Journal of Materials Research* 20 (7): 1802–1807.

63 Bernal, R.A., Ramachandramoorthy, R., and Espinosa, H.D. (2015). Double-tilt in situ TEM holder with multiple electrical contacts and its application in MEMS-based mechanical testing of nanomaterials. *Ultramicroscopy* 156: 23–28.

64 Xu, T.T., Ning, Z.Y., Shi, T.W. et al. (2014). A platform for in-situ multi-probe electronic measurements and modification of nanodevices inside a transmission electron microscope. *Nanotechnology* 25 (22): 225702. (8 pp).

65 Arita, M., Tokuda, R., Hamada, K., and Takahashi, Y. (2014). Development of TEM holder generating in-plane magnetic field used for in-situ TEM observation. *Materials Transactions* 55 (3): 403–409.

66 Lau, J.W., Schofield, M.A., and Zhu, Y. (2007). A straightforward specimen holder modification for remnant magnetic-field measurement in TEM. *Ultramicroscopy* 107 (4–5): 396–400.

6

In-Situ Solid–Liquid Interactions

> *The future is fluid. Each act, each decision, and each development creates new possibilities and eliminates others. The future is ours to direct.*
>
> – Jacque Fresco

Materials' interactions with liquids result in both desired and undesired phenomena that require atomic-level understanding to promote or mitigate them, as needed. Also, the nature of biological samples must be observed in their native (wet) environment. Therefore, efforts to confine samples in liquids have been going on for long time, but the field of making in-situ observations of materials in liquids has exploded in last couple of decades. Nucleation and growth of nanoparticles, corrosion [1, 2], etching [3, 4], galvanic replacement [5], electrochemistry, battery materials, polymer self-assembly, are some of the research areas that have taken advantage of the liquid-phase TEM (LP-TEM) or also known as liquid cell TEM (LC-TEM) and unveiled unprecedented details for atomic-scale mechanisms. Readers are encouraged to check the references in this chapter for further details, especially review articles [6–13], the book edited by Frances Ross [14], and the references therein. This chapter is aimed to make the readers familiar with the fundamental principles for liquid confinement, flow, incorporating external stimuli such as heating, biasing, and the practical aspects of conducting in-situ LC-TEM experiments, a few representative applications, limitations, and concluding summary.

6.1 Historical Perspective

The idea to contain liquid around the sample, originally called "wet cells," is as old as the TEM itself. In the beginning, the motivation to contain water or water vapor within two thin windows was to observe biological samples. Abrams and McBain were first one to construct an enclosed wet cell to successfully observe dynamic events for materials (cement) [15]. The design and performance were further

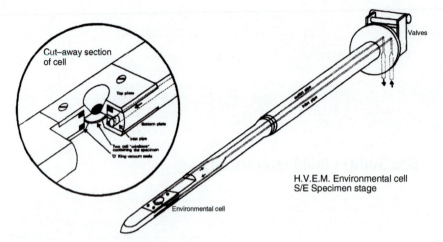

Figure 6.1 Diagrammatic representation of the improved environmental cell specimen stage for the H.V.E.M., with access pipes to the cell from outside the microscope. Source: Adapted from Double et al. [17].

improved to study materials' interactions with water, and first in-situ observations for the reaction of cement with water were reported by Double et al. [16] using a closed cell that was further developed into a flow system by incorporating water inlet and outlet lines (Figure 6.1) [17].

Another approach was to place the apertures in a differentially pumped sample chamber, similar to the one described in Figure 7.1 [18, 19]. This approach allows water vapor to be introduced in the TEM column at low pressure that was enough for water droplets to condense on the sample for short period. Direct water uptake by individual aerosol particles prior to deliquescence and their phase transition was studied at low temperature using a modified liquid nitrogen holder in an ETEM [20, 21]. ETEM was also used to make direct observation of nanoparticles in a liquid medium at room temperature, by using Pickering emulsions as a template to investigate the multiphase interactions and self-assembled structure of nanoparticles at a trichloroethylene–water interface [22]. This work also revealed the detailed self-assembled structure of nanoparticles at a liquid/liquid interface [22]. However, making routine observation using ETEM is difficult as the sample often dehydrates during transfer and condensed water also evaporated from the sample after short duration of observation and ETEM is expensive.

On the other hand, open cell design has been successfully employed for liquids with low vapor pressure, such as ionic liquids. For example, electrochemical lithiation process of SNO_2 nanowire electrode was observed using ionic liquid as electrolyte [23, 24]. Again, such open cells have limited application; therefore, a closed cell with windows presents a better alternative to confine liquid around the sample. The electron transparent windows used in earlier designs were made of amorphous carbon or SiO_2, which (i) could not withstand pressures above 100 mbar and (ii) broke after 10 to 20 minutes of observation period, and (iii) did not meet the mass thickness criteria needed for high-resolution imaging and/or EELS analysis [14, 25]. As explained in Chapter 5, with the advent of easy access to the processing technology used for microelectromechanical systems (MEMS) and microfluidic devices, thin and robust windows made of Si, SiN_x or SiC became available. Ross and Searson reported the observation of Si etching at room temperature using a closed LC [26]. First of this new generation of LCs was constructed using two silicon chips with windows glued together with a spacer that controlled the liquid layer thickness. Electrochemical cells, incorporating electrodes, were introduced soon after for dynamic observation of nucleation and growth of nanoclusters [27].

Original closed cells used sealed static liquid electrolyte, separated by electrode with the possibility to image through windows using the TEM while measuring the electrochemical process [28]. Later, liquid flow was introduced by pumping the liquid using an inlet and outlet to the LC, via tubes led through the sample holder [29] (also included flow control systems [30]). High-angle annular dark-field (HAADF) STEM imaging mode has been used to obtain high-resolution images through the liquid [31, 32]. In the last decade, several commercial companies have developed LC TEM holders incorporating liquid flow, electrical and temperature controls, which have helped to unveil a large number of novel phenomena (see Section 6.6).

6.2 Holder Design and Selection

As mentioned above and explained in Chapter 5, many commercial holders with varied design and capabilities are now available. Moreover, often the vendors are willing to work with researchers to modify the microchips and/or holder design for new applications. Therefore, the choice of the holder is dictated by your project's needs. Sections 6.2.1–6.2.5 describe some of the commercial and modified holders reported so far. The applications Section 6.6 may be used as guide to help choose an appropriate holder and technique.

6.2.1 Closed Cells

Following are two types of closed cells for the containment of static liquid suspensions:

6.2.1.1 Graphene Cells
Sandwiching a sample between two layers of graphene is the simplest type of closed cell, as graphene is the most thin and robust window material. In simple terms, small droplets of liquids with suspended nanoparticles or solutions are encapsulated between two graphene sheets that make a liquid cell. Since graphene is impermeable to most liquids, and the thinnest film possible, these cells have been used to obtain images with unprecedentedly high resolution.

Yuk et al. were first to report fabrication of a graphene liquid cell [33]. They used a large-area single-layer graphene (SLG), grown by chemical vapor deposition, and transferred to a rigid support such as holey amorphous carbon film on quasi-freestanding Quantifoil TEM grids (Figure 6.2a). Nanoparticles or other material of interest was then drop-casted on it (Figure 6.2b). A second SLG sheet was then transferred on top of the first graphene sheet thus covering the pre-deposited

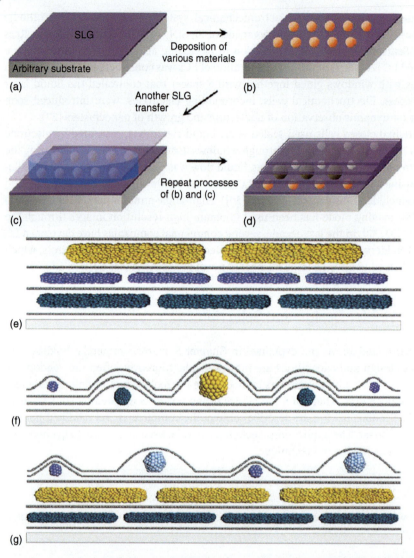

Figure 6.2 Schematic of the process flow for graphene-based LC fabrication. (a) Large-area single-layer graphene (SLG) on an arbitrary substrate. (b) Deposition of various materials by either drop-casting chemically synthesized materials or e-beam evaporation. (c) A second SLG is superimposed on the first graphene layer, thus encapsulating the deposited materials. IPA is used to wet the graphene layers and intercalants to improve graphene–graphene or graphene–intercalant bonding after drying. (d) Repeating steps (b) and (c) results in a multilayer graphene-based superstructure encapsulating different materials. (e–g) Schematics of a graphene sandwich superstructure (GSS), graphene veil superstructure (GVS), and a hybrid superstructure of GSSs and GVSs. Various types of intercalants are represented by different colors and sizes. Source: Yuk et al. [33]/American Chemical Society.

materials (Figure 6.2c,d). A rigid support is needed to prevent folding or tearing of the second graphene layer. They used Isopropanol (IPA) to wet both graphene layers, such that surface tension generated by the evaporation of IPA pulled the graphene layers together to form a physical seal and was termed as "a graphene sandwich superstructure (GSS)" (Figure 6.2e,f).

Despite the lack of liquid flow or possibility to incorporate other stimuli, their report resulted in a significant advancement in the application of graphene LCs. Currently, there are three broad categories of graphene liquid cells [34] depending on the encapsulation technology and windows used (Figure 6.3). One way is to deposit liquid suspension on a thin SiN_x or carbon film and cover it with graphene sheet. Here the base film may have pits etched part of the way to provide wells for liquid (Figure 6.3, Type A) [35]. Another approach is to encapsulate the liquid by placing graphene layer on both sides as shown in Figure 6.3 Type B [33]. Yet another way is to enclose the liquid sample in wells in a perforated thin film and close this film on both sides with graphene (Figure 6.3, Type C) [36].

While constructing a graphene liquid cell requires some strategy and experience, they are designed and made in the lab based on the sample, right before starting the

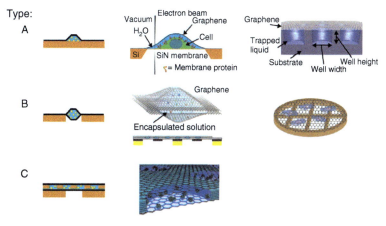

Figure 6.3 Basic types of graphene liquid cells. For type A, a sample is immobilized on a thin film and covered by a sheet of graphene as described in Figure 6.2. Source: Top middle image adapted with permission from Mansfeld et al. [24]. Copyright 2017 American Chemical Society. The membrane may contain liquid pockets. Source: Top right image reprinted with permission from Williamson et al. [27]. Copyright 2016 John Wiley and Sons. The liquid sample is enclosed on both sides by graphene in liquid cell type B and supported by a perforated thin membrane. This type has been used to study growth processes of nanoparticles. The liquid cell is typically placed on a support grid. Source: Middle row middle image reprinted with permission from Sutter et al. [5]. Copyright 2012 AAAS. Also, this type of liquid cell was used to image cells. Source: Middle row right image adapted with permission from Wise et al. [20]. Copyright 2015 American Chemical Society. Another approach, type C, is to enclose a perforated thin membrane at both sides with graphene. The holes in the membrane serve as liquid pockets. Source: Bottom image reprinted with permission from Radisic et al. [28]. Copyright 2018 American Chemical Society. Nano Letters Mini Review. DOI: 10.1021/acs.nanolett.8b01366 [34].

experiment or introduction to the TEM column. Therefore, we must consider that the nature of the liquid as the structural integrity of the specimen depends on the specific properties of the liquid. The amount of liquid should be proportional to the sample amount, which is often small. Also, a constant temperature should be maintained during preparation to avoid temperature effects on the sample. A detailed methodology of making graphene cell can be found in Ref. [34].

6.2.1.2 Microfabricated Window Cell

Graphene windows can be replaced by microfabricated windows and are often preferred as other stimuli such as temperature and electrical bias can be easily incorporated during the microfabrication process, as explained in detail in Chapter 5 and will not be included here.

6.2.2 Limitations of Closed Cells and Need for External Stimuli

The amount and nature of the liquid in the closed cells, described above, remain the same during observation period, i.e. the reaction may be over before the sample is introduced in the microscope unless the process is triggered or stopped by some controlled stimuli, such as heat or bias for the former and freezing for the latter. Another way to overcome this limitation is to introduce the liquid after sample to be studied is already loaded in the column (flow reactors, also see Section 5.4.2).

Radiolysis of liquids, especially water, has fortuitously provided another alternative to trigger the reactions only when irradiated by electron beam, i.e. under observation. Many reactions triggered by electron beam have provided unprecedented information about the nucleation and growth of metal nanoparticles (Section 6.6.1). In Sections 6.2.3–6.2.5, we describe various modifications that are incorporated during the fabrication of LC such that we don't need to depend on the electron beam to initiate the process.

6.2.3 Flow Reactors: Microfluidic Design

This section is a continuation of Section 5.4.2 and provides detailed description of specific flow cells that provide a way to introduce the reactants in the cell after the holder is loaded in the column and the TEM/STEM optics is aligned. This step can also control the reactions triggered by electron beam. As explained in Section 5.4.2, liquid flow in the cell can be achieved by etching microfluidic channels on the Si chips supporting the thin membrane windows [29, 30, 37, 38]. As commercial holders for select applications are readily available, here we describe a couple of examples of "home-built" flow cells that use different strategies to control the flow.

An example of such nanofluidic cell is shown in Figure 6.4a, where the inlet and outlet ports of the microfabricated cell are connected to a syringe pump (Figure 6.4c), outside of the TEM column, through tubing enclosed within a custom-built TEM holder. This permits the flow of benign fluid after TEM alignment, to introduce the reactants and/or mixing of reactants if two or more channels are present. The advantage of such nanofluidic cell is that the liquid pressure, or the amount of liquid, can

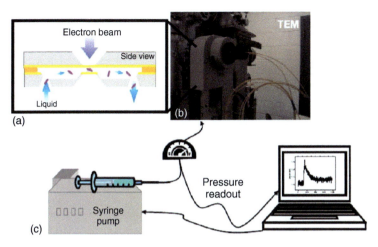

Figure 6.4 (a) Nanofluidic cell design with a feedback loop to control the liquid flow through the inlet and the outlet ports etched into the cell, and a liquid layer thickness defined by the rigid spacer material (orange). (b) Image of the modified TEM holder inserted in the TEM column showing tubes that connect to (c) syringe pump outside the TEM column. The liquid flow rate is computer-controlled with pressure readout (c). Source: Mueller et al. [29]/from American Chemical Society.

be kept constant during operation or exchanged on the fly, and multiple experiments can be run without removing the TEM holder [29].

> Where the flow carries a large quantity of water, the speed of the flow is greater and vice versa.
> – Leonardo da Vinci

Another way to delay the start of the reaction until you are ready to start the observation is to flow the liquid using a built-in reservoir within the cell itself that can be pumped into the region under observation with an on-chip electrochemical pump and burst valve structure [30]. The basic LC structure follows the established geometry and is made of two micromachined silicon chips, with thin SiN_x membranes, bonded together such that it encapsulates fluid sample but allows electron beam to transmit through (Figure 6.5a). The thickness of the liquid layer is defined by a patterned spacer layer that also provides a narrow channel with reservoirs at each end (Figure 6.5b). A combination of valve and pump structures is integrated on the chip to control the flow (Figure 6.5c). The valve action is achieved by patterning a hydrophobic barrier crossing the channel near its center (Figure 6.5c, red dashed line) that divides the channel into two chambers. Liquid samples are loaded in the right chamber and are prevented from flowing into the left chamber by this hydrophobic barrier. The electrical contacts on one side of the functional chip are used to drive flow within the central channel by applying voltage through the interconnects. Application of voltage generates the gas in the chamber that forces the liquid to flow across the barrier until it reaches the hydrophilic layer on the other

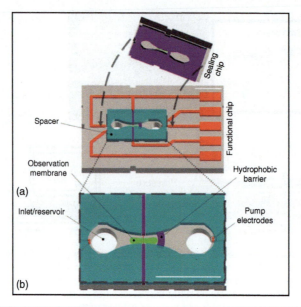

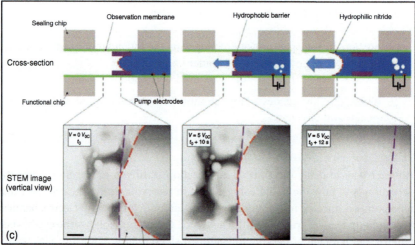

Figure 6.5 (a) Schematic illustration of the unassembled LC device showing the electrical interconnects (red lines), the cell (green section), and the top sealing chip with windows (light green). (b) A zoomed view where the electron-transparent TEM membrane (light green center), reservoirs with pump electrodes (red dots) and fluidic channel can be seen. Scales = 1 mm. (c) Schematic cross-sectional view of the observable region of the device in the initial state (left), during pumping (center), and after release of the burst valve (right). (bottom) Dark-field Z-contrast STEM view of the hydrophobic barrier during pump actuation. The liquid on the right is initially pinned in place until voltage is applied to the pump electrodes. The liquid front bulges toward the left until it finally fills the chamber in the last pane. Scales = 10 μm. Source: Menon et al. [30]/from IOP Publishing.

side and is drawn into the left chamber through capillary action (Figure 6.5c). In this design, the reactants are kept in separate chambers with closed microfluidic valves until TEM observation conditions are established, and flow is triggered subsequently using integrated pumps. The capillary burst valves utilize a hydrophobic region within a hydrophilic chamber to stop liquid flow until some forward pressure is applied using electrochemical pumps placed on microfluidic devices to create pressure gradient within the channel from the vapors produced by an electrolysis process [30].

A number of other flow reactors, including one with high sample refreshing rates (over 1 kHz) that can be used for femtosecond electron diffraction collection, have also been reported [39].

6.2.4 Electrochemical Cell: Biasing

The first liquid electrochemical cell, reported by Williamson et al. [27], was used for biasing via electrodes. It was made by gluing two Si wafers, face to face. Viewing window was obtained by coating each Si wafer with 100 nm Si_3N_4 and selectively etching from the back to leave a 10 nm window, similar to the process described in Section 5.4 (Figure 6.6a). On the lower wafer, a patterned ring of SiO_2 maintained a distance of 0.5 to 1 µm between the two wafers. A polycrystalline Au was deposited across the viewing window and over a via as working electrode that was connected through the wafer electrode to an external contact. Two reservoirs were etched in upper wafer and liquid electrolyte was introduced with a syringe and flowed between the viewing windows by capillary action (Figure 6.6a). The body of the cell was sealed by gluing sapphire lids over the holes in the spacers. Counter electrode was made of Au wire and placed in one of the reservoirs and copper wire in second reservoir was placed as reference electrode. The assembly was covered by sapphire lids using a heat-curing epoxy (Figure 6.6b) [27].

This setup was successfully used to measure the real-time growth of Cu clusters over a large area (several mm^2) and at a spatial resolution of around 5 nm. The nucleation and growth or dissolution thereafter of Cu nanoparticles were controlled by applying current through the electrodes [27]. Quantitative analysis showed the relationship between the current and cluster growth to be comparable to ex-situ results such that simple models could be applied to understand the key processes involved in nucleation and cluster growth [27].

In-situ electrochemical TEM/STEM has since been applied to gain fundamental understanding of electron, charge, and mass transfer mechanisms, and reaction kinetics for batteries, fuel cells, and super capacitors [40]. The structure evolution of electrochemical deposits has also been explored by several research groups. Examples include, but not limited to, synthesis of Cu_2O cubes [41], growth mechanisms of lead dendrites as deposited on the gold electrodes immersed in an aqueous solution of lead nitrate under an applied potential [42], aggregative growth rather than 3D island growth of Pd particles [43], Li deposits [44], etc.

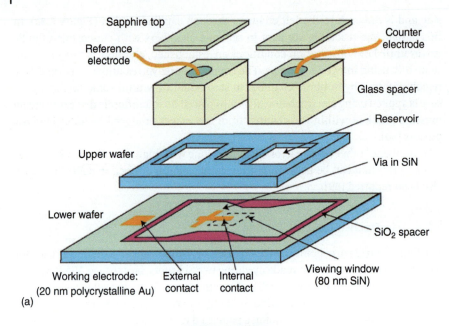

Figure 6.6 The liquid cell. (a) Components of the cell. The viewing window is enlarged for clarity. (b) Photograph of a two-electrode cell with an optical micrograph of the viewing window. Source: Williamson et al. [27]/from Springer Nature.

6.2.5 Heating in Liquids

Heat is another stimulus that can be used to trigger liquid–solid interactions in liquid environment and provide additional control over the reaction process. First and foremost, electron beam could be a source of heating as well as ionization of liquids. (The effect of ionization will be covered further in Section 6.5.2). Fritsch et al. [45] have used thermal expansion coefficients to measure the sample temperature from selected area electron (parallel beam) diffraction patterns for samples in gas environment and in vacuum [46, 47]. They calibrate the expansion in observed d-spacing of Au nanoparticles with temperature using thermal heating (Figure 6.7a). Local sample temperature was then measured using the same

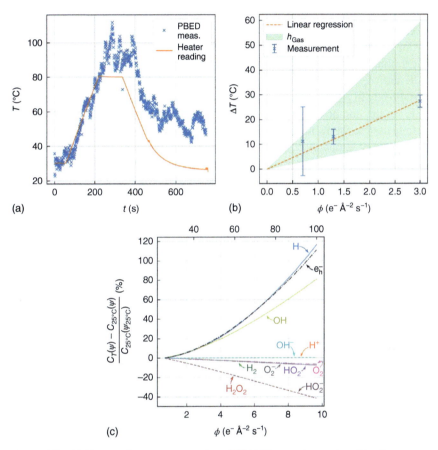

Figure 6.7 (a) Measured temperature using a TEM LC heater (orange) and in-situ temperature measured from selected area (parallel beam) electron diffraction of Au nanoparticles. Particle-based measurements perfectly reproduce the temperature ramp (until 220 seconds) but show an offset related to beam heating effects upon subsequent delay and decline. (b) Increase of specimen temperature, derived from electron diffraction measurements, after five minutes of illumination as a function of electron flux density. Error bars mark the maximum deviation of the {220} and {311} data from the mean value. The green area spans the considerable range of temperature variation modeled for a gaseous environment. (c) Beam heating strongly affects the equilibrium concentrations of water radiolysis products and thus the prospected chemical environment. Assuming a reference temperature of 25 °C, the curves give the percentage change of relevant species at the corresponding dose rate against a conventional radiolysis model neglecting beam-heating effects. The corresponding sample temperature in the observation area is plotted on the top axis. Source: Fritsch et al. [45]/The Royal Society of Chemistry/CC BY 3.0.

procedure as a function of electron dose (Figure 6.7b). The average temperature after five minutes of irradiation was derived from the {220} and {311} diffraction rings and the error bars correspond to the maximum deviation of the measurements from the mean value. They found that the uncertainties increased as electron flux dropped from 0.3 to 0.07 e nm^2 s^{-1} and attributed it to be due the small electron flux density itself that lowered the signal-to-background ratio. However, a direct proportionality between temperature increase and the electron dose rate could be established (Figure 6.7b). They determined that the measured temperature increase is not due to electron interaction with liquid only, but nanoparticles and the surrounding also play an important role [45]. Their measurement also indicated that dose rate and temperature also affected the concentration of radiolysis product (Figure 6.7c).

Alternatively, controlled heating of the liquid can be achieved by placing the LC chips directly in a furnace-based heating holder [48], also known as Joule heating, similar to the one described in detail in Chapter 3. For example, it has been successfully used to follow the agglomeration behavior of Au nanoparticle in graphene LCs at different temperatures with time [33] and to explore the effect of temperature on electrochemical etching of Ni–Pt bimetallic nanoparticles [49]. van Omme et al. have developed a "Stream Liquid Heating Holder," which combines on-chip liquid flow channel with a microheater. The channel enables direct flow over the imaging area and rapid replenishment of the solution inside the nano-cell with simultaneous heating to more than 100 °C [50].

> Liquid cell holder with more than one stimulus such as heating and biasing is also commercially available. MEMS-based microchips allow us to combine more than one stimulus on the same holder.

A capacitively coupled heat feedback technique has also been implemented to control the temperature of the LC from room temperature to 100 °C with an accuracy of ±0.1 °C [51]. Inclusion of nanoscale heating strip or spiral, directly on the microfabricated windows, is now commonly used for the LCs made using microfabricated windows. LC with both biasing and heating filaments is also now commercially available, e.g. (in alphabetic order), DENSsolutions (https://denssolutions.com), Hummingbird Scientific (https://hummingbirdscientific.com), and Protochips (https://www.protochips.com/about-us). Although laser heating is more commonly used in the Dynamic or Ultrafast TEM platforms (Section 1.5.4), it has also been used to control the nucleation and growth of nanoparticles [52]. A detailed discussion of temperature control in LCs and application can be found in Ref. [53]. But as explained above, electron beam effects should always be measured and mitigated (if possible).

6.3 Specimen Design and Preparation

Most often the sample in an LC consists of powder solid and/or liquid, and we must consider them separately as well as a combination of the two. The size of the solid material, whether a reactant or product must fit within the cell windows, whereas particle mobility is desirable for a reaction to proceed, their adherence to windows is needed to obtain high-resolution images [54]. The properties of the liquid also need to be considered, especially its viscosity and affinity to the window material. Most of the sample preparation techniques described in Section 2.3 can be used here except, their compatibility with the window material and LC volume needs to be considered for final selection. Recently, FIB has also been used for preparing a lamella from bulk samples and transferred them to the window of a microchip [55, 56].

6.4 Data Acquisition

Direct electron detection cameras play an important role for imaging in liquids as they enable low-dose imaging and high temporal resolution. Multiple images can be added, after drift correction, to overcome the effects of vibration and motion on resolution and provide a way to reduce the dose rate for individual image but increase SNR for the final image. Although the total dose to the sample does not change, the dose rate for individual image is reduced. Therefore, a balance between the spatial and temporal resolution can be chosen for both the total electron dose and the dose rate during irradiation [57]. This balance must be set between the spatial resolution required, the number of images recorded during the experiment, the temporal resolution, and both the dose and dose-rate tolerance of the sample. Detailed theoretical treatments for various parameters effecting spatial resolution in liquids are available in articles published by de Jonge and collaborators [58, 59].

6.5 Practical Challenges

One of the major challenges for using LC holder in regular TEM is that windows might break and leak the liquid that may damage the column vacuum system. As we explain in Section 6.5.1, the windows must be visually inspected and leak tested before loading the microchip containing the sample in the TEM column. However, it is still possible for windows to break during observations. It is interesting to note that if the cold finger (anticontamination device) surrounding the sample has reached the required temperature, i.e. the liquid N_2 in the reservoir connected to it has stopped bubbling, the liquid leaking out will freeze instantaneously and clog a small crack or hole. Therefore, cooling the cold finger of the TEM is not just desirable to keep good vacuum in the column but a must when using LC holder. Some of the other challenges are explained below.

6.5.1 Sample Loading

As mentioned in Section 2.5.4, LC windows, irrespective of the material, are the weakest point of the cell as they are fragile and can have holes generated during manufacturing or sample loading. Thereby, a visual inspection of the chips under high-magnification optical microscope or SEM is highly recommended to make sure that the windows are completely intact before loading the sample. The microchips should be thoroughly cleaned using appropriate chemical to remove leftover photoresist or other contaminants from the fabrication process.

Normally salt solutions, powders, or nanoparticles suspensions are suited for LC-TEM experiments. Even powder samples are often loaded as suspensions in iso-propanol to obtain an even distribution across the window. As described above, a choice hydrophobic or hydrophilic window can be made depending upon the sample properties and observation conditions. For example, nanoparticles, suspended in water, will stick to the hydrophilic window and will be easier to image at higher resolution compared with moving particles. The static or flow LC holders with two microchips, top and bottom, with windows, are sealed together after loading the liquid. Therefore, loading the liquid sample on the microchips is also a critical step for executing successful experiments. Liquids should be drop-casted using microliter pipettes with precisely known amounts as recommended by the microchip manufacturer. Sudden drop or weight of liquid can break the windows. Entire assembly, for either the static or flow LCs, should be leak tested using a vacuum chamber, before loading in the microscope. Apart from the resolution limits for imaging and EELS, specifications such as the thickness of windows, maximum flow rate, drift rate, etc. should be considered when buying or making an LC holder/microchip cell. EELS can be used to map the liquid layer thickness across the windows and controlled for optimum performance [60]. Sometimes, microchips with thicker window may be desirable, especially if atomic resolution is not required to obtain the needed information. Furthermore, following steps are recommended for loading the microchip in the holder (Khalid Hattar, Sandia National Laboratory, private communication):

- Clean the tip of the holder with an air gun before loading the new chips to remove the residual debris that might be present from previously broken microchip.
- Make sure that the dimensions of the home-built or an older microchip are compatible with sample well of the holder as tolerances are tight and slight errors in dimension will apply stress on the window and break.
- Large thin sections of bulk materials are not commonly used in LC, as they can puncture the thin windows and make the cell unusable. Special attention should be paid when loading FIB-based sample lamella as they can puncture the windows. This is not obvious for those doing lift-out onto Cu or metal half-moon grids as they often mill with larger boxes than necessary to compensate for eucentric errors in the FIB, which if done with a window microchip will result in holes in the windows and in full window failure.
  Leonard and Hellmann have used FIB to prepare lamellar sections of mineral wollastonite, $CaSiO_3$, to observe its dissolution mechanism [61]. Complete details of the sectioning process with FIB, thinning a FIB section from 4 µm to $\approx$ 70 nm electron transparent lamella, transferring it to SiN_x window, can be found in this reference and Section 2.4.6.
- The difference in thermal coefficient of expansion between the window and the support during an experiment needs to be considered with respect to the thickness and x–y dimensions of the window (see Section 5.3). To be safe

(sacrificing resolution), you want the thickest window and the smallest window area. The rapid heating can also result in window failure.

For graphene LCs, first a graphene layer is often placed on a lacy carbon grid followed by a drop of liquid and capped with another graphene sheet. The amount of liquid enclosed between the two sheets is thus quite small, still it is a good idea to leak test the assembly before transferring to the TEM column.

6.5.2 Electron Beam Effects

As mentioned earlier, high-energy electrons can modify both the liquid and nanoparticles and has been widely employed to stimulate nanoparticle growth in liquid solutions, especially when using graphene cells [62]. Most of the liquids dissociate when irradiated with electron beam and hydrogen bubble formation was the first indication of radiolysis of water due to electron beam [63, 64]. The pressure and volume of bubbles have been determined using EELS [65]. A detailed discussion of energy transfer by electrons to liquids can be found in publications by Schneider [66, 67]. A detailed discussion of mechanisms of electron beam effects for various imaging conditions, evaluation, and calibration can also be found in Ref. [53]. A computer program to simulate the electron beam effects is also available and should be used to estimate the electron beam effects [68]. Although exact mechanism of electron interaction with liquid that stimulates the nanoparticle growth is not well established, it has been used to decipher the nucleation and growth of solute particles (also see Section 6.6.1). Several systematic studies of the effect of electron dose rate and total dose on the nucleation process and mitigating the dose affect have been reported [69–72]. Therefore, it is recommended that the electron dose effects must be evaluated for every liquid system under observation to either use electrons as stimuli or to avoid the electron beam effects.

> "In Liquid-Phase Electron Microscopy experiments there is no such thing as no beam effects. Beam effects can either be significant or insignificant and either useful or harmful. Beam damage occurs when the beam effects are significant and harmful, low dose experiments are viable when the beam effects are harmful but insignificant. However, there are many examples where beam effects can be useful – such as electron beam synthesis."
>
> – Joe Patterson, UC Irvine

Experimentally, electron beam on and off observations should be made and reported when using other stimuli such as biasing or heating or both. Recently, cryo-TEM has been used to mitigate the electron beam–induced damage in LC (see Chapter 4 for details). Patterson's group has proposed a multimodal approach to follow reactions in liquids. They named it as "distributed electron microscopy" as the data are collected using multiple imaging modalities, such as liquid TEM/STEM, cryo-TEM/STEM, etc. from samples extracted at different stages of reaction. This collective data set was then used to understand the reaction mechanism and strength of this approach is that the different imaging techniques have complimentary benefits and limitations [73].

Fortunately, such detailed studies have led to understanding the chemistry of the liquids as a function of radiation and have been successfully applied to control the growth and degradation of nanoparticles, and electron beam–induced synthesis of polymers [52, 74, 75]. Moreover, as electron beams are widely used for manufacturing of a number of polymers, LC TEM observations provide an unprecedented opportunity to characterize and understand the formation mechanisms of nanostructured polymers [76] (see Section 6.6.5).

6.5.3 Windows Bulging

Understanding the behavior of liquid within confined space is complicated, especially for graphene cells due to interplay between the vdW interlayers interaction, the layer interaction with the liquid sample, and the bending strain energy of the layers to accommodate encapsulation. However, Ghodsi et al. have used EELS data to measure the thickness of the encapsulated water layer between the two graphene sheets and plasmon peak energy shifts to obtain the density of water [77]. They show that a comparison between the change in water density obtained from a flow LC, assumed to be at normal pressure at RT, and graphene cell with varying water thickness layer can be used to calculate the pressure in the cell. Their results indicate $\approx 12\%$ increase in the density of water under tight graphene encasement, where pressure as high as 400 MPa can be expected. Their estimation of pressure and strain energy is in agreement with the theoretical analysis based on the effect of vdW forces [77]. It is important to note that exact value of electron mean path, λ, depends on the nature of material, electron energy, and the convergence and divergence angles of the microscope electron optics. Therefore Yesibolati et al. have directly measured the inelastic mean free path (λ_{IMFP}) of water as functions of the EELS collection angle (β) at 120 and 300 keV using a monolithic LC holder [78]. This value can be used to make thickness measurements across the window using EELS.

As we know that the bulging of windows is due to the pressure difference on both sides of the windows (see Section 5.3.3), i.e. the pressure inside the cell is much higher than outside in the TEM column (high vacuum), limits the functionality of the entire cell. Experimental measurements (Figure 6.8) by Holtz et al. show windows deflection is as high as 1 μm for 40 μm window and 650 nm for 25 μm window, respectively (Figure 6.8a,b) [79]. They also show that a limited part of viewing window is available where the thickness for useful data collection for core-loss and valence EELS is possible [79]. There has been an effort to reduce the size of the windows; however, as explained earlier (Sections 5.3 and 5.4), it is usually kept above 20 μm due the processing limitations.

Electron scattering from windows also reduces SNR, thereby reducing the effective image resolution and peak intensities of core-loss peaks. Image processing algorithms are being developed to improve SNR that requires further improvements [80].

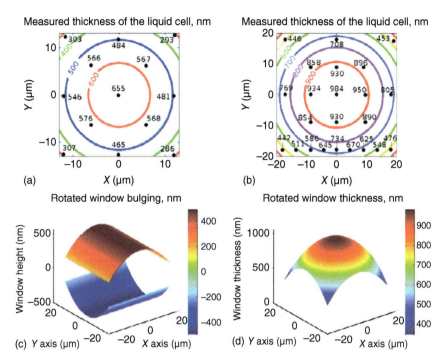

Figure 6.8 Measurement of the variations in liquid thickness due to the bulging of the windows. STEM-EELS measurements (points) with a parabolic fit contours are shown for (a) water between 25-μm wide windows with a 150 nm Au spacer and (b) ethylene glycol between 40 μm wide windows with a 500 nm SU-8 spacer (c): A three-dimensional representation of the bulging of the two rotated SiN_x membranes, assuming each window bows out as an independent orthogonal parabola and (d) the total thickness between the windows from (c). The profile shown in (d) corresponds to the contour lines in (b). Source: Holtz et al. [79]/Cambridge University Press.

6.5.4 Interaction of Sample with Windows

Reactivity of window material with liquid is an important consideration; however, SiN_x windows have been found to be inert for most of liquids. Location of solid particles in liquid with respect to windows is another important consideration, not only for image resolution but also for the mobility of the nanoparticles. Often, nanoparticles will stick to either upper, lower, or both windows, which may or may not be desirable. Whereas it is easier to obtain atomic resolution images from immobile particles, overlap between the particles stuck on both sides creates a problem. Also, for flow reactors, these particles are not being continuously replaced by the new ones. Moreover, the window surface could be made hydrophilic by using pretreatments such as plasma cleaning or flushing empty cell with the fluid to be used for experiments. Hoppe et al. have reported effect of different pretreatments, such as soaking the chips with bovine serum albumin on absorption of liposomes on the window surface [81].

> A pretreatment of the windows is needed to make their surface chemistry compatible to the liquid to be used.
> – Khalid Hattar, Sandia National Lab

6.6 Select Examples of Applications

In the past couple of decades, the integration of liquid flow and other stimuli with LC have greatly expanded the field. Pu et al. have provided a brief history of early accomplishments obtained using various types of LCs in table 1 of their review article [9]. As explained in Chapter 4, cryo-TEM is now often used to monitor reaction steps in liquids and will not be covered further in this chapter. Also, there is large volume of literature published for the imaging of biological and geological applications of the LC [61, 82–84]. Here we concentrate on in-situ and operando measurements for understating materials synthesis and properties in liquid environment.

6.6.1 Nucleation and Growth of Nanoparticles

Pathways of nucleation and growth of metal [52, 85–89], colloidal [90], oxide nanoparticles [91–93], and phase transformations [94] in liquid medium have been intensively investigated. We know that electron beam can trigger changes in samples depending on the materials properties, but it is omnipresent for liquids. As mentioned in Section 6.5.2, ionization of liquids by electron beam has been investigated extensively, and the results can be used to either mitigate or to trigger nucleation and growth of metallic nanoparticles from their solutions. For example, the quantitative measurement of electron dose on the nucleation and growth of Ag and Pt particles has been reported [4, 52, 85], and methods to mitigate these effects have also been explored [95]. One of the advantages to follow the electron beam induced processes is that we can make multiple measurements from the same liquid sample by just moving the electron beam to a new observation area, enabling us to determine the reproducibility of the results as described below.

Liao et al. measured the growth rate of various facets in Pt nanoparticles during electron beam–induced decomposition of Pt(acetylacetonate)$_2$ (20 mg ml^{-1}) in a solvent mixture of oleylamine, oleic acid, and pentadecane [96]. A combination of graphene LC and direct electron detection camera enabled them to distinguish the growth rates of different facets leading to the formation of facet-controlled nanoparticles (Figure 6.9a). They found similar growth rates for low index surfaces until {100} facets stopped growing (Figure 6.9b). Theoretical simulations show that mobility of surfactant legend on (111) surfaces favored their growth, while the low mobility on (100) facet blocked the growth, i.e. the growth rate of facets is controlled by the mobility of the ligands in solution on specific facet. The shape of the nanoparticle continuously changed during the growth as shown in Figure 6.9c,d. These findings shed a light on the nanocrystal shape-control mechanisms and future design for selected shapes by using selective facet–arrested ligand mobility. Growth rates of Ag nanoparticles in the presence of Pt in solution have also been investigated, and the

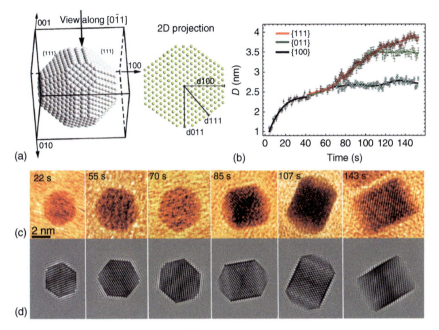

Figure 6.9 The facet development of a Pt nanocube viewed along the [011] axis. (a) The atomic model of a truncated Pt nanocube and its projection along the [011] zone axis. The distances from the crystal center to each of the (100), (011), and (111) facets are highlighted. (b) The measured average distances from the crystal center to each facet as a function of time. Error bars indicate the standard deviation. (c) Sequential images show the growth of the Pt nanocube extracted from a video. (d) Simulated TEM images of the Pt nanoparticle in (c). Source: Liao et al. [96]/from American Association for the Advancement of Science.

anomalous growth behavior is found to be due to catalytic properties of Pt as well as due to electron beam effects [97].

Oxidative dissolution mechanism of Pd nanoparticles, where facet-dependent etching rate converted a cubic nanoparticle to a nearly spherical shape [3], and of Cu nanoparticles have been reported [98]. Similar oxidative shape transformation and selective oxidation of Ga–In alloy nanoparticles in water as a function of temperature, instead of electron beam, have also been performed [99]. In summary, this technique has been used to follow the coalescence mechanism of metallic and bimetallic nanoparticles [100, 101], cluster formation [102, 103], self-assembly [104, 105], formation of novel structures such as needles, pillars, stars, 2D shapes [106] and other shape-controlled particles [107].

The ability to control liquid flow in the cell increased the opportunity to investigate the effect of the liquid chemistry, especially the pH value, on the formation of nanoparticles. For example, Pd particles were observed to switch the growth mode from three-dimensional island to clusters when HCl was added to the solution [43]. Similarly, addition of a capping agent (tri-n-octylphosphine [TOP]) resulted in the growth sub 3 nm Pd nanoparticles [108]. Also, a flow cell provided a means to change the composition of the liquid during observation and has led to understand the nucleation and growth mechanisms of materials other than metal, such as PbS [52]. Joule or laser heating, instead of electron beam, has also been used to understand the nucleation and growth of a number of other nanoparticles, e.g. oscillatory growth behavior of metallic Bi [48], and non-metallic hexagonal polymorph of BeO nanoparticles [91]. LC-TEM has also enabled us to understand the molecular self-assembly mechanism of small molecules and polymers [109].

6.6.2 Corrosion/Oxidation

Corrosion is a destructive process that degrades/alters the properties of materials, especially for metals and alloys, due to their interaction with the environmental conditions over time, even at room temperature [110]. Corrosion of iron is simple example from our everyday life. Therefore, practical applications of materials are limited by the rate of its corrosion in a specific environment and understanding the process provides us a control to mitigate or reduce it. There are many kinds of corrosion and multiple methods have been developed over the years to understand and control them. LC TEM provides a platform to characterize and measure the corrosive effects of solutions. Aqueous corrosion can be described as electrochemical oxidative process initiated by either the oxygen dissolved in water, by proton, or by water itself.

However, the chemistry of solution can be changed in a flow reactor to follow the corrosion by other radicals, such as Cl^- or Br^-. Shan et al. followed the etching mechanist of Pd@Pt cubic nanoparticles by Br-containing electrolyte [111]. They found two pathways, depending on the particle morphology, for etching to proceed. For example, on a regular, defect-free cube (Figure 6.10a), the etching of the Pd core started at the corners forming four voids, marked by white arrows in Figure 6.10b, and moved inward with time until most of the Pd was etched out after 60 seconds,

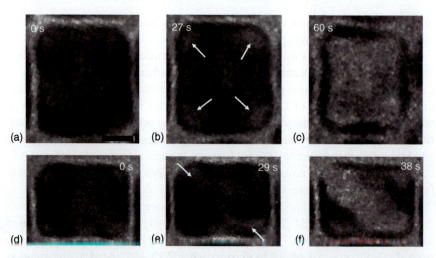

Figure 6.10 Etching process of regular and corner defected cubes. Time sequential TEM micrographs showing the etching process of internal Pd atoms in(a–c) a single regular and (d–f) corner defected Pd@Pt cube, respectively. Scale bars in all panels are 5 nm. Source: Shan et al. [111]/from Springer Nature, CC BY 4.0.

6.6 Select Examples of Applications | 193

leaving a Pt shell behind (Figure 6.10c). It is an indirect etching process and is similar to galvanic etching. However, surface defects exposed Pd in direct contact with the electrolyte and provided a starting point for direct etching as is the case for the nanoparticle shown in Figure 6.10d. Although the etching started at all four corners, the etching rate from upper-left and lower-right corners, marked by white arrows in Figure 6.10e, was much faster and the two voids coalesced, forming a nanochannel after 38 seconds (Figure 6.10f). They also reported that the location of defects at the surface, instead of corner, also resulted in providing a start location for etching [111].

Other examples include the onset and development of corrosion pits in Cu and Al thin films by NaCl solution [112], electrochemical etching of Mn and S from stainless steel under various flow conditions [113]. As described in Section 5.5.1, redesigned microchips or four-detector configuration has made it possible to collect EDS data in liquids [114].

6.6.3 Galvanic Replacement Reactions

Galvanic replacement of one metal by other is one of the methods to produce hollow metallic nanoparticles. For example, the transient stages of galvanic replacement of Ag by Au in Ag nanoparticles have been elucidated using LC-TEM, where voids are formed inside the Ag nanocubes [115]. Such hollow structures are technologically important for catalysis, energy conversion, and medicine. Chee et al. have followed the galvanic replacement by Au in an Ag nanocube at room temperature and 90 °C. First, the nanocube is encapsulated by Au before the Ag starts to leach out (Figure 6.11 $t - t_0 = 5.6$ seconds). Next, two voids are formed at 9.6 seconds, marked

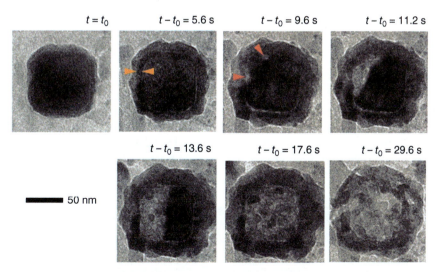

Figure 6.11 Time series of in-situ TEM images, extracted from a video, showing the morphological evolution of an Ag nanocube during galvanic replacement at 90 °C. Orange arrows indicate the inner shell with darker contrast, which should be the Au layer. Red arrows indicate observed void nucleation. Here, t_0 indicates the start of the recording time. Source: Chee et al. [115]/from Springer Nature, CC BY 4.0.

by red arrows in Figure 6.11 $t - t_0 = 9.6$ seconds and grow to form a single larger void upon merging (Figure 6.11 $t - t_0 = 11$ seconds). The void continues to grow as more Ag is leached out of the Ag nanocube (Figure 6.11 $t - t_0 = 13.6$ seconds) until a hollow Au shell is formed (Figure 6.11 $t - t_0 = 17.6$ seconds). With continued observation, large pores appear in the walls, which lead to degradation of the entire structure (Figure 6.11 $t - t_0 = 29.6$ seconds) that could be attributed to electron beam effect.

In summary, the nanocube became hollow via the nucleation, growth, and coalescence of voids during the galvanic replacement with gold (Au) ions at different temperatures. Here Kirkendall effect is a direct result of nanoscale process that occurs in conjunction with galvanic replacement and can be controlled by controlling the reaction parameters [115].

These are particularly interesting results as the silver–palladium nanocages were found to have similar morphologies as those obtained without electron beam irradiation (ex-situ control experiments) such that the fundamental mechanisms observed with TEM could be directly applied to large-scale synthesis. They also show that scavengers can be added to the aqueous solution to identify the role of radicals generated via radiolysis by high-energy electrons in modifying galvanic reactions. It is important to note that such scavengers can provide a way to minimize electron beam effects to achieve reaction conditions similar to ex-situ processes [5].

6.6.4 Growth of Core–Shell Nanoparticles

Synthesis of core–shell nanoparticles from solutions is a simple way to coat seed particles of one element (core) by another (forming a shell). Jungjohann et al. [116] have explored the mechanisms and conditions to coat Au seed particles with Pd via deposition of Pd^0, generated by the reduction of chloropalladate complexes by radicals generated during the radiolysis of aqueous solution by the high-energy electron beam [116]. They show that the size and shape of the seed particles determined the morphology of the deposited shell. For example, a continuous, uniform shell was observed to deposit up to large thicknesses by Pd monomer incorporation for Au nanoparticles with size 5 nm or smaller (Figure 6.12). On the other hand, for larger (15, 30 nm) particles, uneven deposit was observed as Pd growth accelerated at corners and asperities, with suppressed deposition on facets leading to different shapes and morphologies [116].

Figure 6.12 Growth of core–shell structure: Image series showing the Pd growth on an individual seed of 5 nm Au nanoparticle in 10 μM aqueous $PdCl_2$ solution over 92.4 seconds. The electron dose per image is 150 e^- Å^{-2} (scale bar: 5 nm). Source: Jungjohann et al. [116]/from American Chemical Society.

Another example of core–shell nanoparticle growth was reported by Liang et al. [117], where they observed the formation of bimetallic core (Fe_3Pt) covered by an oxide shell (Fe_2O_3). They achieved the growth of bimetallic alloy particles followed by the oxide shell in the electron beam–induced reactions by controlling the Fe to Pt ratio in the precursor solution [117]. They found that growth of the core stopped after Pt was depleted from the solution and the growth kinetics of the iron oxide shell did not change substantially after the Fe–Pt alloy core stopped to grow. A heteroepitaxy of Fe_3Pt [101] (core)||α-Fe_2O_3 [111] (shell) was observed in most of the nanoparticles, while a strain relaxation caused the formations of a polycrystalline iron oxide shell. They propose that Pt atoms catalyzed the reduction of Fe ions in the solution to form the Fe_3Pt alloy core, and a direct precipitation of iron oxide after Pt was depleted, resulting in the core–shell nanostructure formation [117]. Formation of Au–Pd core–shell nanoparticles under kinetically and thermodynamically controlled reaction conditions has also been reported [118].

A number of other studies for understanding the electrochemical behavior of liquids have also been reported, e.g. growth of metallic Ni [119] and Ag [4, 120] nanoparticles; growth of lead dendrites [42, 121]; water splitting [122, 123], etc.

6.6.5 Soft Nanomaterials Analyzed by In-Situ Liquid TEM

Polymers and block copolymer materials have applications ranging from being components of hybrid materials to drug delivery systems. Therefore, understanding of their formation and functionality is important toward designing novel materials for specific applications. For example, aggregation of amphiphilic block copolymers forms micelles that are ideal drug nanocarriers. Parent et al. have reported the evolution, growth, collision, and fusion mechanism of amphiphilic polymeric micelles in water [124]. In-situ observations revealed formation, growth, and motion of multiple micelles with time as captured by recording a video (Figure 6.13a). Quantitate analysis of the motion distance and directions suggested that micelles move in stochastic fashion in the confined water space, reflecting fractional Brownian motion with time. Such motion often results in collision and fusion of two micelles, marked by red and blue arrow in Figure 6.13b ($t = 0$ seconds). High-magnification images extracted from the video show the diffusion, collision, and fusion of two micelles into one in 90 seconds (marked by black arrow) (Figure 6.13b). However, non-fusion collisions were also observed in 2D images that could be attributed to Coulombic repulsion between them. The frequency of sticking, slipping, and sliding of the micelles on the membrane, i.e. the motion, increased with the dose rate, but the hopping distance remained constant over all doses. The quantitative analysis of LC-TEM data, combined with in silico simulations, presents a unique view of solvated soft matter nano-assemblies as they form and grow in time and space [124].

The mechanism of the micelle formation and encapsulation of target nanomaterials to prevent their fusion was revealed by capturing the dynamics of micelle formation and their encapsulation of gold nanoparticles (NPs) in an aqueous solution [125]. Li et al. report that a micelle, with a hydrophobic and rigid core

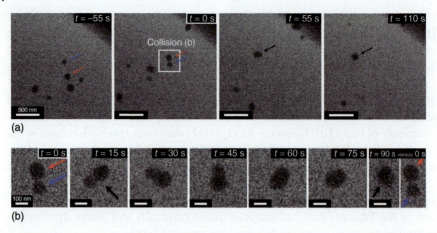

Figure 6.13 (a) Micelle–micelle fusion event captured by LCTEM. (a) Single frames (time-lapse) extracted from an in-situ LCTEM video (6 e$^-$ Å^{-2} s^{-1}) starting at $t = -55$ seconds (prior to collision). The video frame where collision between the red and blue micelles first occurs is referred to as "$t = 0$ seconds" (start of fusion-relaxation process). Frames labeled as $t = 55$ and 110 seconds show relaxation yielding a single assembly (black-labeled micelle). (b) Magnified time-lapsed images starting at $t = 0$ seconds, showing the fusion-relaxation process between the red and blue micelles. Source: Parent et al. [124]/ from American Chemical Society.

surrounded by a corona of hydrophilic blocks, which extend into the solution, is formed when the amphiphilic block copolymers aggregate and rearrange. They found these micelles to be stable against coalescence, and that the encapsulation of hydrophobic NPs is a self-limiting process. The gradual adsorption of block copolymers forms a polymeric shell around the NPs and the growth stops when the NPs are completely covered by the adsorbed copolymers [125].

The formation and evolution mechanism of a membrane for vesicle from polymer-rich liquid droplets that can affect the assembly process has also been reported. The amphiphilic self-assembly was proposed to proceed by liquid–liquid phase separation [126]. By studying the motion of polymer nanoparticles, formed by direct polymerization of an oxaliplatin analogue, in water points toward key design parameters that enable this LC-TEM characterization approach for organic nanomaterials [127]. The insights from these observations are of fundamental importance for the design of biocompatible soft materials.

Other examples include, but not limited to, gaining mechanistic insight into nanoscale assembly dynamics of thermos-responsive polymers [128], fragmentation mechanism of block copolymer micelle at 170 °C [129], formation of polymeric nanomaterials as function of temperature in LC [130].

Mobility of free-floating nanoparticles should follow a Brownian path. In-situ observation of unhindered, free-floating, Au particle was made using STEM imaging and quantified to measure the diffusion coefficient of Au nanoparticles in water [131]. Recently, Clark et al. have used double graphene liquid cell, consisting of a central molybdenum disulfide monolayer separated by hexagonal boron nitride spacers from the two enclosing graphene windows to follow mobility of Pt adatoms in an aqueous salt solution [132]. The STEM measurements also allowed to quantify electron beam effects on the mobility and diffusivity of the nanoparticles.

6.6.6 Quantitative Electrochemical Measurements

Since the first measurements of Cu nanoparticle nucleation and growth, made using a home-built LC holder [27], researchers have been interested in understanding electrochemical reactions [133, 134]. The effect of the chemistry of deposition solution (electrolyte) on the morphology of electrodeposit has also been evaluated by adding phosphates in the electrolyte that leads to the formation of copper phosphate complexes in solution [135]. Availability of commercial liquid flow cell with microfabricated electrodes moved the field further into making quantitative measurements of electrochemical parameters, including methods to differentiate between effect of electron beam and biasing [133]. Examples of some of the early results are described below.

Quantitative in-situ STEM measurements of electrochemical parameters, such cyclic voltammetry (CV), choronoamperometry (CA), and electrochemical impedance spectroscopy (EIS), were reported by Unocic et al. using a commercial holder [136]. The electrochemical cell used in their experiments consisted of silicon microchip device, microfabricated with a three-electrode configuration with a glassy carbon working electrode and platinum reference and counter electrodes. All measurements were performed using a standard one electron transfer redox couple $[Fe(CN)_6]^{3-/4-}$-based electrolyte.

Two CV data were acquired at a 1000 mV s^{-1} scan rate with scan limits of −200 to +200 mV and an electrolyte flow rate of 10 µl min^{-1} for both scans (Figure 6.14a). Measured response has the shape characteristic of a diffusion-controlled reaction of species dissolved in solution. A number of experiments were performed by varying flow and scan rates to fully understand the redox behavior of the solution. Furthermore, they determined the diffusion coefficient of electroactive species as well as the diffusion-controlled electrochemical reactions using CA. EIS was used to determine the electrochemical measurement of the cell and to confirm the redox behavior as determined using CA measurements. The experimentally measured diffusion coefficients of the $[Fe(CN)_6]^{4-}$ analyte, extracted from CA measurements, were consistent with reported values and CV measurements. EIS curves show (Figure 6.14b,c) the effect of mass transport limit that results in a larger resistance for high potentials and low flow rates. The high frequency response is controlled by the cell geometry and thus is relatively unaffected by the increase in mass transport [136, 137].

Similar operando monitoring of the charge and discharge process in a Na–O$_2$ battery, using liquid electrochemical cell, revealed a solution-mediated nucleation of NaO$_2$ cubes [138]. Furthermore, the subsequent oxidation of NaO$_2$ was observed to proceed via a solution mechanism. Based on the in-situ observations, it was proposed that the poor cyclability of Na–O$_2$ batteries is due the formation of shell of a parasitic product through the chemical reaction at the interface between the growing NaO$_2$ cubes and the electrolyte [138]. Cyclic voltammetry measurements using Ni-rich layered oxides electrodes have also been reported [139].

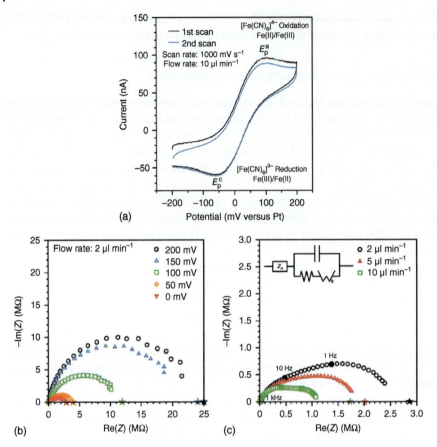

Figure 6.14 (a) Cyclic voltammograms of the first two scans from a glassy carbon microband electrode within a 2 mM $K_3Fe(CN)_6$ and 2 mM $K_4Fe(CN)_6$ in a 1 M KCl/10 mM NaCN electrolyte. Cyclic voltammetry (CV) scans were acquired at a scan rate of 1000 mV s^{-1} and at an electrolyte flow rate of 10 μl min^{-1}. (b, c) Electrochemical impedance spectroscopy Nyquist plots of $[Fe(CN)_6]^-$ on the glassy carbon microband electrode at constant flow rate of 2 μl min^{-1} with varied potential (b) and constant potential of 0 V with varied flow rate (c). Inset in (c) is the equivalent circuit model used to fit the impedance data from (a). The Warburg element here is the general finite diffusion element from ZView. Source: Unocic et al. [136]/Cambridge University Press.

The prospect of conducting novel investigative in-situ electrochemistry experiments has been realized by measuring various electrochemical parameters and couples the sinusoidal potential/current perturbations with simultaneous characterization of dynamic events using STEM imaging, EDS, EELS, and/or electron diffraction [49, 50, 114, 140]. It has also been used for electron beam-induced lithography, mass transport, growth of dendritic structures on battery electrodes [42, 50, 140, 141], etc.

6.6.7 Battery Research

Electrochemical measurements described above can be applied to understand the nanoscale mechanisms of charge and discharge phenomenon in batteries.

Development of new battery materials requires a fundamental understanding of the three basic steps that control the operation and degradation during the battery's lifetime: intercalation, alloying, and conversion to power [142, 143]. Although open cells, using ionic liquids, have been successfully employed to follow charge and discharge phenomenon for Li batteries [23, 144], LC-TEM has provided the much-needed platform to obtain in-situ and operando localized information of solid–electrolyte interface and other micro-to-nanosized components [144–147]. Therefore, battery research has become a big part of in-situ LC-TEM research for the direct observations to unveil the charge and discharge mechanisms in unprecedented details [144, 148–152]. Such information can then be used to design the materials system with improved properties such as durability for large number of cycles.

6.6.7.1 Open Cell

In-situ TEM measurements have impacted the understanding of charge/discharge phenomenon for several battery systems. First groundbreaking observations were made by Huang et al. using ionic liquids in an open cell configuration [23]. Figure 6.15 shows their setup (a) and lithiation mechanism into single nanowire, such as SnO_2 [153]. Here, they use ionic liquid (with very low vapor pressure) as an electrolyte. The cell consisted of single nanowire as the electrode under observation, vacuum-compatible lithium bis(trifluoromethanesulfonyl)imide (LiTFSI) dissolved in a hydrophobic 1-butyl-1-methylpyrrolidium (P14) TFSI (P14TFSI) as ionic liquid, and $LiCoO_2$ as the counter electrode [154].

This setup made it possible to collect nanoscale information of Li-ion insertion mechanism in SnO_2 during initial charging, where large irreversible volume expansion is due to the formation of Li_2O. The Li_2O did not participate in the discharging process and only the Li_xSn nanoprecipitates were found to be active [23]. The failure mechanism could be attributed to the volume change in the SnO_2 nanowires upon lithiation. Although the cell could be cycled several times, the number of cycles was nowhere near the desired real-life operation of Li ion batteries. But this technique

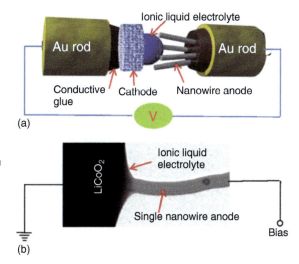

Figure 6.15 (a) Schematic drawing showing the experimental setup of the open-cell approach using ionic liquid as the electrolyte, (b) TEM image showing a working nanobattery in TEM column, where the single nanowire anode can be imaged during charge and discharge of this nanobattery. Source: Huang et al. [23]/from American Association for the Advancement of Science.

opened the door to investigate lithiation mechanism in other relevant anode materials such as Si [155, 156], Ge [157], Al_2O_3 [158], ZnO [159], etc. More information and references can be found in recent publications that review the recent progress of using in-situ TEM to study individual nanostructures in battery materials using an open-cell design, including for anode and cathode materials in lithium ion batteries, and Li–S batteries [148, 151, 160].

6.6.7.2 Closed Liquid Cell

A closed or flow cell with biasing contacts is ideally suited for investigating the mechanism and performance of battery systems [161]. It is important to note that a flow cell may be used to introduce the liquid in the cell and then seal it such that it acts as a closed cell.

Here we provide an example of monitoring Li ion transport kinetics and degradation mechanisms in $LiFePO_4$ particles during the charge/discharge cycles obtained by combining TEM and EFTEM imaging with voltage cycles, using a commercial LC with modified microfabricated electrode on a chip, reported by Holtz et al. [162]. They utilized the valency-loss peak at 5 eV in low-loss EELS data, which is unique to $FePO_4$ and not to $LiFePO_4$, to acquire EFTEM images. Slow nucleation during transformation, indicating particle-by-particle lithiation, was observed in the 5 eV window images, centered at 5 eV peak of $FePO_4$, acquired as function of time during voltage cycling (Figure 6.16). The structural change in one cluster of particles, corresponding to the voltage profile (Figure 6.16a), with time, marked by black arrows, corresponds to the de-lithiation process. As expected, lithiated ($LiFePO_4$) regions are dark and de-lithiated ($FePO_4$) regions are bright in EFTEM

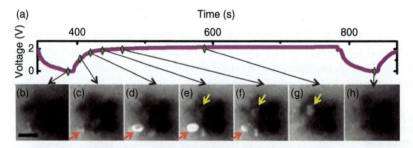

Figure 6.16 Temporal evolution of a $LiFePO_4$/$FePO_4$ cluster during one charge/discharge cycle. The voltage profile is shown in (a), corresponding to the second imaged cycle (fourth cycle after assembly). (b) EFTEM image acquired using 5 eV window around 5 eV valence-loss peak (from 2.5 to 7.5 eV) of $FePO_4$ represents the completely discharged state. (c–h) EFTEM images acquired at different time during voltage cycling (marked by black arrows). At the bottom of (c), we see the emergence of core–shell structure (d) that becomes clearer with dark core surrounded by bright shell, (e–g) the core partially disappears, and bright shell reduced in size before disappearing. The de-lithiating core–shell regions are marked by red arrows. More regions of bright $FePO_4$, marked by yellow arrows, develop in (e–g), and the particle returns to the discharged state in (h), where it is darker. In general, images taken for the charged state (d–g) have more bright regions than the images taken in the discharged state, (a) and (h). Red arrows indicate particles delithiating by a core–shell pathway, and yellow arrows indicate delithiation propagating from left to right through the particle. The scale bar is 200 nm. Source: Holtz et al. [162]/American Chemical Society.

images and development of a core–shell structure, marked by red arrows in Figure 6.16c,d, is observed. A number of white patches, marked by yellow arrows, develop while the bright core–shell structure disappears (Figure 6.16c–g). Fully charged cluster (Figure 6.16h) turned dark again. Interestingly, they found that formation and disappearance of de-lithiated particles have also been observed by ex-situ experiments and in the regions not irradiated by electron beam during their experiments. Therefore, using EFTEM imaging, the lithiated state of a $LiFePO_4$ electrode and surrounding aqueous electrolyte was monitored in real time, with nanoscale resolution, during electrochemical charge. These images revealed the phase transition and Li transport characteristics between $LiFePO_4$ and $FePO_4$ [162].

Large number of reports has since been published to understand the cyclic behavior of the lithiation process and degradation mechanisms. These reports include, but are not limited to, quantification of the solid–electrolyte interface (SEI) growth kinetics during lithiation of graphite and the dynamic self-healing nature of the SEI with changes in cell potential [147]; Electrochemical lithiation and de-lithiation of Au anodes in a commercial $LiPF_6$/EC/DEC electrolyte for lithium ion batteries [163]; charging/discharging and rate-dependent capacity degradation of Li-ion batteries [164–166]. Moreover, recently in-situ liquid cell observations have also revealed that the fluoride-rich SEI yields a denser Li structure that is particularly amenable to uniform stripping, thus suppressing lithium detachment and isolation [167, 168]. Additionally, the particle and dendritic growth at electrode and electrolyte interfaces have revealed the reasons for the loss of activity for Li ion batteries [169]. These findings highlight the importance of operando measurements to appropriately tailor the SEI, improving the cyclic performance of lithium batteries and expand the field of advanced rechargeable batteries [170, 171].

> Let a drop of wine fall into a glass of water; whatever be the law that governs the internal movement of the liquid, we will soon see it tint itself uniformly pink and from that moment on, however we may agitate the vessel, it appears that the wine and water can separate no more. All this, Maxwell and Boltzmann have explained, but the one who saw it in the cleanest way, in a book that is too little read because it is difficult to read, is Gibbs, in his Principles of Statistical Mechanics.
> – Henri Poincaré

6.7 Limitations

Despite all the efforts made for developing LC-TEM technique and reported success stories, it still has limited applications. Most of them arise from the challenges outlined in Section 6.5 and outlined below:

- **Widow rupture:** The thin electron transport windows are known to break during observation that can in principle deteriorate the column vacuum and adversely affect the TEM functionality in the long run. However, it has been shown that

(i) the amount of liquid is too small to cause appreciable harm and (ii) the liquid freezes at the break point(s) to clog the hole/cracks immediately if the anticontamination device is being used.

- **Electron beam:** The chemistry of liquids is modified by electron beam while using closed or flow cells. Whereas the liquid affected by the interaction of high-energy electron may be continuously replaced by fresh batch in a flow cell, the effects cannot be ignored. However, these effects have been thoroughly investigated, and methods to mitigate have been established. On the other hand, electrons can be used as stimuli for the nucleation and growth of nanoparticle from the salt solutions and understanding the industrial application of electron sources for polymer synthesis. In summary, electron bean effects are a mixed bag of positive and negative, depending on the process to be observed and questions to be answered.
- **Pressure limit:** Although windows can withstand reasonable pressure, but the windows bulging due to the pressure difference between the cell interior and exterior, the total path length increases (Section 6.5.3) that decreases SNR for both EELS and imaging. Alternative design of the microchips and working with lower pressures have been reported to overcome this limitation.
- **HREM imaging and spectroscopy:** The windows and the liquid increase the electron path length to reduce the spatial resolution, except for the graphene liquid cell. Moreover, the thickness also reduces the SNR for obtaining good quality of EELS data. The thickness and shape of the microchip also affect the EDS data collection. Zaluzec's group has worked with Protochips, Inc. to change the shape of the chip such that the EDS signal from sample is not blocked by the walls of the cell (Figure 5.11).
- **Electrode thickness:** The thickness of electrode hinders observation of electrode–electrolyte interface, where the electrochemical reaction is occurring. Some efforts to reduce the thickness have been reported, but it is not clear if it has any effect on the reaction process.
- **Experimental constraints:** LC-TEM is considered as an example of "lab-on-chip," implying lab-scale experiments can be performed using this technique. However, not all the actions, such as mixing and stirring, are possible. Combining other stimuli such as heating is also not readily available.

Overall, LCs provide an invaluable platform to follow liquid–solid interactions under other stimuli such as heating and biasing. Microfabricated chips have revolutionized the field and made it possible to incorporate other stimuli for measuring mechanical or magnetic properties.

6.8 Take-Home Messages

Commercial availability of LC-TEM holders has made it possible to observe solid–liquid interaction in unprecedented details and relate them to the properties. Although originally used for imaging biological materials in their native state, they are now routinely used for materials science applications. Direct visualization of nanoparticle nucleation and growth from solutions, particle agglomeration, and electrode and electrolyte interactions are some of the exemplary applications. Understanding and measuring the electrochemistry of battery materials is one the valuable applications. Moreover, electron beam–induced material modification is also being explored to understand the polymerization process for industrial applications.

Electron beam effects, window bulging and compatibility of liquid with LC materials including windows must be considered while planning an in-situ LC experiments. Windows are fragile and break easily, therefore, careful loading procedures should be established and followed. EDS data collection requires special consideration due to (a) possible restriction on holder tilt-angle posed by objective pole-piece gap and (b) restriction on the line of site to the detector by microchip side-walls.

References

1 Hayden, S.C., Chisholm, C., Grudt, R.O. et al. (2019). Localized corrosion of low-carbon steel at the nanoscale. *npj Materials Degradation* 3 (1): 17.
2 Kosari, A., Zandbergen, H., Tichelaar, F. et al. (2019). Application of in situ liquid cell transmission electron microscopy in corrosion studies: a critical review of challenges and achievements. *Corrosion* 76 (1): 4–17.
3 Jiang, Y., Zhu, G., Lin, F. et al. (2014). In situ study of oxidative etching of palladium nanocrystals by liquid cell electron microscopy. *Nano Letters* 14 (7): 3761–3765.
4 Sun, M., Tian, J., and Chen, Q. (2021). The studies on wet chemical etching via in situ liquid cell TEM. *Ultramicroscopy* 231: 113271.
5 Sutter, E., Jungjohann, K., Bliznakov, S. et al. (2014). In situ liquid-cell electron microscopy of silver–palladium galvanic replacement reactions on silver nanoparticles. *Nature Communications* 5 (1): 4946.
6 Dwyer, J.R. and Harb, M. (2017). Through a window, brightly: a review of selected nanofabricated thin-film platforms for spectroscopy, imaging, and detection. *Applied Spectroscopy* 71 (9): 2051–2075.
7 Liao, H.-G. and Zheng, H. (2016). Liquid cell transmission electron microscopy. *Annual Review of Physical Chemistry* 67 (1): 719–747.
8 de Jonge, N. and Ross, F.M. (2011). Electron microscopy of specimens in liquid. *Nature Nanotechnology* 6 (11): 695–704.
9 Pu, S., Gong, C., and Robertson, A.W. (2020). Liquid cell transmission electron microscopy and its applications. *Royal Society Open Science* 7 (1): 191204.
10 Ross, F.M. (2015). Opportunities and challenges in liquid cell electron microscopy. *Science* 350 (6267): aaa9886.
11 Bharda, A.V. and Jung, H.S. (2019). Liquid electron microscopy: then, now and future. *Applied Microscopy* 49 (1): 9.
12 Ross, F.M., Wang, C., and de Jonge, N. (2016). Transmission electron microscopy of specimens and processes in liquids. *MRS Bulletin* 41 (10): 791–803.

13 Wu, H., Friedrich, H., Patterson, J.P. et al. (2020). Liquid-phase electron microscopy for soft matter science and biology. *Advanced Materials* 32 (25): 2001582.

14 Ross, F.M. (2017). Liquid cell electron microscopy. In: *Advances in Microscopy and Microanalysis* (ed. M.I.P. Calarco), 407. Cambridge: Cambridge University Press.

15 Abrams, I.M. and McBain, J.W. (1944). A closed cell for electron microscopy. *Journal of Applied Physics* 15 (8): 607–609.

16 Double, D.D. (1973). Some studies of the hydration of Portland cement using high voltage (1 MV) electron microscopy. *Materials Science and Engineering* 12 (1): 29–34.

17 Double, D.D., Hellawell, A., Perry, S.J., and Hirsch, P.B. (1978). The hydration of Portland cement. *Proceedings of the Royal Society of London. A. Mathematical and Physical Sciences* 359 (1699): 435–451.

18 Ruska, E. (1942). Beitrag zur uebermikroskopischen Abbildungen bei hoeheren Drucken. *Kolloid Zeitschrift* 100: 212–219.

19 Swann, P.R. and Tighe, N.J. (1972). Performance of differentially pumped environmental cell in the AE1 EM7. Proceedings of the Fifth European Congress, Manchester, England, UK, NSA-27-005367.

20 Wise, M.E., Biskos, G., Martin, S.T. et al. (2005). Phase transitions of single salt particles studied using a transmission electron microscpe with an environmental cell. *Aerosol Scienec and Technology* 39: 849–856.

21 Wise, M.E., Martin, S.T., Russell, L.M., and Buseck, P.R. (2008). Water uptake by NaCl particles prior to deliquescence and the phase rule. *Aerosol Science and Technology* 42 (4): 281–294.

22 Dai, L.L., Sharma, R., and Wu, C.-Y. (2005). Self-assembled structure of nanoparticles at a liquid–liquid interface. *Langmuir* 21 (7): 2641–2643.

23 Huang, J.Y., Zhong, L., Wang, C.M. et al. (2010). In situ observation of the electrochemical lithiation of a single SnO_2 nanowire electrode. *Science* 330 (6010): 1515–1520.

24 Mansfeld, U., Hoeppener, S., and Schubert, U.S. (2013). Investigating the motion of diblock copolymer assemblies in ionic liquids by in situ electron microscopy. *Advanced Materials* 25 (5): 761–765.

25 Butler, P. and Hale, K. (1981). In situ gas–solid reactions, in practical methods in electron microscopy. In: *Experimental Microscopy*, vol. 9 (ed. A.M. Glauert), 239–309. North Holland Co.

26 Ross, F.M. and Searson, P.C. (1995). *Dynamic Observation of Electrochemical Etching in Silicon*. IOP Publishing Ltd.

27 Williamson, M.J., Tromp, R.M., Vereecken, P.M. et al. (2003). Dynamic microscopy of nanoscale cluster growth at the solid–liquid interface. *Nature Materials* 2: 532–536.

28 Radisic, A., Vereecken, P.M., Searson, P.C., and Ross, F.M. (2006). The morphology and nucleation kinetics of copper islands during electrodeposition. *Surface Science* 600 (9): 1817–1826.

29 Mueller, C., Harb, M., Dwyer, J.R., and Miller, R.J.D. (2013). Nanofluidic cells with controlled pathlength and liquid flow for rapid, high-resolution in situ imaging with electrons. *The Journal of Physical Chemistry Letters* 4 (14): 2339–2347.

30 Menon, V., Denoual, M., Toshiyoshi, H., and Fujita, H. (2019). Self-contained on-chip fluid actuation for flow initiation in liquid cell transmission electron microscopy. *Japanese Journal of Applied Physics* 58 (9): 090909.

31 Jungjohann, K.L., Evans, J.E., Aguiar, J.A. et al. (2012). Atomic-scale imaging and spectroscopy for in situ liquid scanning transmission electron microscopy. *Microscopy and Microanalysis* 18 (3): 621–627.

32 de Jonge, N., Poirier-Demers, N., Demers, H. et al. (2010). Nanometer-resolution electron microscopy through micrometers-thick water layers. *Ultramicroscopy* 110 (9): 1114–1119.

33 Yuk, J.M., Kim, K., Alemán, B. et al. (2011). Graphene veils and sandwiches. *Nano Letters* 11 (8): 3290–3294.

34 Textor, M. and de Jonge, N. (2018). Strategies for preparing graphene liquid cells for transmission electron microscopy. *Nano Letters* 18 (6): 3313–3321.

35 Dahmke, I.N., Verch, A., Hermannsdörfer, J. et al. (2017). Graphene liquid enclosure for single-molecule analysis of membrane proteins in whole cells using electron microscopy. *ACS Nano* 11 (11): 11108–11117.

36 Kelly, D.J., Zhou, M., Clark, N. et al. (2018). Nanometer resolution elemental mapping in graphene-based TEM liquid cells. *Nano Letters* 18 (2): 1168–1174.

37 Jonge, N.D., Peckys, D.B., Kremers, G.J. et al. (2009). Electron microscopy of whole cells in liquid with nanometer resolution. *Proceedings of the National Academy of Sciences* 106 (7): 2159.

38 Ring, E.A. and de Jonge, N. (2010). Microfluidic system for transmission electron microscopy. *Microscopy and Microanalysis* 16 (5): 622–629.

39 Petruk, A.A., Allen, C., Rivas, N. et al. (2019). High flow rate nanofluidics for in-liquid electron microscopy and diffraction. *Nanotechnology* 30 (39): 395703.

40 Unocic, R.R., Jungjohann, K.L., Mehdi, B.L. et al. (2020). In situ electrochemical scanning/transmission electron microscopy of electrode–electrolyte interfaces. *MRS Bulletin* 45 (9): 738–745.

41 Arán-Ais, R.M., Rizo, R., Grosse, P. et al. (2020). Imaging electrochemically synthesized Cu_2O cubes and their morphological evolution under conditions relevant to CO_2 electroreduction. *Nature Communications* 11 (1): 3489.

42 Sun, M., Liao, H.-G., Niu, K., and Zheng, H. (2013). Structural and morphological evolution of lead dendrites during electrochemical migration. *Scientific Reports* 3 (1): 3227.

43 Yang, J., Andrei, C.M., Chan, Y. et al. (2019). Liquid cell transmission electron microscopy sheds light on the mechanism of palladium electrodeposition. *Langmuir* 35 (4): 862–869.

44 Leenheer, A.J., Jungjohann, K.L., Zavadil, K.R. et al. (2015). Lithium electrodeposition dynamics in aprotic electrolyte observed in situ via transmission electron microscopy. *ACS Nano* 9 (4): 4379–4389.

45 Fritsch, B., Hutzler, A., Wu, M. et al. (2021). Accessing local electron-beam induced temperature changes during in situ liquid-phase transmission electron microscopy. *Nanoscale* 3: 2466–2474.

46 Niekiel, F., Kraschewski, S.M., Müller, J. et al. (2017). Local temperature measurement in TEM by parallel beam electron diffraction. *Ultramicroscopy* 176: 161–169.

47 Winterstein, J.P., Lin, P.A., and Sharma, R. (2015). Temperature calibration for in situ environmental transmission electron microscopy experiments. *Microscopy and Microanalysis* 21 (6): 1622–1628.

48 Xin, H.L. and Zheng, H. (2012). In situ observation of oscillatory growth of bismuth nanoparticles. *Nano Letters* 12 (3): 1470–1474.

49 Tan, S.F., Reidy, K., Klein, J. et al. (2021). Real-time imaging of nanoscale electrochemical Ni etching under thermal conditions. *Chemical Science* 12 (14): 5259–5268.

50 Sasaki, Y., Yoshida, K., Kawasaki, T. et al. (2021). In situ electron microscopy analysis of electrochemical Zn deposition onto an electrode. *Journal of Power Sources* 481: 228831.

51 Denoual, M., Menon, V., Sato, T. et al. (2018). Liquid cell with temperature control for in situ TEM chemical studies. *Measurement Science and Technology* 30 (1): 017001.

52 Evans, J.E., Jungjohann, K.L., Browning, N.D. et al. (2011). Controlled growth of nanoparticles from solution with in situ liquid transmission electron microscopy. *Nano Letters* 11 (7): 2809–2813.

53 Dillon, S.J. and Chen, X. (2017). Temperature control in liquid cells for TEM. In: *Liquid Cell Electron Microscopy* (ed. F.M. Ross), 127–139. Cambridge: Cambridge University Press.

54 Parent, L.R., Bakalis, E., Proetto, M. et al. (2018). Tackling the challenges of dynamic experiments using liquid-cell transmission electron microscopy. *Accounts of Chemical Research* 51 (1): 3–11.

55 Hammad Fawey, M., VSK, C., Reddy, M.A. et al. (2016). In situ TEM studies of micron-sized all-solid-state fluoride ion batteries: preparation, prospects, and challenges. *Microscopy Research and Technique* 79 (7): 615–624.

56 Lee, H., OFN, O., Kim, G.-Y. et al. (2021). TEM sample preparation using micro-manipulator for in-situ MEMS experiment. *Applied Microscopy* 51: 8.

57 Abellan, P., Woehl, T.J., Parent, L.R. et al. (2014). Factors influencing quantitative liquid (scanning) transmission electron microscopy. *Chemical Communications* 50 (38): 4873–4880.

58 de Jonge, N. (2018). Theory of the spatial resolution of (scanning) transmission electron microscopy in liquid water or ice layers. *Ultramicroscopy* 187: 113–125.

59 de Jonge, N., Houben, L., Dunin-Borkowski, R.E. et al. (2019). Resolution and aberration correction in liquid cell transmission electron microscopy. *Nature Reviews Materials* 4 (1): 61–78.

60 Wu, H., Su, H., RRM, J. et al. (2021). Mapping and controlling liquid layer thickness in liquid-phase (scanning) transmission electron microscopy. *Small Methods* 5 (6): 2001287.

61 Leonard, D.N. and Hellmann, R. (2017). Exploring dynamic surface processes during silicate mineral (wollastonite) dissolution with liquid cell TEM. *Journal of Microscopy* 265 (3): 358–371.

62 Egerton, R.F. (2013). Control of radiation damage in the TEM. *Ultramicroscopy* 127: 100–108.

63 Grogan, J.M., Schneider, N.M., Ross, F.M. et al. (2014). Bubble and pattern formation in liquid induced by an electron beam. *Nano Letters* 14 (1): 359–364.

64 Bae, Y., Kang, S., Kim, B.H. et al. (2021). Nanobubble dynamics in aqueous surfactant solutions studied by liquid-phase transmission electron microscopy. *Engineering* 7 (5): 630–635.

65 Taverna, D., Kociak, M., Stéphan, O. et al. (2008). Probing physical properties of confined fluids within individual nanobubbles. *Physical Review Letters* 100 (3): 035301.

66 Schneider, N.M., Norton, M.M., Mendel, B.J. et al. (2014). Electron–water interactions and implications for liquid cell electron microscopy. *The Journal of Physical Chemistry C* 118 (38): 22373–22382.

67 Schneider, N.M. (2017). Electron beam effects in liquid cell TEM and STEM. In: *Liquid Cell Electron Microscopy* (ed. F.M. Ross), 140–163. Cambridge: Cambridge University Press.

68 Demers, H., Poirier-Demers, N., Couture, A.R. et al. (2011). Three-dimensional electron microscopy simulation with the CASINO Monte Carlo software. *Scanning* 33 (3): 135–146.

69 Alloyeau, D., Dachraoui, W., Javed, Y. et al. (2015). Unravelling kinetic and thermodynamic effects on the growth of gold nanoplates by liquid transmission electron microscopy. *Nano Letters* 15 (4): 2574–2581.

70 Park, J.H., Schneider, N.M., Grogan, J.M. et al. (2015). Control of electron beam-induced Au nanocrystal growth kinetics through solution chemistry. *Nano Letters* 15 (8): 5314–5320.

71 Woehl, T.J., Jungjohann, K.L., Evans, J.E. et al. (2013). Experimental procedures to mitigate electron beam induced artifacts during in situ fluid imaging of nanomaterials. *Ultramicroscopy* 127: 53–63.

72 Prabhudev, S. and Guay, D. (2020). Probing electrochemical surface/interfacial reactions with liquid cell transmission electron microscopy: a challenge or an opportunity? *Current Opinion in Electrochemistry* 23: 114–122.

73 Wu, H., Li, T., Maddala, S. et al. (2021). Studying reaction mechanisms in solution using a distributed electron microscopy method. *ACS Nano* 15 (6): 10296–10308.

74 Woehl, T.J. and Abellan, P. (2017). Defining the radiation chemistry during liquid cell electron microscopy to enable visualization of nanomaterial growth and degradation dynamics. *Journal of Microscopy* 265 (2): 135–147.

75 Donev, E.U.B., Matthew, and Hastings, J.T. (2017). Nanoscale deposition and etching of materials using focused electron beams and liquid reactants.

In: *Liquid Cell Electron Microscopy* (ed. F.M. Ross), 291–315. Cambridge: Cambridge University Press.

76 Gibson, W. and Patterson, J. (2021). Liquid phase electron microscopy provides opportunities in polymer synthesis and manufacturing. *Macromolecules* 54 (11): 4986–4996.

77 Ghodsi, S.M., Sharifi-Asl, S., Rehak, P. et al. (2020). Assessment of pressure and density of confined water in graphene liquid cells. *Advanced Materials Interfaces* 7 (12): 1901727.

78 Yesibolati, M.N., Laganá, S., Kadkhodazadeh, S. et al. (2020). Electron inelastic mean free path in water. *Nanoscale* 12 (40): 20649–20657.

79 Holtz, M.E., Yu, Y., Gao, J. et al. (2013). In situ electron energy-loss spectroscopy in liquids. *Microscopy and Microanalysis* 19 (4): 1027–1035.

80 Marchello, G., De Pace, C., Duro-Castano, A. et al. (2020). End-to-end image analysis pipeline for liquid-phase electron microscopy. *Journal of Microscopy* 279 (3): 242–248.

81 Hoppe, S.M., Sasaki, D.Y., Kinghorn, A.N. et al. (2013). In-situ transmission electron microscopy of liposomes in an aqueous environment. *Langmuir* 29 (32): 9958–9961.

82 Ring, E.A., Peckys, D.B., Dukes, M.J. et al. (2011). Silicon nitride windows for electron microscopy of whole cells. *Journal of Microscopy* 243 (3): 273–283.

83 Peckys, D.B. and Niels, D.J. (2017). Liquid stem for studying biological function in whole cells. In: *Liquid Cell Electron Microscopy* (ed. F.M. Ross), 334–355. Cambridge: Cambridge University Press.

84 Igami, Y., Tsuchiyama, A., Yamazaki, T. et al. (2021). In-situ water-immersion experiments on amorphous silicates in the MgO–SiO$_2$ system: implications for the onset of aqueous alteration in primitive meteorites. *Geochimica et Cosmochimica Acta* 293: 86–102.

85 Zheng, H.M., Smith, R.K., Jun, Y.W. et al. (2009). Observation of single colloidal platinum nanocrystal growth trajectories. *Science* 324 (5932): 1309–1312.

86 Longo, E., Avansi, W., Bettini, J. et al. (2016). In situ transmission electron microscopy observation of Ag nanocrystal evolution by surfactant free electron-driven synthesis. *Scientific Reports* 6 (1): 21498.

87 Lu, Y., Wang, K., Chen, F.R. et al. (2016). Extracting nano-gold from HAuCl$_4$ solution manipulated with electrons. *Physical Chemistry Chemical Physics* 18 (43): 30079–30085.

88 Cheng, Y., Tao, J., Zhu, G. et al. (2018). Near surface nucleation and particle mediated growth of colloidal Au nanocrystals. *Nanoscale* 10 (25): 11907–11912.

89 Dong, M., Wang, W., Wei, W. et al. (2019). Understanding the ensemble of growth behaviors of sub-10-nm silver nanorods using in situ liquid cell transmission electron microscopy. *The Journal of Physical Chemistry C* 123 (31): 21257–21264.

90 Kim, B.H., Yang, J., Lee, D. et al. (2018). Liquid-phase transmission electron microscopy for studying colloidal inorganic nanoparticles. *Advanced Materials* 30 (4): 1703316.

91 Wang, L., Liu, L., Chen, J. et al. (2020). Synthesis of honeycomb-structured beryllium oxide via graphene liquid cells. *Angewandte Chemie International Edition* 59 (36): 15734–15740.

92 Zhou, Y., Powers, A.S., Zhang, X. et al. (2017). Growth and assembly of cobalt oxide nanoparticle rings at liquid nanodroplets with solid junction. *Nanoscale* 9 (37): 13915–13921.

93 De Yoreo, J.J. (2016). In-situ liquid phase TEM observations of nucleation and growth processes. *Progress in Crystal Growth and Characterization of Materials* 62 (2): 69–88.

94 Tao, J., Nielsen, M.H., and De Yoreo, J.J. (2018). Nucleation and phase transformation pathways in electrolyte solutions investigated by in situ microscopy techniques. *Current Opinion in Colloid & Interface Science* 34: 74–88.

95 Zhu, G., Reiner, H., Cölfen, H., and De Yoreo, J.J. (2019). Addressing some of the technical challenges associated with liquid phase S/TEM studies of particle nucleation, growth and assembly. *Micron* 118: 35–42.

96 Liao, H.G., Zherebetskyy, D., Xin, H. et al. (2014). Facet development during platinum nanocube growth. *Science* 345 (6199): 916–919.

97 Ge, M., Lu, M., Chu, Y., and Xin, H. (2017). Anomalous growth rate of Ag nanocrystals revealed by in situ STEM. *Scientific Reports* 7 (1): 16420.

98 Ahmad, N., Wang, G., Nelayah, J. et al. (2018). Driving reversible redox reactions at solid–liquid interfaces with the electron beam of a transmission electron microscope. *Journal of Microscopy* 269 (2): 127–133.

99 He, J., Shi, F., Wu, J., and Ye, J. (2021). Shape transformation mechanism of gallium–indium alloyed liquid metal nanoparticles. *Advanced Materials Interfaces* 8 (6): 2001874.

100 Liao, H., Cui, L., Zheng, H., and Whitelam, S. (2012). Imaging of Pt_3Fe nanwire growth in liquids by in situ TEM. *Microscopy and Microanalysis* 18 (S2): 1092–1093.

101 Wang, H., Zhou, X., Huang, Y. et al. (2020). Interactions of sub-five-nanometer diameter colloidal palladium nanoparticles in solution investigated via liquid cell transmission electron microscopy. *RSC Advances* 10 (57): 34781–34787.

102 Wu, Y., Chen, X., Li, C. et al. (2019). In situ liquid cell TEM observation of solution-mediated interaction behaviour of Au/CdS nanoclusters. *New Journal of Chemistry* 43 (32): 12548–12554.

103 Li, C., Chen, X., Liu, H. et al. (2018). In-situ liquid-cell TEM study of radial flow-guided motion of octahedral Au nanoparticles and nanoparticle clusters. *Nano Research* 11 (9): 4697–4707.

104 Tan, S.F., Raj, S., Bisht, G. et al. (2018). Nanoparticle interactions guided by shape-dependent hydrophobic forces. *Advanced Materials* 30 (16): 1707077.

105 Lee, W.C., Kim, B.H., Choi, S. et al. (2017). Liquid cell electron microscopy of nanoparticle self-assembly driven by solvent drying. *The Journal of Physical Chemistry Letters* 8 (3): 647–654.

106 Yang, J., Zeng, Z., Kang, J. et al. (2019). Formation of two-dimensional transition metal oxide nanosheets with nanoparticles as intermediates. *Nature Materials* 18 (9): 970–976.

107 Asghar, M.S.A., Inkson, B.J., and Möbus, G. (2020). In situ formation of 1D nanostructures from ceria nanoparticle dispersions by liquid cell TEM irradiation. *Journal of Materials Science* 55 (7): 2815–2825.

108 Abellan, P., Parent, L.R., Al Hasan, N. et al. (2016). Gaining control over radiolytic synthesis of uniform sub-3-nanometer palladium nanoparticles: use of aromatic liquids in the electron microscope. *Langmuir* 32 (6): 1468–1477.

109 Rizvi, A., Mulvey, J.T., Carpenter, B.P. et al. (2021). A close look at molecular self-assembly with the transmission electron microscope. *Chemical Reviews* 121 (22): 14232–14280.

110 Song, Z. and Xie, Z.-H. (2018). A literature review of in situ transmission electron microscopy technique in corrosion studies. *Micron* 112: 69–83.

111 Shan, H., Gao, W., Xiong, Y. et al. (2018). Nanoscale kinetics of asymmetrical corrosion in core-shell nanoparticles. *Nature Communications* 9 (1): 1011.

112 Chee, S.W., Pratt, S.H., Hattar, K. et al. (2015). Studying localized corrosion using liquid cell transmission electron microscopy. *Chemical Communications* 51 (1): 168–171.

113 Schilling, S., Janssen, A., Zaluzec, N.J., and Burke, M.G. (2017). Practical aspects of electrochemical corrosion measurements during in situ analytical transmission electron microscopy (TEM) of austenitic stainless steel in aqueous media. *Microscopy and Microanalysis* 23 (4): 741–750.

114 Zaluzec, N.J., Burke, M.G., Haigh, S.J. et al. (2014). X-ray energy-dispersive spectrometry during in situ liquid cell studies using an analytical electron microscope. *Microscopy and Microanalysis* 20 (2): 323–329.

115 Chee, S.W., Tan, S.F., Baraissov, Z. et al. (2017). Direct observation of the nanoscale Kirkendall effect during galvanic replacement reactions. *Nature Communications* 8 (1): 1224.

116 Jungjohann, K.L., Bliznakov, S., Sutter, P.W. et al. (2013). In situ liquid cell electron microscopy of the solution growth of Au–Pd core–shell nanostructures. *Nano Letters* 13 (6): 2964–2970.

117 Liang, W.-I., Zhang, X., Zan, Y. et al. (2015). In situ study of $Fe_3Pt-Fe_2O_3$ core–shell nanoparticle formation. *Journal of the American Chemical Society* 137 (47): 14850–14853.

118 Tan, S.F., Bisht, G., Anand, U. et al. (2018). In situ kinetic and thermodynamic growth control of Au–Pd core–shell nanoparticles. *Journal of the American Chemical Society* 140 (37): 11680–11685.

119 Chen, X., Noh, K.W., Wen, J.G. et al. (2012). In situ electrochemical wet cell transmission electron microscopy characterization of solid–liquid interactions between Ni and aqueous $NiCl_2$. *Acta Materialia* 60 (1): 192–198.

120 Chen, X., Zhou, L., Wang, P. et al. (2012). A study of nano materials and their reactions in liquid using in situ wet cell TEM technology. *Chinese Journal of Chemistry* 30 (12): 2839–2843.

121 White, E.R., Singer, S.B., Augustyn, V. et al. (2012). In situ transmission electron microscopy of lead dendrites and lead ions in aqueous solution. *ACS Nano* 6 (7): 6308–6317.

122 Rodriguez Manzo, J.A., Salmon, N.J., and Alsem, D.H. (2017). In situ TEM observation of water splitting. *Microscopy and Microanalysis* 23 (S1): 936–937.
123 Lu, Y., Yin, W.-J., Peng, K.-L. et al. (2018). Self-hydrogenated shell promoting photocatalytic H_2 evolution on anatase TiO_2. *Nature Communications* 9 (1): 2752.
124 Parent, L.R., Bakalis, E., Ramírez-Hernández, A. et al. (2017). Directly observing micelle fusion and growth in solution by liquid-cell transmission electron microscopy. *Journal of the American Chemical Society* 139 (47): 17140–17151.
125 Li, C., Tho, C.C., Galaktionova, D. et al. (2019). Dynamics of amphiphilic block copolymers in an aqueous solution: direct imaging of micelle formation and nanoparticle encapsulation. *Nanoscale* 11 (5): 2299–2305.
126 Ianiro, A., Wu, H., MMJ, v.R. et al. (2019). Liquid–liquid phase separation during amphiphilic self-assembly. *Nature Chemistry* 11 (4): 320–328.
127 Patterson, J.P., Proetto, M.T., and Gianneschi, N.C. (2015). Soft nanomaterials analysed by in situ liquid TEM: towards high resolution characterisation of nanoparticles in motion. *Perspectives in Science* 6: 106–112.
128 Korpanty, J., Parent, L.R., Hampu, N. et al. (2021). Thermoresponsive polymer assemblies via variable temperature liquid-phase transmission electron microscopy and small angle X-ray scattering. *Nature Communications* 12 (1): 6568.
129 Early, J.T., Yager, K.G., and Lodge, T.P. (2020). Direct observation of micelle fragmentation via in situ liquid-phase transmission electron microscopy. *ACS Macro Letters* 9 (5): 756–761.
130 Scheutz, G.M., Touve, M.A., Carlini, A.S. et al. (2021). Probing thermoresponsive polymerization-induced self-assembly with variable-temperature liquid-cell transmission electron microscopy. *Matter* 4 (2): 722–736.
131 Yesibolati, M.N., Mortensen, K.I., Sun, H. et al. (2020). Unhindered Brownian motion of individual nanoparticles in liquid-phase scanning transmission electron microscopy. *Nano Letters* 20 (10): 7108–7115.
132 Clark, N., Kelly, D.J., Zhou, M. et al. (2022). Tracking single adatoms in liquid in a transmission electron microscope. *Nature* 609 (7929): 942–947.
133 Han, C., Islam, M.T., and Ni, C. (2021). In situ TEM of electrochemical incidents: effects of biasing and electron beam on electrochemistry. *ACS Omega* 6 (10): 6537–6546.
134 Mehdi, B.L., Stevens, A., Qian, J. et al. (2016). The impact of Li grain size on Coulombic efficiency in Li batteries. *Scientific Reports* 6 (1): 34267.
135 Zhao, J., Sun, L., Canepa, S. et al. (2017). Phosphate tuned copper electrodeposition and promoted formic acid selectivity for carbon dioxide reduction. *Journal of Materials Chemistry A* 5 (23): 11905–11916.
136 Unocic, R.R., Sacci, R.L., Brown, G.M. et al. (2014). Quantitative electrochemical measurements using in situ ec-S/TEM devices. *Microscopy and Microanalysis* 20 (2): 452–461.
137 Sacci, R.L., Black, J.M., Balke, N. et al. (2015). Nanoscale imaging of fundamental Li battery chemistry: solid-electrolyte interphase formation and preferential growth of lithium metal nanoclusters. *Nano Letters* 15 (3): 2011–2018.

138 Lutz, L., Dachraoui, W., Demortière, A. et al. (2018). Operando monitoring of the solution-mediated discharge and charge processes in a Na–O_2 battery using liquid-electrochemical transmission electron microscopy. *Nano Letters* 18 (2): 1280–1289.

139 Hou, J., Freiberg, A., Shen, T.-H. et al. (2020). Charge/discharge cycling of $Li_{1+x}(Ni_{0.6}Co_{0.2}Mn_{0.2})_{1-x}O_2$ primary particles performed in a liquid microcell for transmission electron microscopy studies. *Journal of Physics Energy* 2 (3): 034007.

140 Beker, A.F., Sun, H., Lemang, M. et al. (2020). In situ electrochemistry inside a TEM with controlled mass transport. *Nanoscale* 12 (43): 22192–22201.

141 Park, J.H., Steingart, D.A., Kodambaka, S. et al. (2017). Electrochemical electron beam lithography: write, read, and erase metallic nanocrystals on demand. *Science Advances* 3 (7): e1700234.

142 Yousaf, M., Naseer, U., Li, Y. et al. (2021). A mechanistic study of electrode materials for rechargeable batteries beyond lithium ions by in situ transmission electron microscopy. *Energy & Environmental Science* 14 (5): 2670–2707.

143 Fan, Z., Zhang, L., Baumann, D. et al. (2019). In situ transmission electron microscopy for energy materials and devices. *Advanced Materials* 31 (33): 1900608.

144 Wang, C.-M. (2015). In situ transmission electron microscopy and spectroscopy studies of rechargeable batteries under dynamic operating conditions: a retrospective and perspective view. *Journal of Materials Research* 30 (3): 326–339.

145 Wang, C.M., Yan, P., Zhu, Z. et al. (2017). Multimodal and in-situ chemical imaging of critical surfaces and interfaces in advanced batteries. *Journal of Surface Analysis* 24 (2): 141–150.

146 Ghosh, C., Singh, M.K., Parida, S. et al. (2021). Phase evolution and structural modulation during in situ lithiation of MoS_2, WS_2 and graphite in TEM. *Scientific Reports* 11 (1): 9014.

147 Unocic, R.R., Sun, X.G., Sacci, R.L. et al. (2014). Direct visualization of solid electrolyte interphase formation in lithium-ion batteries with in situ electrochemical transmission electron microscopy. *Microscopy and Microanalysis* 20 (4): 1029–1037.

148 Wu, J., Fenech, M., Webster, R.F. et al. (2019). Electron microscopy and its role in advanced lithium-ion battery research. *Sustainable Energy & Fuels* 3 (7): 1623–1646.

149 Wang, C.M., Xu, W., Liu, J. et al. (2010). In situ transmission electron microscopy and spectroscopy studies of interfaces in Li ion batteries: challenges and opportunities. *Journal of Materials Research* 25 (8): 1541–1547.

150 Yuan, Y., Amine, K., Lu, J. et al. (2017). Understanding materials challenges for rechargeable ion batteries with in situ transmission electron microscopy. *Nature Communications* 8 (1): 15806.

151 Wu, X., Li, S., Yang, B., and Wang, C. (2019). In situ transmission electron microscopy studies of electrochemical reaction mechanisms in rechargeable batteries. *Electrochemical Energy Reviews* 2 (3): 467–491.

152 Zhang, C., Firestein, K.L., JFS, F. et al. (2020). Recent progress of in situ transmission electron microscopy for energy materials. *Advanced Materials* 32 (18): 1904094.

153 Huang, J.Y., Zhong, L., Wang, C.M. et al. (2010). In situ observation of the electrochemical lithiation of a single SnO_2 nanowire electrode. *Science* 330 (6010): 1515.

154 Wang, C.M., Xu, W., Liu, J. et al. (2010). In situ transmission electron microscopy and spectroscopy studies of interfaces in Li ion batteries: challenges and opportunities. *Journal of Materials Research* 25 (8): 1541–1547.

155 Liu, X.H., Wang, J.W., Huang, S. et al. (2012). In situ atomic-scale imaging of electrochemical lithiation in silicon. *Nature Nanotechnology* 7 (11): 749–756.

156 Ghassemi, H., Au, M., Chen, N. et al. (2011). Real-time observation of lithium fibers growth inside a nanoscale lithium-ion battery. *Applied Physics Letters* 99 (12): 123113.

157 Liu, X.H., Huang, S., Picraux, S.T. et al. (2011). Reversible nanopore formation in Ge nanowires during lithiation-delithiation cycling: an in situ transmission electron microscopy study. *Nano Letters* 11 (9): 3991–3997.

158 Liu, Y., Hudak, N.S., Huber, D.L. et al. (2011). In situ transmission electron microscopy observation of pulverization of aluminum nanowires and evolution of the thin surface Al_2O_3 layers during lithiation–delithiation cycles. *Nano Letters* 11 (10): 4188–4194.

159 Kushima, A., Liu, X.H., Zhu, G. et al. (2011). Leapfrog cracking and nanoamorphization of ZnO nanowires during in situ electrochemical lithiation. *Nano Letters* 11 (11): 4535–4541.

160 Zou, R., Cui, Z., Liu, Q. et al. (2017). In situ transmission electron microscopy study of individual nanostructures during lithiation and delithiation processes. *Journal of Materials Chemistry A* 5 (38): 20072–20094.

161 Gu, M., Parent, L.R., Mehdi, B.L. et al. (2013). Demonstration of an electrochemical liquid cell for operando transmission electron microscopy observation of the lithiation/delithiation behavior of Si nanowire battery anodes. *Nano Letters* 13 (12): 6106–6112.

162 Holtz, M.E., Yu, Y., Gunceler, D. et al. (2014). Nanoscale imaging of lithium ion distribution during in situ operation of battery electrode and electrolyte. *Nano Letters* 14 (3): 1453–1459.

163 Zeng, Z., Liang, W.-I., Liao, H.-G. et al. (2014). Visualization of electrode–electrolyte interfaces in $LiPF_6$/EC/DEC electrolyte for lithium ion batteries via in situ TEM. *Nano Letters* 14 (4): 1745–1750.

164 Kushima, A., Koido, T., Fujiwara, Y. et al. (2015). Charging/discharging nanomorphology asymmetry and rate-dependent capacity degradation in Li–oxygen battery. *Nano Letters* 15 (12): 8260–8265.

165 Zeng, Z., Zhang, X., Bustillo, K. et al. (2015). In situ study of lithiation and delithiation of MoS_2 nanosheets using electrochemical liquid cell transmission electron microscopy. *Nano Letters* 15 (8): 5214–5220.

166 Liu, P., Han, J., Guo, X. et al. (2018). Operando characterization of cathodic reactions in a liquid-state lithium-oxygen micro-battery by scanning transmission electron microscopy. *Scientific Reports* 8 (1): 3134.

167 Gong, C., Pu, S.D., Gao, X. et al. (2021). Revealing the role of fluoride-rich battery electrode interphases by operando transmission electron microscopy. *Advanced Energy Materials* 11 (10): 2003118.

168 Sacci, R.L., Dudney, N.J., More, K.L. et al. (2014). Direct visualization of initial SEI morphology and growth kinetics during lithium deposition by in situ electrochemical transmission electron microscopy. *Chemical Communications* 50 (17): 2104–2107.

169 He, K., Bi, X., Yuan, Y. et al. (2018). Operando liquid cell electron microscopy of discharge and charge kinetics in lithium-oxygen batteries. *Nano Energy* 49: 338–345.

170 Chen, C.-Y., Tsuda, T., Oshima, Y. et al. (2021). In situ monitoring of lithium metal anodes and their solid electrolyte interphases by transmission electron microscopy. *Small Structures* 2 (6): 2100018.

171 Xie, J., Li, J., Mai, W., and Hong, G. (2021). A decade of advanced rechargeable batteries development guided by in situ transmission electron microscopy. *Nano Energy* 83: 105780.

7

In-Situ Gas–Solid Interactions

> *The gaseous condition is exemplified in the soiree, where the members rush about confusedly, and the only communication is during a collision, which in some instances may be prolonged by button-holing.*
> – James Clerk Maxwell

Air is the most common example of gaseous environment that surrounds us all the time, and rusting of iron is its most prevalent effect, making gas–solid interaction an integral part of our life. It will be desirable to be able to determine the rusting mechanism at atomic scale so that we can find ways to prevent it. Of course, TEM is an excellent tool for such characterization if we can follow the reaction in air, especially in humid air, instead of vacuum. In other words, we need to develop techniques to confine gases around the sample, like the liquids as described in Chapter 6. Moreover, gaseous environment plays an import role in synthesis and functioning of a large number of materials, especially at high temperatures. Therefore, following the gas–solid interactions is highly desirable to understand and control catalytic reactions, synthesis of nanomaterials, materials degradation, to name a few. Several review articles and books have been published that describe various methods to confine gas environment around the sample and their applications to follow gas–solid interactions in TEM [1–5]. In this chapter, we will review the historical aspect of techniques developed for gas confinement, current status, emphasize on the challenge of designing experiments, some representative applications, followed by limitations.

7.1 Historical Perspective

The technical challenge for observing samples in gas or hydrated environment is to confine the gas environment around the sample without deteriorating the vacuum in the microscope column, especially in the gun chamber as the electron guns require 10^{-6} mbar or better vacuum to operate. Also, the electron scattering from the gas molecules contributes to the image intensities and reduces the signal-to-noise ratio

(SNR) in the image. Materials community has always been interested in observing the effect of temperature and environment on the structure and thereby on the properties of solids. Therefore, the pursuit to be able to observe structural changes in-situ at the atomic level during a chemical reaction is as old as the advent of TEM itself [2, 6–11]. In early days, a specially modified microscope to control the environment around the sample was called as "controlled atmosphere" TEM. The simplest way to isolate the sample and its environment from TEM column is to use electron transparent windows above and below the sample in a TEM holder and seal them before inserting in the microscope column. This approach was originally designed for liquids as explained in Chapter 6. While window holders are quite effective to contain liquid, it is challenging to replace air with any specific gas without incorporating a flow system. Moreover, integration of a heating element to observe gas–solid interactions at desired temperature posed another challenge. The electron scattering from the thick amorphous windows deteriorated the image contrast, making high-resolution observation difficult as explained in Chapter 5 (Figure 5.2) [9].

Incorporation of differential pumping within the TEM column to confine the gas or water vapor within the sample stage area was an alternative method developed in 1930s and 1940s [12, 13]. To achieve this, a physical cell, with apertures to reduce gas flow rate from the sample area to the column, was built and connected to additional pumping above and below the sample stage area [7, 8]; hence, the term environmental cell or E-cell was also used for this gas containment design. The differential pumping design continued to be developed and used as it allowed (i) the gas flow and (ii) use of heating holders and thus enabled us to observe samples in the gas environment at elevated temperature. Introduction of high-voltage TEMs helped in advancing this field of research due to improved spatial resolution and larger pole-piece gap to accommodate the physical cell with large enough gap between the differential apertures to allow high tilting range for sample holders [9, 10]. It was also assumed that high-energy electrons (1 MeV), with greater penetrating power, should be able to address the problem of increased gas pathlength [10]. But the high-voltage microscopes had major practical problems: most of the materials damaged by the high-energy electron beam; the resolution limit, after installation of E-cell, was still not suitable for atomic-level imaging; and finally, these microscopes required special rooms with high ceiling, low vibrations and proved to be expensive to maintain.

The technical improvements in the pole-piece design of medium voltage (300 to 400 keV) microscopes in the late 1990s revitalized the interest in designing aperture-type (differential pumping system) E-cells column for gas introduction in the sample area. [14–17].These medium-voltage microscopes have pole-piece gap large enough (5 to 9 mm) to accommodate the cell but low enough C_s to obtain near-atomic-resolution images. The radiation damage problem is also reduced due to the reduced electron energy (300 keV instead of 1000 keV).

During this period, the efforts to integrate a differential pumping system with TEM continued toward building a physical enclosure for gas confinement to be placed between the two objective pole pieces [18]. Such an enclosure was built by Gatan Inc. for Shell Inc. that was transferred to Arizona State University and modified to fit in a Philips 430 TEM [16]. Figure 7.1a describes the design principle and the placement of a physical E-cell between the upper and lower objective pole pieces.

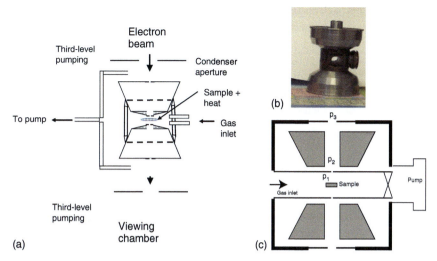

Figure 7.1 (a) Schematic of a differentially pumped physical cell placed between the objective pole pieces of a Phillips 430 TEM. (b) Photograph of the physical cell. Source: Adapted from Sharma and Weiss [16]/from John Wiley & Sons. (c) Schematic of a double-aperture differentially pumped environmental cell for a JEOL 4000 EX [19].

First set of apertures were part of the physical cell (Figure 7.1b), and the second set of apertures were inserted within the bores of both top and bottom pole-pieces. With careful design, it was possible to obtain image resolution of 0.28 nm at high temperatures [16]. Incorporation of Gatan imaging filter (GIF) provided a possibility to image using only elastically scattered electrons to mitigate the effect of inelastic scattering from the gases on image formation. Moreover, GIF was also used to obtain chemical information using electron energy-loss spectroscopy (EELS) [20].

The disadvantages of this physical enclosure were that that the body of the E-cell reduced the space available for specimen tilt, use of objective aperture and EDS detector. Therefore, the entire assembly had to be dismounted for regular TEM applications. Robertson et al. used an alternative enclosure for JEOL 4000 EX where the objective pole-piece block was modified to place plates with a set of apertures (Figure 7.1c). The larger pole-piece gap of JEOL 4000 EX (15 mm) allowed the double tilt heating, straining, and straining with heating holder to be used [19]. In 1996, Boyes and Gai modified the column of a CM30 microscope such that no physical enclosure was needed to confine gases around sample at pressures up to 50 mbar [17]. To control the gas leak rate away from the sample area, they used four apertures, which were installed on the both ends of the upper and lower pole-pieces and the gases were pumped out through bores drilled in the pole pieces (see Figure 7.5). This design was adapted and improved by FEI (now ThermoFisher) and became commercially available. Moving the differential pumping apertures in the pole pieces mitigated the need for a larger pole-piece gap and the regular TEM operation, including access to the sample tilt and EDS detector, remained unaffected.

During last couple of decades, the development of microfabrication techniques to make microchips for window holders contributed to advancement of the application of closed cell approach and exploded the field of in-situ TEM/STEM observations of gas–solid interaction (see also Chapters 5 and 6) [21–23].

> Gases are distinguished from other forms of matter, not only by their power of indefinite expansion so as to fill any vessel, however large, and by the great effect heat has in dilating them, but by the uniformity and simplicity of the laws which regulate these changes.
> – James Clerk Maxwell

7.2 Current Strategies

Current motivation for confining the gas around sample is the same as in the early days, i.e. we want to observe the gas–solid interactions, preferably as a function of temperature and/or other stimuli at nanoscale. Our gas confinement strategy should allow us to follow chemical, morphological, and structural changes as a function of sample temperature and gas pressure. Sometimes effect of other stimuli, such as mechanical, magnetic, and/or electrical is also need to be explored. Both the window and differential pumping method to confine gases around the sample are commercially available and currently used in combination of other stimuli. In this section, we will look at their design, function, and experimental considerations for each individually and discuss their pros and cons.

7.2.1 Window Holders

General principle of gas confinement around sample is based on the objective that most of the TEM-based techniques, such as electron diffraction, TEM/STEM imaging (bright-field and dark-field), EDS and EELS, should remain unaffected (Figure 7.2). Window holders designed for containment of liquids, described in Chapters 5 and 6, can also be used to confine gaseous environment around the sample. Depending on the thickness and subsequent rupture strength of the windows, it can be used to study gas–solid interactions at gas pressures higher than 1 atm [24, 25]. As explained in the Chapters 5 and 6, the material used to make the windows should be: (i) electron-transparent; (ii) sturdy enough to tolerate high gas pressure; and (iii) weakly diffracting so as not to interfere with the electrons scattered from the sample (image formation). In the early days, amorphous carbon or silica was used to make windows as they best satisfied these conditions. JEOL made first window holder for liquid flow using amorphous carbon windows commercially available [26]. However, the fragile nature of thin carbon windows used made it very difficult and time-consuming to conduct successful experiments.

With the development of microelectromechanical systems (MEMS), windowed holders became generally available at a fast pace [22, 23, 27]. The design of a window gas cell is identical to a liquid flow cell (Chapters 5 and 6), i.e. the gaseous environment is confined between thin windows on a microchip with channels etched for gas inlet and outlet that are connected to a gas handling system outside the column (instead of liquid) through the specimen rod. The samples are loaded

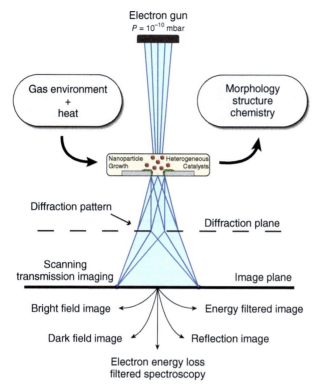

Figure 7.2 A flow chart explaining the concept of enclosing sample in a gas or liquid environment without compromising the vacuum in the rest of the microscope column or characterization functions/capabilities.

directly on one of the windows and the microchips with windows and other stimuli are often called "lab on a chip." Currently amorphous SiN_x is commonly used window material and the design requirements for windowed gas-cell holders are also very similar to the LC holders as discussed in Chapters 5 and 6. Here we will concentrate on the technical details that are specific to gas confinement around the samples.

Windowed E-Cells or nano-reactors are generally incorporated in a side-entry TEM holder. The basic principle of one of such window microchip design is shown in Figure 7.3. It consists of two Si chips with gas inlet and outlet channels fabricated on the bottom chip (Figure 7.3a) and windows containing spiral heating element with Au contacts (Figure 7.3b) There are two parts of the holder with a modified (a) sample well, where the microchip with windows is loaded (Figure 7.3c), and the rod. The sample is sandwiched between two windows that are aligned (Figure 7.3d) and sealed after loading the sample [21]. This particular set-up has been further developed and is now commercially available (DENSsolutions Inc.).

Another design based on similar principles with amorphous SiN_x windows supported on a thin ceramic membrane, but using O-rings to seal the top and bottom Si chips instead of glue, has also been reported [22, 28]. The ceramic membrane is used to heat the sample with fast heating and cooling rates ($10^6 \, °C \, s^{-1}$)

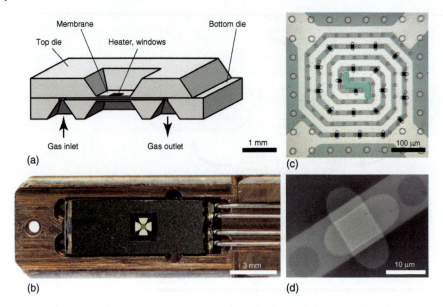

Figure 7.3 Illustration of the nanoreactor device. (a) Schematic cross section of the nanoreactor. (b) Optical image of the TEM holder with the integrated nanoreactor and the four electrical contacts. (c) Optical close-up of the nanoreactor membrane. The bright spiral is the Pt heater. The small ovaloids are the electron-transparent windows. The circles are the SiO_2 spacers that define the minimum height of the gas channel. (d) A low-magnification TEM image of a pair of superimposed 10 nm thick windows. Their alignment creates a highly electron-transparent square through which high-resolution TEM imaging can be performed. Source: Creemer et al. [21]/from ELSEVIER.

and is commercially available from Protochips Inc. Similar gas-cell holders are also available from Hummingbird Scientific. Further changes, such as depositing small thin circular windows on a thicker SiN_x membrane with circular geometry, are reported to withstand pressures up to 4.5 bar for H_2 [29, 30]. Similar to liquid cell, monolithic gas cell, microfabricated as a single chip, instead of two, has been reported [31]. This single-chip nanoreactor contains pillars that separate shallow channel (0.5 μm) with very thin windows (15 nm) and a resistive heater and has been shown to maintain high resolution (0.15 nm) and a pressure up to 14 bar can be achieved at 660 °C [31]. However, their application is limited to powders as samples can only be introduced as suspensions and is only available at lab scale.

Thickness of the sample holder tip (Figure 5.1), dependent on the objective pole-piece gap of the specific to TEM to be used, is crucial for sample tilting. Therefore, the tip of the side-entry TEM holders is around 2 mm thick, and the entire assembly, windows, washers, and specimen clamp must fit in it. Currently, most of the commercially available holders use 30 to 50 nm SiN_x thick windows, made using cleanroom technology, with a gap of around 30 nm. Overall, the gas path length should be below 50 μm for high-resolution imaging.

Browning's group at Pacific Northwest National Laboratory has reported an alternative design that allows the samples to be loaded on 3 mm grids to be sandwiched between two commercially available SiN_x grids [32]. They included a laser beam line through the holder rod for heating the sample locally along with gas inlet and outlet lines. The main advantage was the ability to use 3 mm ion milled discs or focused ion beam (FIB) lamella, thereby allowing to observe semiconducting thin film samples. Later, methods to prepare and transfer FIB lamellae on microchips of commercial gas-cell holders were developed [33], and this holder has not been used a lot.

7.2.1.1 Incorporation of Other Stimuli

Gas–solid interactions are invariably affected by high temperature; therefore, all of the window gas cells, reported so far, are equipped with heaters and are routinely used to follow the gas–solid interactions as a function of temperature. In principle, multiple contacts can be integrated on a microchip during fabrication, as described in Chapter 5, and inclusion of other stimuli such as biasing is possible.

7.2.1.2 Specimen Design and Preparation

As explained in Chapter 6 (Section 6.5.1), microchips or sample grids should always be visually examined and cleaned before loading them in the holder. The fabricated chips with windows and heating elements are best suited for powder samples that can be either dry-loaded or loaded in suspension and dried. For dry loading, the powder is sprinkled carefully on the membrane without coating the contacts. Powders can also be suspended by adding small amount to ethanol or iso-propanol and sonicating the solution for 5 to 10 minutes. General steps outlined in Section 6.5.1 for loading liquids on microchips should be followed to drop-cast the suspension carefully on the microchips using micro-pipettes and keeping the contacts clean. Usually, one drop of suspension is enough depending on concentration of powder in suspension.

Thin films or thin electron transparent sections from bulk materials can be prepared using FIB and step-by-step process of transferring FIB lamella on a gas-cell or liquid cell window can be found in Refs. [33, 34] and on YouTube video (https://www.youtube.com/watch?v=uXJI5GmdGCw).

7.2.1.3 Practical Challenges for Gas-Cell Holders

Just like other in-situ techniques described so far, designing experiments using window gas cell holders is challenging. Some of the challenges are similar to the one described in Section 6.5, following are specific to designing experiments using window gas cell holders.

Flow Rate and Pressure Equilibrium The gas inlet and outlet are etched on the microchip during fabrication and are connected through small capillary tubes from the microchip to the external end of the sample rod. The connection from the holder rod to the gas source (manifold or gas cylinder) is made using thin flexible tubes (PEEK) to flow the gas in and out of the windows containing sample. We know that the flow rate is dependent on the nature of the gas as the volumetric flow

$$V_F = \pi \Delta P r^4 / 8\eta l \tag{7.1}$$

where V_F is volumetric flow, ΔP is pressure difference between the external gas container and in the cell, r is radius of the tube, η is the viscosity, and l is the length of the tube. From Eq. (7.1), we can infer the following:

- The volumetric flow rate is controlled by the diameter and length of the tube.
- The pressure difference will determine the time it will take to achieve equilibrium pressure in the cell.
- Both the flow rate and equilibrium conditions depend on the viscosity or density of the gas.

We could have viscous, molecular, or turbulent flow, depending on the pressure in the cell, which in turn determines the mean free path between the gas molecules in the cell. In the viscous regime, also known as continuous flow, there are frequent collisions between gas molecules, but less frequently with the walls of the vessel. In this case, the mean free path of the gas molecules is significantly shorter than the dimensions of the flow channel. In the case of viscous flow, a distinction is made between laminar and turbulent flow. In laminar mode, the gas flow remains smooth in the same displaced layers that are constantly parallel to each other. If the flow velocity increases, these layers are broken up and the gas molecules run into each other in a completely disordered way and is termed turbulent flow. It is important to note, that the mean free path also determines the residence time of gas molecules on the solid and thereby affects the reaction rates.

Because the pressure difference between the cell and the source is large and flow rate is very small, it takes time to obtain equilibrium pressure in the cell. As the length and diameter of capillary in the holder and the PEEK tubing is fixed, we can control this time by reducing the pressure in the external container such that ΔP is small.

Reactivity of Gases and Sample with Windows We are usually interested in following gas–solid interaction at elevated temperatures, i.e. above RT. Therefore, here we need to pay attention that:

- Gases to be used do not etch the window material at high temperature,
- Sample does not react with the window material at temperatures to be used, i.e. there are no side reactions.
- Windows can tolerate the pressure increase due to heating. Unlike liquids, gases don't freeze and plug the ruptured windows.

Although SiN_x windows are most robust and inert, they have been reported to crystallize above 800 °C in oxygen environment [35] and stochiometric film corrode above 1400 °C in oxidative environment [36]. Therefore, it is important to research the chemical and physical aspects of the windows pertinent to the gas and solid to be used for specific experiments. Such information can be obtained from online handbooks, materials safety data sheets (MSDS), etc.

7.2.1.4 Review of Benefits and Limitations of Gas-Cell Holders

Introduction of window gas-cell holders and their commercial availability have motivated a large number of researchers to make in-situ observations of gas–solid interactions at nanoscale. One of the main advantages is that the cost of buying a TEM holder is much less than buying a modified TEM (environmental TEM [ETEM]) dedicated for observing gas–solid interactions. Moreover, it is easier and less expansive to modify a TEM holder than the TEM column. Microchips also allow us to combine more than one stimulus, such as heating and/or biasing.

Another advantage is that windows can withstand high gas pressures, higher than an atmosphere such that realistic reaction conditions, especially for catalytic processes, can be achieved [30]. Thus gas cells or nanoreactors bring us closer to overcome the so called "pressure gap," i.e. the gas pressure in the nanoreactor (cell) compared with the realistic pressure used in a commercial reactor. Also, the gas path length can be minimized by reducing the gap between the two windows without appreciable loss of spatial resolution. Moreover, by knowing the volume of the nanoreactor and gas pressure, the collision frequency and thereby residence time of gas molecules on the sample surface can be estimated. Therefore, may be possible to make operando measurements by connecting a GC–MS system to gas outlet to quantitatively measure the gas composition before and during the reaction. However, the density of the sample loaded on the chip, the gas flow rate within the cell, the temperature gradient and heat dissipation are still far from the reactor conditions.

The high gas pressure also has its drawbacks, such as increased electron path length leading to the reduced SNR from electron scattering by the gas molecules. Also, the effect of ionization of gas molecules by electrons is not properly evaluated yet. Another limitation is that the tilt of the sample is restricted as the total thickness of the holder tip is increased and objective pole-piece gap for most of the high-resolution TEMs is around 5 mm or less. Some of these limitations can be addressed by using ETEM/ESTEM open cell configuration.

7.2.2 Environmental Microscopes (Open Cell)

Although an ETEM/ESTEM is now commercially available, it is imperative to understand the designing principles behind the particular configuration to be used for its optimum performance. There are three types of open cell configurations that are currently being used: First is the modified UHV-TEM, where a small amount of gas or vapors can be leaked in the sample chamber; the second is by using a gas injection holder; and the third is modified column with differential pumping. All of them are commercially available, whereas the design of a ultrahigh vacuum (UHV) and differential pumping column follows similar principles, gas-injection holder, developed by Hitachi, is still at developing stage as discussed below.

7.2.2.1 ETEM Combined with Gas Injection Sample Holder

Currently, two types of modified TEM columns for gas introduction in the sample area, commonly known as environmental TEM (ETEM), are based on differentially pumping concept [8]. Hitachi has developed gas injection holder that is used in a modified TEM column with two stage differential pumping system installed above and below the sample chamber. The gas injection system is part of a resistive heating holder, developed by Kamino et al. [37] that is made of tungsten spiral wire of

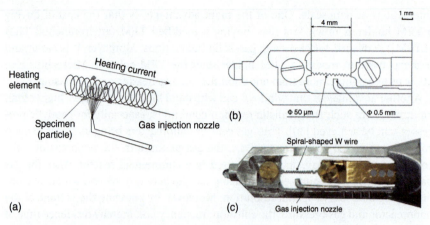

Figure 7.4 (a) Design of a spiral heating element with gas injection system, (b) incorporated in a TEM specimen holder. (c) Photograph of the tip of the holder. Source: Kishita et al. [38]/from Oxford University Press.

the order of 25 μm (Figure 7.4a). The gas is introduced via a nozzle (0.5 mm inner diameter) near the sample (Figure 7.4b,c) and pressure decreases rapidly away from the sample making it possible to have high gas pressure near the sample [38]. The wire is resistively heated by current supplied by a battery. The pressure in the sample chamber is measured using pressure gauges and is automatically saved. The pressure drop away from the gas nozzle and the dynamics of gas flow in ETEM have been estimated using computational fluid dynamic (CFD) simulations [39]. This microscope is commercially available and being used by several research groups. The advantage of this design is that higher pressure can be achieved near the sample without filling the entire sample chamber with gas environment. The inaccuracy in both the temperature and the pressure measurement due to the large gradients in the system leads to significant inaccuracy in measuring kinetics of an individual sample. Also, the applications in oxidizing environment are not compatible with the W wire as heating element as it will oxidize at high temperatures [40].

7.2.2.2 Differentially Pumped TEM

As mentioned above, another way to confine gas around the sample region is by using the differentially pumping system first proposed by Swann and Tighe [8]. In this design, two sets of apertures are placed above and below the sample between the objective pole-pieces. Various gases can then be introduced around the sample at controlled pressures (Figure 7.5a). The gases leaking through the first set of apertures are differentially pumped using a turbo molecular pump (TMP) (first level of pumping) (Figure 7.5b), while the second set of apertures further reduces the leak rate of the gas, not pumped out, into the microscope column, The column between the upper pole piece and condenser aperture above the sample and between the lower pole piece and differential aperture above the viewing area is pumped using a second line connected to another TMP (second level of pumping) such that column vacuum

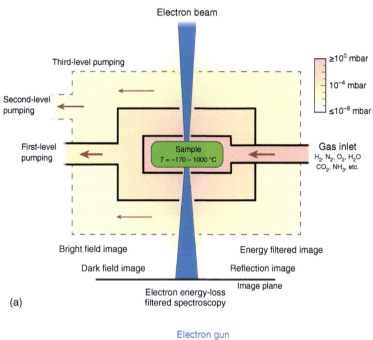

(a)

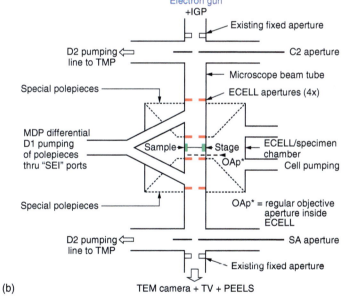

(b)

Figure 7.5 (a) General principle of designing a differentially pumped (DP) TEM column, as proposed by Swann and Tighe [8] consists of a set of apertures that are placed above and below the sample area to reduce the gas flow from sample chamber to the rest of the column and pumped by magnetically levitated turbo molecular pumps (TMP). The pressure limit is controlled by the differential pumping aperture size and pumping speed.
(b) Schematic of the modifications of objective pole pieces and the column to incorporate the DP system as marked. Source: Gai et al. [3]/Springer Nature.

is maintained at 10^{-6} mbar level or better (Figure 7.5b). Following factors are used to design a differentially pumped system:

1. The gas leak is controlled by the size of pumping apertures and pumping speed.
2. The rate of gas leak above the sample, i.e. toward the gun chamber is more important since the life of the gun depends on the vacuum around it, for example, a field-emission gun require 10^{-10} mbar for best performance.
3. On the other hand, the aperture below the sample is information limiting as it is located in diffraction plane and controls the number diffracted beam that can be used for TEM image formation. It also determines the possibility to collect HAADF-STEM as the diffracted beams at high angles may be blocked by this aperture.

Let us consider the leak rates through the apertures. Assuming that the leak rate beyond the aperture is proportional to the bore area and the pressure difference, in the molecular flow regime (a few mbar of gas pressure in cell), the conductance is given by

$$Q = t \cdot d^2 \sqrt{(T/M)} \qquad (7.2)$$

where Q is the conductance, t is the aperture plate thickness, d is the diameter of the aperture, T is temperature in $°K$, and M is molecular weight in $g\,mol^{-1}$, and the leak rate R_L through the apertures is given by:

$$R_L = Q(P_1 - P_2) = (t \cdot d^2 \sqrt{(T/M)})(P_1 - P_2) \qquad (7.3)$$

From these equations we can understand that the pressure in the cell is indirectly proportional to the leak rate through the apertures, i.e. higher pressure in the cell can be achieved by reducing the diameter of the apertures (Figure 7.6a) [19]. However, it is not feasible to reduce the aperture size beyond specific values due to practical constraints; the electron scattering from the edges of a small upper aperture will result in deforming the images, and reducing the size of lower aperture will reduce the angular contribution of the diffracted beam that is used in image formation. On the other hand, the achievable pressure does not improve appreciably beyond

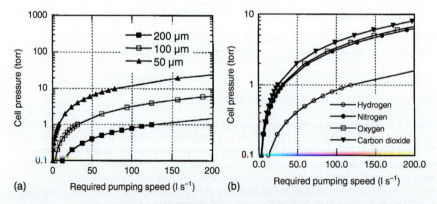

Figure 7.6 (a) Pumping speed required to maintain a particular cell pressure of air as a function of aperture diameter. (b) Pumping speed as a function of cell pressure for different gases. Source: Robertson and Teter [19]/John Wiley & Sons.

a certain value with pumping speed for each aperture size (Figure 7.6a). Therefore, a compromise between the aperture size and achievable pressure in the cell must be made. Usually, the size of first aperture in the upper pole piece is smaller than the one on the lower pole piece, but the difference must not disturb the balance of pumping rate as the pumping lines from above the top and below the lower apertures are joined and connect to the same TMP.

The pumping speed needed to achieve and maintain a specific pressure is also function of molecular weight of the gas, for example, the achievable pressure for H_2 in the sample area will be much less than for N_2, O_2, or CO_2 at the same pumping speed (Figure 7.6b) [19]. This principle is also used for UHV TEMs where base pressure of better than 10^{-9} mbar is required to image clean surfaces [41]. However, UHV-TEM has been used to observe a number of CVD processes, such as Ge island and Si nanowire growth on clean Si surfaces by introducing small amount of gas/vapor, $\sim 10^{-5}$ mbar around the sample [42–44].

> You can consider the ETEM to be the world's most expensive chemical vapor deposition system.
>
> – Eric Stach, University of Pittsburg

An ETEM thus can be considered as a flow reactor without compromising functionalities of the TEM, such as diffraction, imaging, electron spectroscopies, and use of modified holders for external stimuli (heating, cooling, biasing, indenting, etc.). A large number of researchers are utilizing this setup since it became commercially available through FEI (now ThermoFisher) reporting groundbreaking understanding of gas–solid interactions, especially in the field of catalysis [45, 46].

Higher gas pressure in the sample region can, in principle, be achieved by modifying the pumping stages and their location. For example, Hitachi has incorporated two-stage pumping above the sample chamber to keep high vacuum condition in the electron gun chamber, while only one-stage pumping below the specimen area. However, higher gas pressure increases the value to t/λ and thereby degrades the image intensity, depending on the nature of the gas. Moreover, the image resolution approximately defined by signal-to-noise ratio (SNR), known as Rose criteria, [47] which decreases due to the contribution from the random electron scattering by gas molecules as discussed in Section 7.4.2.

7.3 Gas Manifold Design and Construction

Notwithstanding the choice of microscope or E-cell holders available, a gas handling system is essential part of the setup. It could be commercially available as part of the microscope or holder purchase, or custom-built, but follow the same design principle. Building your own system can be beneficial as it provides you flexibility to add/remove ports or entirely redesign it. A sample of design of the setup used for Titan 300 FEG ETEM, at National Institute of Standards and Technology (NIST), is shown in Figure 7.7. The central piece of the setup is a gas manifold, which needs to be large enough to have space for as many inlet ports as you envision and one outlet

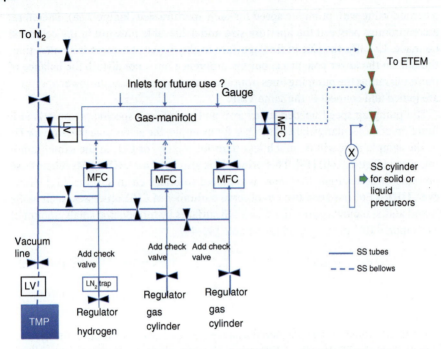

Figure 7.7 Design of gas manifold used at NIST-ESTEM lab, with flow rates controlled by mass flow controller (MFC). Entire system is controlled using LabView.

port and small enough to fill and evacuate quickly to exchange gases. It is a good idea to have one inlet for nitrogen for purging the mixing chamber, gas lines, and ETEM column, when changing from one gas to another and an independent TMP. The entire assembly can fit on a cart that can be placed near the TEM. You should keep in mind that full-size cylinders in the microscope room are fire hazard and will not be allowed, but small-size cylinder or lecture bottles may be acceptable.

> Appropriate gas monitors should be installed in the TEM room before starting gas–solid TEM experiments.
> – Seiji Takeda, Osaka University, Japan

The MFCs enable the flow control into the gas manifold to maintain certain back pressure at the inlet and mix gases in specific ratio. The flow system can be controlled using LabView. A detailed description of another design for a gas mixing system can be found in Ref. [48].

7.4 Practical Aspects of Performing Experiments in Gas Environment

Most of the practical challenges described in Chapter 2 are applicable here and will not repeated, but the challenges specific to gas–solid reaction are described in more detail. We expect that most of the readers of this book are trained to use the TEM as a characterization technique, but it is easy to forget that the sample space in the ETEM column or gas cell holder or both are practically a chemical reactor when used to observe gas–solid interactions. It is important to remember that the gas reacting with your sample can also react with other components such as grid and/or support material, heating and/or thermocouple wires, etc. Most of the vendors may not be willing to share the chemical composition of the internal components used, but they will let you know if the proposed gas environment is safe to be used at desired temperature. Moreover, the experimental setup must be clean and all sources of possible contamination should be removed, for example,

1. MEMS-based grids may have some debris left from the clean room processing that can be removed using the air duster. It is also recommended to plasma clean them before loading the sample and heat them under a lamp after loading the sample to remove adsorbed water.
2. The gas handling system can be a source of contamination should be cleaned by baking it occasionally. It is important to make sure that the inlet from the gas manifold to the microscope is closed during the baking and is pumped out using its own turbo pump.
3. Cross-contamination from the gases can be avoided by purging the gas lines and the specimen chamber by N_2 a couple of times between the experiments before changing gases. Purging with hot H_2 is helpful in removing water from the metal surfaces.
4. TEM specimen chamber should be baked (if possible) or plasma cleaned (if possible) overnight as often as needed.
5. Similarly, the flexible tubing used for nanoreactors should be cleaned between the gaseous environment changes.

In the past, the spatial resolution deteriorated from the noise and mechanical vibrations from the TMPs traveling to the TEM column through the pumping lines. The availability of noise and vibration-free magnetically levitated TMP (Maglev TMP) has practically mitigated this problem [49]. However, a user must pay attention to any outer connections made to the TEM column remain vibration-free. The wires connecting the sample holder to a stimulus controller (e.g. heater) or gas lines often will bring vibrations to the sample. Best practice is to check the image resolution before connecting wires or lines to the holder and clamp them to the column if needed. Some of the other practical challenges we face in performing experiments and interpreting the results are described below.

7.4.1 Electron Beam Effects

As mentioned in almost every chapter of this book, electron beam effects are omnipresent but could vary depending on the nature of sample and environment. Whereas radiolysis is most commonly observed for liquids (Chapter 6), ionization of gases is a matter of concern and can change the kinetics of the reaction. For example, multiwalled carbon nanotubes (MWCNTs) were observed to damage faster in gaseous environment (N_2 or O_2) during electron irradiation than in vacuum [50]. The surface adsorption of ionized gas species is found to be responsible to activate the carbon oxidation. CeO_2 reduction by electron beam is another example, where it is reported to reduce as a function electron beam dose [51] that can be mitigated by introducing oxygen in the sample chamber [52]. This observed mitigation is in fact the result of surface adsorption of ionized O species on CeO_2 particles that replace the lattice O vacancies generated by the interaction of high energy electrons. Similar control on CeO_2 reduction by electron beam has also been demonstrated by heating the sample above 700 °C where the rate of intake of environmental oxygen by the particles overrides the oxygen depletion by the electron beam [53]. Surface oxidation of Pd nanoparticles, induced by electron beam irradiation in low O_2 partial pressure, has also been reported [54].

Therefore, it is recommended to evaluate the electron beam effects by making measurements (i) by blanking the beam, (ii) as a function of electron dose, (iii) check the sample regions after removing the stimuli (i.e. at RT in vacuum) away from the in-situ observation area (not radiated by electrons).

The ionization of O_2 by electron beam and subsequent adsorption on the sample surface have been further demonstrated by in-situ conductivity measurements of CuO and SnO_2 nanowires by Steinhauer et al. [55]. Here the resistivity measurements made using $I-V$ curves collected at 300 °C in 20 mbar of oxygen with beam on and beam off conditions showed a decrease of resistivity for CuO nanowires and an increase for SnO_2 nanowires. They performed controlled experiments by moving the 250 nm diameter electron beam a few μm away from the nanowire to eliminate direct contribution of the beam to the sample. However, it continued to create ionized species in the environment. It is interesting to note that despite the sample not being directly irradiated, both the electron beam dose and oxygen pressure influenced the resistivity (Figure 7.8). Moreover, it also went up and down as electron

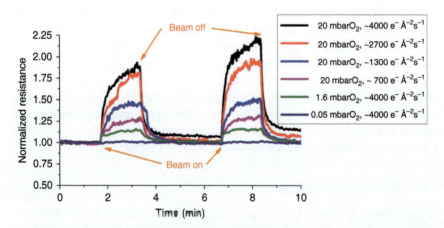

Figure 7.8 Observed increase in resistance of the single SnO_2 nanowire device operated at 300 °C due to electron–gas interactions at different O_2 pressures and electron dose rates. Source: Steinhauer et al. [55]/AIP Publishing LLC. Arrows mark the on and off events for remotely placed electron beam.

beam was turned on or off (Figure 7.8). Their experiment clearly demonstrated that adsorbed ionized O species are changing the resistivity of the nanowire, a property that makes these nanowires suitable as sensors for chemisorbed species [55].

7.4.2 Gas Pressure and Resolution

It is essential to keep in mind that the interaction of gas molecules with the sample is controlled by the flow regime of the system. The sample chamber falls in a molecular flow regime under low pressure ($<10^{-3}$ mbar) conditions, where the intermolecular interactions are limited due to large gas path length. Most of the time ETEM is operated under gas pressure of a few mbar bringing the chamber into viscous flow regime that can be further divided into two categories: laminar or turbulent. As explained earlier, in laminar flow, the mean free path between the gas molecules is shorter and intermolecular collisions of gas molecules are frequent but infrequent with the walls of the container. Turbulent flow regime is generally not achieved under differentially pumped systems. High gas pressures in the ETEM specimen chamber are desirable in order to observe the gas–solid interactions under realistic conditions. However, the inelastic scattering from the gas molecules results in reducing the image intensity or SNR, thereby reducing effective image resolution.

Systematic investigations of image intensity for Ar, H_2, He, N_2, and O_2, i.e. show a loss in the intensity as a function of gas pressure with ETEM operated at 80 or 300 keV (Figure 7.9a,b) and is greater for heavier gases as measured at the CCD camera [56]. However, the decrease in intensity is also much larger at 80 keV (Figure 7.9a) than 300 keV (Figure 7.9b), due to lower penetration power of low-energy electrons. Furthermore, the normalized intensity-loss plots for N_2 at different operating voltages further confirmed a greater intensity reduction at lower operating voltage (Figure 7.9c) [57].

Zhu and Browning have reported the reduction in SNR for lacey carbon films to be less pronounced for ESTEM than for ETEM images. They compared the effect of gas pressure for TEM and STEM imaging for different gases as a function of pressure [58]. They show that under low-angle annular dark-field (LAADF) conditions, the SNR remains above the Rose criterion for high resolution up to 10 mbar of N_2. They assessed the contributions of electron scattered by gas molecules by calculating the elastic and inelastic scattering angles for the two imaging conditions (Figure 7.9d). We note that the total scattering angles by gas molecules in ESTEM mode are below 25 mrad, which can be avoided using appropriate annular detector. However, all the electrons scattered within the DP aperture (around 71 mrad for Titan ESTEM) are used for TEM imaging. They conclude that LAADF imaging could be used at high gas pressures without loss of resolution, but image acquisition speed is much lower for STEM than for TEM imaging, thereby reducing the temporal resolution.

In principle, effective image resolution in gas environments is a function of the nature of the gas (mass), electron beam energy, and the dose rate. We expect the resolution to deteriorate as the SNR decreases due to increased gas pressure, however, Jinschek and Helveg report that the signal attenuation caused by the gas molecule is

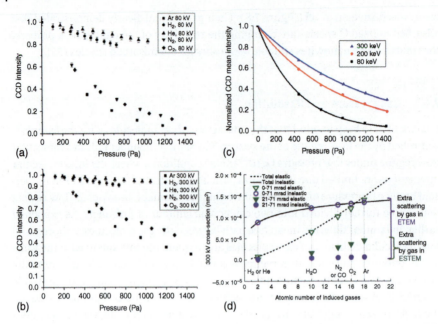

Figure 7.9 (a) Intensity recorded on the CCD camera as a function of gas pressure for 80 kV and (b) 300 kV. Source: Hansen and Wagner [56]/Cambridge University Press. (c) Normalized image intensity (without a specimen) measured on a pre-GIF Ultrascan CCD camera, plotted as a function of gas pressure for Ar at three different acceleration voltages, the data points have been fitted to exponential functions. Source: Wagner et al. [57]/with permission of Elsevier. (d) Theoretically calculated cross sections of elastically and inelastically scattered electrons at different scattering angular ranges as a function of the atomic number of the object the electrons interact with at 300 kV. Different types of gases are marked based on their atomic number. Source: Zhu et al. [58]/John Wiley & Sons.

also a function of electron dose [4]. They show that high electron dose induces specimen vibrations that smear out the electrostatic potential, thereby reducing the lateral resolution. Therefore, low-dose imaging provides a way to retain atomic resolution even in 19.3 mbar of N_2 [4].

7.4.3 Sample Temperature and Cell Pressure

Measuring temperature of the sample at nanoscale in gaseous environment is somewhat more challenging than in vacuum as the heat conducted away from the sample by the gases also needs to be considered. The problem is less severe in the MEMS-based heating holders where the gas flow rate, pressure, and volume can be measured accurately for theoretical calculations. Also, the change in gas volume with increased temperature can be measured from the shifts of plasmon peaks in low-loss EELS signal [59]. EELS can also be used to measure the thickness (t/λ) change of the cell as function of pressure [60]. Moreover, the pressure in the cell can also be estimated from the pressures measured at the gas inlet and outlet of a flow reactor and a combination of the two should result in reliable values.

The situation with ESTEM is more complicated as the exact values for the diameter and length of the pumping line in the column are not known. However, there is space to install a pressure gauge for approximate measurements that can be further calibrated against known reaction processes. For example, Wise et al. calibrated the pressure gauge located at the inlet of a Tecnai ETEM, using the known values of relative humidity needed for the deliquescence of different aerosol particles [61]. Zhao et al. have used CFDs to model the gas pressure distribution, pressure drop away from the inlet, and gas flow for different inlet pressures for the gas injection holders [39]. Either of these methods can be used to calibrate the pressure gauge used in the sample chamber.

Temperature calibration methods, described in Chapter 3 may not be applicable as it will require measuring temperature in various gases as a function of pressure. However, some of them may be used for in-situ temperature measurement under the experimental condition. For example, plasmon shift in EEL spectrum can be directly related to the temperature of a single nanoparticle [62, 63]. On the other hand, reliable chemical engineering models that account for heat transfer (by conduction in the grid material, by the gas phase, and radiation from the grid surface), as mass transport processes within the ETEM cell have been proposed [59, 64, 65]. Mortenson et al. simulated temperature distribution across the sample region in an ESTEM (open cell) using CFDs and estimated the temperature and velocity field, pressure distribution, and the temperature variation for a furnace-type heating holder. They report that although conduction was prominent way of heat transfer, radiation became an important factor at higher temperature and scaled with the emissivity of grid material [64].

Vincent et al. proposed a 3-D model to evaluate the gas pressure and temperature distribution within the sample chamber of an ESTEM [65]. They developed a finite element model combining fluid dynamics, heat transfer, multi-component mass transport, and chemical reaction engineering to determine the gas and temperature profiles. Their model accounts for the inflows of reactant gas mixtures and the property of reaction products and determines steady-state solutions for an open cell ETEM setup and found that both the temperature and pressure are nearly uniform in the region where sample is located or in the hot zone. Experimental measurements, using partial oxidation of CO over Ru/SiO_2 catalyst, matched with their simulations. Both groups have reported conduction to be the main path for heat transfer, but radiation dominated at temperatures above 400 °C. It is also possible to redesign the MEMS-based heating holders to accurately measure the temperature in gaseous environment as recently reported by Li et al. [66].

7.4.4 Anticontamination Device

Most microscopes are equipped with an anticontamination plate, located in the sample area that is kept at liquid nitrogen temperature to improve the vacuum by condensing the water vapor on the plate. However, it cannot be used if gases with boiling point higher than liquid nitrogen (−195.8 °C) are to be used as they will condense on the plate too. Therefore, anticontamination device can only be used in hydrogen (−253 °C), helium (−269 °C), and maybe oxygen (−182.9 °C) environments.

7.5 Select Examples of Applications

Since the 1990s, as modified ETEM/ESTEM and window gas holders became commercially available, researchers have taken full advantage of their capability to explore the gas–solid interactions for atomic-scale understanding of chemical reactions. The number of publications escalated with the advent of windowed gas cell TEM holders that are commercially available and much cheaper than a modified column. From early on, the ETEM had made significant contributions to the field of catalysis where the activity and reactivity of catalytic nanoparticles are controlled by the size, shape, temperature, and gas environment. Therefore, we need to explore the relationship between these parameters during the functioning of a catalyst. It is beyond the scope of this chapter to refer to this large body of available literature, but we include a few noteworthy research outcomes as examples here. References to some of the books and review articles related to heterogeneous catalysis are also provided at the end of the chapter [67–72].

7.5.1 Effect of Gas Environment on Catalyst Nanoparticles

Observations of catalytic nanoparticles under gaseous environment (near reaction conditions) were some of the first applications of the ETEM [2, 45]. It is obvious that study of catalytic process is the most suitable application as (i) catalysts particle are nanometer in size, i.e. require TEM to observe them and (ii) they affect the thermodynamics and kinetics of the reaction but remain unchanged, i.e. pre- and post-mortem examination does not provide information about their contribution to the reaction process. Therefore, it is not surprising that the first commercial ETEM was installed at a chemical company (Haldor Topsoe in Denmark). As mentioned above, there is large body of literature available in this field, here we will look at some select examples that made unprecedented revelations or settled an ongoing debate about the role of catalyst nanoparticles for a specific reaction.

First atomic-resolution images showing the shape change of Cu nanoparticles supported on ZnO, a catalyst system used for methanol synthesis and hydrocarbon conversion, were reported by Hansen et al. [45]. A reversible shape change is observed as the gaseous environment around the sample was changed from reducing to oxidizing. The faceted particle in H_2 at 200 °C changed to almost spherical shape as H_2O was added to the environment and became faceted again in reducing environment as H_2O was replaced by CO. It is interesting to note the surface termination of ZnO support also changed with the environmental change indicating that both the interfacial energy and the surface energy of the catalyst are changing due to gas adsorption. Similar changes were also observed for Cu nanoparticles supported

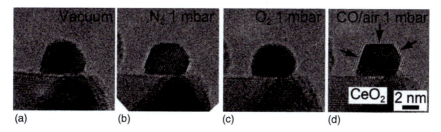

Figure 7.10 Au nanoparticle supported on CeO_2 (a) in vacuum, (b) in pure N_2 gas (1 mbar), (c) in pure O_2 gas (1 mbar) and (d) during CO oxidation (1 vol% CO/air at 1 mbar). Source: Uchiyama et al. [73]/from John Wiley & Sons.

on silica indicating that gas adsorption is dominant effect for inducing the shape change [45].

Similarly, a gold nanoparticle supported on CeO_2 is reported to change from somewhat faceted in vacuum (Figure 7.10a) and in 1 mbar N_2 (Figure 7.10b) to round in 1 mbar of O_2 (Figure 7.10c) but highly faceted in CO/air environment (Figure 7.10d) at room temperature [73]. As expected, N_2 (inert) gas molecules do not interact or adsorb on Au nanoparticle surfaces and result in negligible shape change. On the other hand, adsorption of reactive gases (O_2 and CO/air) on the particle surface changes the surface energy that induces the shape change [74].

Recently, Vendelbo et al. used a gas cell nanoreactor, with outlet connected to a mass spectrometer, to follow CO oxidation on Pt catalyst using a mixture of $CO:O_2$:He with ratio of 4.2% : 21.0% : 74.8% at 727 K (454 °C) [75]. This setup allows to follow the reaction in nearly operando mode as the composition of outlet gas was monitored for reaction kinetics. Interestingly, they found that surface of the Pt nanoparticle oscillated between round and faceted shape that could be directly correlated to the oscillation in the outlet gas composition. The amount of CO and O_2 increased or decreased in a periodic manner that was opposite to the CO_2 amount, i.e. the CO_2 concentration increased as CO was oxidized and decreased if no reaction was taking place. This periodic change was directly related to the observed oscillation of the nanoparticle shape change from round, when CO was being oxidized to faceted to round when it was not. Their detailed modeling agreed with the observed relationship between the particle shape change and indicated that the gas absorption sites on the catalyst particle play an important role [75].

Atomic positions in the surface layer of catalyst nanoparticles have also been reported to change under reactive gas environment. Examples include, but are not limited to, Pt nanoparticle under reducing or oxidizing environments [76], dynamic structural evolution of the Pt/CeO_2 catalyst at varying levels of activity for CO oxidation [77], fluxional cation behavior on nanoparticle surfaces related to oxygen vacancy creation/annihilation at surface sites [78], 3D dynamics of excited Co–Mo–S nanocrystals [79], atomic-scale manipulation of the active Au–TiO_2 interface during CO oxidation [80], dynamic behavior of CuZn nanoparticles under oxidizing and reducing conditions [81], and hydrogen-induced faceting of Pt nanocrystals [82].

7.5.2 Carbon Nanotube (CNT) Growth

Carbon nanotubes (CNTs) or nanofibers were first reported by Baker as the deactivation source for transition metal catalysts using 1 MeV differentially pumped ETEM and proposed a vapor–liquid–solid growth (VLS) as growth mechanism [83]. However, Helveg et al. [46] were first to show the atomic-scale growth mechanism of carbon nanofibers, denouncing the long-standing VLS growth model proposed by Baker. Catalytic chemical vapor deposition (C-CVD) has been a commonly used method for CNT synthesis and ETEM became a most suitable platform to investigate their growth mechanisms. A vapor–solid–solid (VSS) mechanism was further confirmed by a number of other groups, both for multiwalled and single-walled carbon nanotubes (MWCNTs or SWCNTs) growth [84–88]. ETEM observations further revealed that the catalyst particles converted to a metastable carbide phase (Figure 7.11a,b) [84] and the fluctuation between carbide and metal phase resulted in the CNT growth [89]. Atomic-scale time-resolved images revealed that the graphene to metal surface adhesion forces determine the encapsulation or

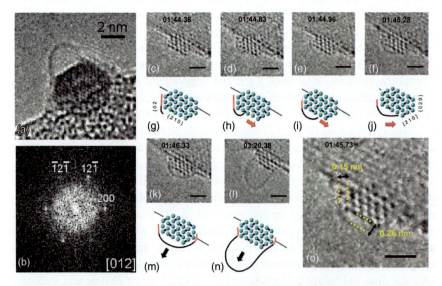

Figure 7.11 (a) A snapshot of a nanoparticle, extracted from a video during SWCNT growth and (b) the corresponding Fourier transform, showing the extra diffraction spots from the Fe-carbide (cementite, Fe_3C) viewed along the [012] direction. Source: Yoshida et al. [84]/from American Chemical Society. (c–o) In-situ time-resolved ESTEM observation of SWNT nucleation and growth. (c, d, e, f, k, l) A series of images extracted from a video recorded during SWCNT growth. (e, f, g, h, k, l) Corresponding atomic models. The red lines indicate the stronger adhesion between graphene and metal on the two Co_2C (020) surfaces, and the black line shows slightly lifted graphene from {210} surface that results in the formation of a cap and growing SWCNT. The arrows are guiding the growth directions. (o) Snapshot showing the average distances between the growing structure and the {020} and {210} catalyst surfaces before the nanotube lift-off. Scale bars are 1 nm. Source: Picher et al. [87]/from American Chemical Society.

SWCNT lift-off phenomenon [87]. Furthermore, time-resolved images, extracted from a video (Figure 7.11c–o), reveal the nucleation of graphene sheet anchored on the {020} facet of a Co_2C nanoparticle (Figure 7.11c,g) that detaches slightly as it moves over to {210} surface (Figure 7.11d,h), continues to grow to the other side of the particle (Figure 7.11e,i) and attaches on {020} surface (Figure 7.11f,j) before a complete lift-off started the growth (Figure 7.11k–n). These observations further confirmed the VSS growth mechanism that helped researcher in their quest for finding catalysts that stay solid at high temperature for selective synthesis of chirality-controlled SWCNTs [90, 91].

7.5.3 Nanowire Growth

Nanowires can be grown with or without a catalyst. While catalytic CVD has been used to grow Si, Ge, SiGe, and III–V nanowires, metal and metal oxide nanowire are synthesized without a catalyst. Direct visualization of Si nanowire growth using catalytic CVD process in a UHV-TEM led to the understanding of growth mechanisms and kinetics for both the VLS and VSS mechanisms [44, 92, 93]. For example, the kinetics and nanowire morphology were observed to change as small amount of oxygen was mixed with disilane during VLS growth of Si nanowires using Au as catalyst (Figure 7.12a). A tapering Si nanowire, due to migration of Au from the droplet, was found to stop as 6.67×10^{-7} mbar of oxygen was introduced at 17 014 seconds while maintaining the disilane pressure constant, and the growth continued at the same rate. On the other hand, the growth was found to stop at 17 239 seconds (Figure 7.12a) in decreasing oxygen pressure as the Au migrated from the tip of growing nanowire [94]. These observations suggested a way to grow longer wires with the same diameter. Furthermore, Madras et al. have reported that the growth directions of the wire are controlled by the growth temperature (Figure 7.12b). Their temperature resolved observation of a single nanowire growth provided a practical route for growing, straight, angled, kinked, or spiral nanowires. Kikkawa et al. measured the growth rates of various Si nanowires growing between 365 and 495 °C and found the activation energy to be 230 kJ mol^{-1} and critical diameter for growth to be around 1.7 nm [100].

In-situ observations also revealed the growth mechanism of Si nanowires via VSS mechanism using Pd as a catalyst instead of Au (Figure 7.12c–f), where a ledge was observed to nucleate (marked by white arrow in Figure 7.12c) at the interface of $PdSi_2$ and Si nanowire and continued to grow (Figure 7.12d,e). The nanowire continued to grow as new ledges nucleated and moved across the interface at different speeds as shown for four ledges in Figure 7.12f [99]. Similar mechanism has been reported for Si nanowire growth catalyzed using Cu [101].

Apart from Si nanowires, in-situ observation for the growth of other materials, such as Ge [92], III–V [102–104], Si–Ge heterostructures [105], CuO [106, 107], etc. has also been reported. A detailed discussion on controlling the growth nanowires can be found in a review article by Ross [108].

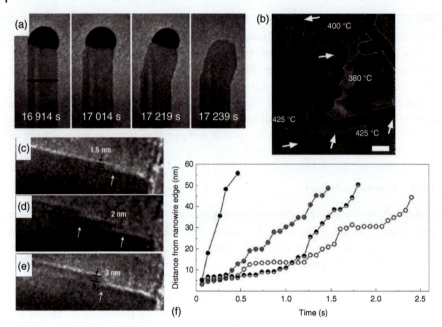

Figure 7.12 Effect of increasing oxygen pressure on Si wire growth kinetics: (a) Series of images extracted from a video showing the effect of reducing the oxygen pressure on wire after 16 914 seconds of Si nanowire growth at 600 °C using 1.33×10^{-6} mbar of disilane and 2.66×10^{-7} mbar of O_2. The O_2 was switched off at 17 014 seconds and the Au droplet at the tip of the wire started to shrink (17 219 seconds) and the growth stopped as the droplet disappeared (17 329 seconds). The scale bar is 50 nm. Source: Kodambaka et al. [94]/from American Chemical Society. (b) Images extracted from the end of an in-situ TEM video of a Si NW grown using 3.3 mbar of disilane acquired along the [110] azimuth. The white arrows delineate segments grown at the indicated temperatures. The NW grows straight at 425 °C with the first segment along [111] and the last along [112]. The segment grown at 400 °C changes its growth axis from one to another crystallographic direction at localized kinks. The segment grown at 380 °C changes its growth direction smoothly and continuously along random directions, following a wormy trajectory. The scale bar is 250 nm. Source: Madras et al. [98]. (c–e) Ledge-flow-controlled catalyst interface dynamics. (c) ETEM image sequence of growing Si nanowire at ≈ 560 °C in $\approx 1.2 \times 10^{-2}$ mbar of Si_2H_6 (scale bar 10 nm). (d, e) Various ledge configurations at the Pd silicide/SiNW interface during disilane exposure. (f) Measured step edge position versus time for four independent ledges of roughly 1.7 nm height in projection. Source: Hofmann et al. [99]/from Springer Nature.

7.5.4 Electron-Beam-Induced Deposition

Electron-beam-induced deposition (EBID) is commonly used in SEM mode and contributed to the development of FIB microscope to pattern metal nanostructures. However, the sizes of the deposits were limited to tens of nanometers due to (i) probe size and (ii) by contributions from secondary electrons. Therefore, it became worthwhile to try to deposit smaller structures using a TEM where sub-nanometer probe is available in STEM mode and thin samples can constraint the contribution from secondary electrons [109, 110]. Original idea was to be able to pattern sub nanometer structures by taking advantage of 0.1 nm probe of the FEG microscopes [111]. However, it was soon realized that deposited nanostructures were limited to 1 nm due to (a) insufficient beam current in small probes and (b) the secondary electron contribution to the decomposition of precursor molecules could not be completely mitigated [95].

7.5.5 REDOX Reactions

Oxidation and reduction (redox) reactions are a part of our everyday life. For example, corrosion of iron is due to oxidation from oxygen in air and humidity is known to expedite the process, or oxidation of Al is retarded by formation of self-limiting oxide surface layer. Redox reactions are also a critical part of heterogeneous catalysis where metal nanoparticles undergo oxidation/reduction cycles under reactive environments. However, direct observation of the atomic-scale mechanisms for the redox processes stayed elusive for long time. Both the gas-reaction cells and modified TEM columns have been critical in elucidating the morphological and atomic-scale mechanism of several redox reaction in recent years. For example, Yang et al. presented an alternative scheme [96] to the prevalent classical theory of Cabrera–Mott that predicted a uniform growth of passivating metal oxide film due to a field-enhanced ionic transport mechanism [97]. They used a modified UHV-TEM to prepare clean Cu surface and the leaked 0.133 mbar of oxygen at 350 °C. Instead of layer-by-layer oxide film formation, predicted by Cabrera and Mott, they observed growth of islands that coalesced with time to form a passivating film. Therefore, they presented a simple phenomenological explanation that the self-limiting oxidation is due to the coalescence of islands, which "switches-off" the surface diffusion route and requires much slower bulk diffusion for further oxidation. These observations have been vital for growing oxide nanowires where lower oxygen bulk diffusion than surface results in one-dimensional growth [106, 107]. In-situ observations have also been critical in understanding the mechanism and the kinetics of other catalytically active nanoparticles such as Fe, Ni, Pt, Co, Pd, etc. [72, 112, 113].

On the other hand, ESTEM observation has also provided direct evidence of existing hypotheses. For example, mechanism of the formation of voids during reduction of metal oxide nanoparticles, also known as the Kirkendall effect, has been investigated [114]. Original model was suggested for the alloys, where the diffusion rate of metals atoms resulted in the void formation as a function of temperature, especially at the grain boundaries. In-situ TEM observations have shown that the model can be applied to oxidation of pure metals (Ni, Pd, etc.), or alloys (Cu–Cr, Cu–Al, etc.) as well to the reduction of metal oxides. For example, Ni metal nanoparticles on SiO_2, formed by in-situ reduction in H_2 (Figure 7.13a), adapt a hollow structure upon oxidation in a mixture CH_4 and O_2 (Figure 7.13b) [115]. Here the oxidation of metal starts at the surface of a nanoparticle with metal oxide nucleating at multiple sites. When the cation diffusion is faster than the oxygen diffusion through the oxide layer, metal atoms from the core of the nanoparticle diffuse toward the surface

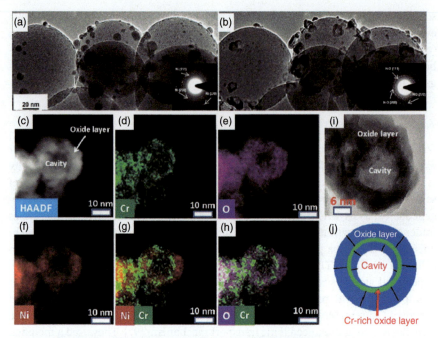

Figure 7.13 In-situ ETEM images and electron diffraction patterns (inset at bottom right corner) of Ni/SiO$_2$: (a) In presence of 100 Pa of H$_2$ at 400 °C; (b) from the same region in the presence of 100 Pa of mixture of CH$_4$ and O$_2$ in 2 : 1 ratio at 400 °C. Source: Chenna et al. [115]/from John Wiley & Sons. STEM-HAADF image and EDS elemental maps showing the spatial distribution of Ni, O, and Cr in Ni$_5$Cr nanoparticle after the high-temperature oxidation. (c) STEM-HAADF image revealing a hollow structured oxide particle. (d) Cr map. (e) O-map. (f) Ni map. (g) Ni–Cr-map. (h) O-Cr map. (i) HRTEM image showing hollow structure. (j) Schematic drawing of the hollow structure and the trapping of Cr-rich oxide layer. A comparison of the dimensions of the cavity and the Cr rich oxide layer clearly shows that the Cr-rich oxide layer is embedded in the nickel oxide. Source: Wang et al. [116]/from Springer Nature.

and react with oxygen to increase the oxide layer thickness. It is interesting to note that the size of the cavity is the same as original Ni particle [115]. Void formation in Ni films during oxidation has also been reported and explained as a result of combination of different mechanisms such as Ni^{2+} diffusion along NiO grain boundaries, self-diffusion of Ni^{2+} and Ni vacancies, and the formation of transverse cracks in the NiO film [117]. Metal diffusion also leaves vacancies behind that then coalesce to form voids.

Similar process has also been reported for alloys where oxidation of one of the metals is faster than the other (Figure 7.13c–j). Wang et al. have shown that for Ni$_5$Cr alloy, Cr oxidation starts at room temperature and forms a shell around the nanoparticle (Figure 7.13d) [116]. However, oxidation of the Ni, from the core in the original particle, starts at 375 °C and diffused to the surface, leaving a cavity behind. Based on the STEM-EELS maps, (Figure 7.13d–h) showing the distribution of Cr, Ni, and O in the nanoparticle, a model (Figure 7.13j) of the resulting nanoparticle could be constructed [116].

A slightly different mechanism has be reported for the oxidation of Ni$_4$Al alloys single crystal posts where the initial oxidation started at multiple sites resulting in the formation of an oxide shell covering a cavity. The cavity formation led to a faceted metal surface and oxidation mechanism changed from multi-sites to atomic ledge migration on the faceted metal surface [118].

Moreover, EELS has been used for qualitative and quantitative measurement of oxidation states of transition metal atoms. We know that the fine-edge structure of EELS peaks for various elements changes as function of their chemical environment, i.e. coordination number. For example, the chemical state of C, amorphous, diamond, graphitic, or carbide, can be distinguished from the near-edge structure and the onset energy of the C peak and provides a qualitative information about the bonding states [119]. Moreover, the white line ratio (L_2, L_3 or M_4, M_5) for certain transition metals varies with their oxidation state [119], and methods to use them for quantitative measurements have been developed [120, 121]. Therefore, in-situ EELS has been used for quantitative measurements of oxidation states during redox reaction of oxides of Ce, Fe, Cr, Mn, etc. [26, 51, 53, 122]. Daulton et al. used a closed cell to follow the microbial reduction of chromium using Cr L_2, L_3 peak ratios [26]. CeO$_2$, an important material due to its ability to go through multiple redox cycles and is used as three-way catalyst support for automobile catalytic convertors and anode material for solid oxide fuel cells. In-situ measurements of Ce white-line ratio (M_4, M_5), combined with atomic resolution images, have revealed unpresented information about redox mechanism and its oxygen storage capacity [53, 123]. Core-loss EELS has also been applied to correlate the valence state of Co nano-porous structure during reduction [124].

Other examples of following redox reactions using imaging and EELS include, but are not limited to, following dynamical changes in the valence states and structure of NiCu$_3$ nanoparticles during redox cycling [125], strong metal support interactions between metal nanoparticles and oxide support under redox conditions [126], structure dynamics at the Cu$_2$O/Cu interface in oxidizing and reducing environment [107, 127], redox cycling of a Ni–ScSZ cermet fuel cell anode [128], redox reactions, and ionic conductivity in solid oxide fuel cells (SOFCs) [129].

7.5.6 Gas Adsorption Sites

Adsorption of gas molecules on catalyst surface may be a fundamental step for a chemical reaction to proceed as evidenced by the shape change of catalyst particles (Figure 7.10). Ideally, specific planes of a catalyst surface may have high adsorption for the reactants and low for the products such that products leave the surface and reactant adsorption sites become continuously available. Therefore, the gas adsorption sites can be the direct indicator of active sites for a catalytic reaction. However, the low scattering gas molecules on metal catalyst surface do not generate high enough contrast in TEM images. Yoshida et al. used the surface reconstruction of Au/CeO$_2$ nanoparticles in CO, as an indicator of adsorption sites [130]. They show that the distance between two top {100} surface planes and the inter-atomic distance between atoms within the plane change from 0.20 and 0.29 nm, respectively, to 0.25 nm for both (Figure 7.14a).

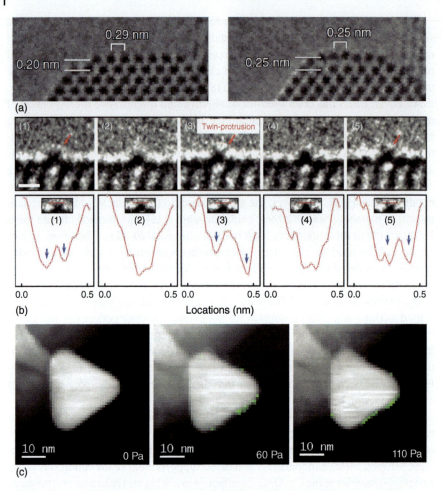

Figure 7.14 (a) High-resolution image of Au{100}-hex reconstructed surface under catalytic conditions. (left) In vacuum, the distance between the topmost and the second topmost {100} surface layers and the adjoining Au atomic columns on the topmost surface layer is measured to be 0.20 and 0.29 nm, respectively, as for crystalline bulk gold. (Right) In the reaction environment both the average distance of the adjoining Au atomic columns and the interplanar distance changed to 0.25 nm. Source: Yoshida et al. [130]/from American Association for the Advancement of Science. These changes in positions of the Au atomic columns correspond well to those of the Au{100}-hex reconstructed surface structure.
(b) (top row) Sequential ETEM images acquired in the mixed gas environment (1:1 ratio of CO and H_2O vapor; total environment pressure: 5 mbar; temperature: 700 °C), viewed in the [010] direction. Enlarged ETEM images show the dynamic structural evolution of the Ti row indicated by red arrows. Scale bar, 0.5 nm. (bottom row) Intensity profiles along the lines across the Ti rows in top row images. Blue arrows denote intensity valleys corresponding to the twin protrusions. au, arbitrary units. Source: Yuan et al. [131]/from American Association for the Advancement of Science. (c) Carbon maps, representing CO adsorption sites on an Au nanoprism (left) in vacuum (middle) in 60 Pa and (right) 110 Pa of CO. After removing the environmental CO contribution to the C K-edge intensity, the STEM-EELS map is overlaid on an ADF image. The green contrast shows the normalized C K-edge intensity integrated from 285 to 315 eV. Source: Yang et al. [132]/from Springer Nature.

Furthermore, evidence of reactive sites on TiO_2-1×4 reconstructed surfaces has recently been revealed using high-resolution TEM and STEM imaging [80]. The reconstructed (001) surface of anatase phase provides ordered four-coordinated titanium atoms for real-time monitoring in reactive environment with H_2O molecules dissociating and reacting on the catalyst surface. A twin-protrusion observed on Ti sites was assigned to adsorbed water (Figure 7.14b). Dynamic changes occurring in these structures during water–gas shift reaction were further visualized at a molecular level [80].

Recently, EELS chemical maps have also been used to directly identify the gas adsorption sites on metal nanoparticles. For example, Yang et al. have exposed shape-controlled Au nanoprisms on TiO_2 to increasing pressure of CO and acquired STEM-EELS maps, using C K-edge as signature for CO adsorption (Figure 7.14c) [132]. It was interesting to note that CO adsorption sites were confined to the edges, indicated by presence of C, instead of surface planes (marked as green lines on the ADF image in Figure 7.14c) that was confirmed by acquiring a tilt series.

7.6 Review of Benefits and Limitations

Instrument developments (both for TEM holders and columns) have made it possible to examine and measure the gas–solid interaction routinely. A combination of TEM-related techniques, such as electron diffraction, imaging (low magnification, high resolution, dark-field, STEM, etc.) EELS and EDS, provides unprecedented information about the morphological, structural, and chemical changes happening when of the nanomaterials are subjected to high temperatures in specific gas environment. Operando conditions can be achieved with careful design of sample, TEM holder, and/or incorporating a mass spectrometer to the gas outlet [75, 133]. EELS has also been used to measure the gas composition [27, 134]. As described above, there are two main configurations, closed cell and open cell, that are now available for gas–solid experiments at elevated temperatures. Closed cell is the part of a modified TEM holder while open cell is a part of modified TEM column. Below is comparison of their performance is given in Table 7.1:

> If everything seems under control, you're just not going fast enough.
> – Mario Andretti

However, just like any other in-situ TEM experiments, following dynamic gas–solid interactions do not have universal applications. Following are main constraints:

1. Compatibility of TEM holder, gas lines, and internal parts with gases to be used: For example, some of the corrosive gases such as halogens, SO_2, and H_2S can destroy the internal parts. Closed cell holders may be modified to tolerate some of these gases.
2. **Temperature limit:** The reaction temperature limit of currently available holders is 1300 °C that makes measurements of ceramic materials, requiring higher reaction temperature, not possible.

Table 7.1 Closed versus open cell configurations.

Functions	Closed cell	Open cell
Pressure limit	>1 bar	<20 mbar
Heating holders	Heating on the microchip only	All types of heating holders are usable
Cryo	Not possible	Possible by using LN_2 or other cryo holders
Image resolution	May be compromised by window thickness and gas path length	Nearly the same as TEM
EELS	Affected by window thickness and gas pressure	Unaffected
EDS	Restricted but in-situ is possible	Possible but not in-situ
STEM performance	Possible	Limited by the differential aperture size
Experimental design flexibility	Limited by the design of the holder	Flexible, other stimuli such as optical, mechanical, biasing holders can be used
Measuring gaseous products	A mass spectrometer on the gas outlet can easily measure the gas composition	Requires special modifications such as to increase the sample size and gas volume for using EELS or mass spectrometer

3. **Chemical analysis:** It is obvious from Table 7.1 that closed cell holders limit the possibility to use EDS and EELS to some extent due to contributions from windows to follow chemical changes but are possible with careful design and modification of the holder. STEM EELS is a possible solution but has longer acquisition time.
4. **Operando conditions:** Ideally, we would want to measure the reactants and products of the gas–solid interactions under reactor conditions. However, the achievable pressure, precise measurement of pressure and the temperature in both closed and open cell limit our capability to achieve this goal for catalytic process. Moreover, measuring gaseous reaction products is further limited when using an open cell configuration. On the other hand, growth kinetics for nanotubes, nanowires, and functionality of electrochemical and indentation processes, can be measured.

7.7 Take-Home Messages

- Nanoscale understanding of the gas–solid interaction is possible using either a gas-cell holder or modified TEM column. However, it is important to keep in mind that we are using TEM as a nanoreactor and should design our experiments with the same care as for a regular reactor.

- Gas-cell holders (nanoreactors) are commercially available, capable of withstanding pressures higher than 1 bar, and can be used in any TEM, but the microchips need to be handled carefully as the windows are fragile, also windows dampen the SNR and thereby effective resolution of TEM images, and only the stimuli incorporated on the chip can be used. Also, the sample tilt is limited that reduces SNR of EDS signal due to detector geometry.
- Modified column microscopes (ETEM) are also commercially available but are much more expensive than a gas-cell holder but other holders with different stimuli such as heating, cryo, biasing, straining, magnetic, optical, can be used. Also, all the functions of the TEM, i.e. tilting, diffraction, atomic-resolution TEM imaging, EELS, EDS, are maintained. However, the achievable gas pressure and STEM imaging are limited.

References

1 Butler, P. and Hale, K. (1981). In situ gas-solid reactions. In: *Practical Methods in Electron Microscopy. Experimental Microscopy*, vol. 9 (ed. A.M. Glauert), 239–309. North Holland Publishing Company.
2 Gai, P.L. (1999). Environmental high resolution electron microscopy of gas-catalyst reactions. *Topics in Catalysis* 8: 97–113.
3 Gai, P.L., Sharma, R., and Ross, F.M. (2008). Environmental (S)TEM studies of gas–liquid–solid interactions under reaction conditions. *MRS Bulletin* 33 (2): 107–114.
4 Jinschek, J.R. and Helveg, S. (2012). Image resolution and sensitivity in an environmental transmission electron microscope. *Micron* 43 (11): 1156–1168.
5 Hansen, T.W. and Wagner, J.B. (ed.) (2016). *Controlled Atmosphere Transmission Electron Microscopy: Principles and Practice*. Springer: Switzerland.
6 Heide, H.G. (1962). Electron microscopic observation of specimen under controlled gas pressure. *Journal of Cell Biology* 13 (1): 147–152.
7 Butler, E.P. and Hale, K.F. (1981). High temperature microscopy. In: *Dynamic Experiments in the Electron Microscope* (ed. A.M. Glauert), 1–457. Amsterdam/New York/Oxford: North Holland Publishing Company.
8 Swann, P.R. and Tighe, N.J. (1972). Performance of differentially pumped environmental cell in the AE1 EM7. *Proceedings of the Fifth European Congress, Manchester, England, United Kingdom*.
9 Swann, P.R. (1972). High voltage microscope studies of environmental reactions. In: *Electron Microscopy and Structure of Materials* (ed. G. Thomas, R. Fulrath and R.M. Fisher), 878. Berkeley: University of California Press.
10 Flower, M.J. (1973). High voltage electron microscopy of environmental reactions. *Journal of Microscopy* 97: 171–190.
11 Baker, R.T.K., Barber, M.A., Harris, P.S. et al. (1972). Nucleation and growth of carbon deposits from the Ni catalyzed decomposition of acetylene. *Journal of Catalysis* 26: 51–62.
12 Marton, L. (1935). Electron microsocpy of biological objects. *Nature* 133: 911.

13 Abrams, I.M. and McBain, J.W. (1944). A closed cell for electron microscopy. *Journal of Applied Physics* 15 (8): 607–609.
14 Doole, R.C., Parkinson, G.M., and Stead, J.M. (1991). High resolution gas reaction cell for the JEM 4000. *Institute of Physics Conference Series* 119: 157–160.
15 Lee, T.C., Robertson, I.M., and Birnbaum, H.K. (1990). An in situ transmission electron microscope deformation study of the slip transfer mechanisms in metals. *Metallurgical Transactions A* 21 (9): 2437–2447.
16 Sharma, R. and Weiss, K. (1998). Development of a TEM to study in situ structural and chemical changes at atomic level during gas solid interaction at elevated temperatures. *Microscopy Research and Technique* 42 (4): 270–280.
17 Boyes, E.D. and Gai, P.L. (1997). Environmental high resolution electron microscopy and applications to chemical science. *Ultramicroscopy* 67 (1): 219–232.
18 Lee, T.C., Dewald, D.K., Eades, J.A. et al. (1991). An environmental cell transmission electron microscope. *Review of Scientific Instruments* 62 (6): 1438–1444.
19 Robertson, I.M. and Teter, D. (1998). Controlled environment transmission electron microscopy. *Microscopy Research and Technique* 42 (4): 260–269.
20 Sharma, R. (2005). An environmental transmission electron microscope for in situ synthesis and characterization of nanomaterials. *Journal of Materials Research* 20 (7): 1695–1707.
21 Creemer, J.F., Helveg, S., Hoveling, G.H. et al. (2008). Atomic-scale electron microscopy at ambient pressure. *Ultramicroscopy* 108 (9): 993–998.
22 Allard, L.F., Overbury, S.H., Bigelow, W.C. et al. (2012). Novel MEMS-based gas-cell/heating specimen holder provides advanced imaging capabilities for in situ reaction studies. *Microscopy and Microanalysis* 18 (4): 656–666.
23 Mele, L., Konings, S., Dona, P. et al. (2016). A MEMS-based heating holder for the direct imaging of simultaneous in-situ heating and biasing experiments in scanning/transmission electron microscopes. *Microscopy Research and Technique* 79 (4): 239–250.
24 Parkinson, G.M. (1989). High resolution, in-situ controlled atmosphere transmission electron microscopy (CTEM) of heteroheneous catalysts. *Catalysis Letters* 2: 303–307.
25 Parkinson, G.M. (1991). Controlled environment transmission electron microscopy (CTEM) of catalysis. *Institute of Physics Conference Series* 119: 151–156.
26 Daulton, T.L., Little, B.J., Lowe, K., and Jones-Meehan, J. (2001). In situ environmental cell-transmission electron microscopy study of microbial reduction of chromium(VI) using electron energy loss spectroscopy. *Microscopy and Microanalysis* 7: 470.
27 Creemer, J.F., Helveg, S., Kooyman, P.J. et al. (2010). A MEMS reactor for atomic-scale microscopy of nanomaterials under industrially relevant conditions. *Journal of Microelectromechanical Systems* 19 (2): 254–264.

28 Pérez Garza, H.H., Morsink, D., Xu, J. et al. (2017). MEMS-based nanoreactor for in situ analysis of solid–gas interactions inside the transmission electron microscope. *Micro & Nano Letters* 12 (2): 69–75.

29 Alan, T., Yokosawa, T., Gaspar, J. et al. (2012). Micro-fabricated channel with ultra-thin yet ultra-strong windows enables electron microscopy under 4-bar pressure. *Applied Physics Letters* 100 (8): 081903–081904.

30 Yokosawa, T., Alan, T., Pandraud, G. et al. (2012). In-situ TEM on (de)hydrogenation of Pd at 0.5–4.5 bar hydrogen pressure and 20–400 °C. *Ultramicroscopy* 112 (1): 47–52.

31 Creemer, J.F., Santagata, F., Morana, B. et al. (2011). An all-in-one nanoreactor for high-resolution microscopy on nanomaterials at high pressures. *2011 IEEE 24th International Conference on Micro Electro Mechanical Systems*, Cancun, Mexico (23–27 January 2011). IEEE.

32 Mehraeen, S., JT, M.K., Deshmukh, P.V. et al. (2013). A (S)TEM gas cell holder with localized laser heating for in situ experiments. *Microscopy and Microanalysis* 19 (2): 470–478.

33 Straubinger, R., Beyer, A., and Volz, K. (2016). Preparation and loading process of single crystalline samples into a gas environmental cell holder for in situ atomic resolution scanning transmission electron microscopic observation. *Microscopy and Microanalysis* 22 (3): 515–519.

34 Lee, H., Beyer, A., Volz, K. et al. (2021). TEM sample preparation using micro-manipulator for in-situ MEMS experiment. *Applied Microscopy* 51 (1): 8–8.

35 Boniface, M., Plodinec, M., Schlögl, R. et al. (2020). Quo Vadis micro-electro-mechanical systems for the study of heterogeneous catalysts inside the electron microscope? *Topics in Catalysis* 63 (15): 1623–1643.

36 Kim, H.-E. and Moorhead, A.J. (1990). High-temperature gaseous corrosion of Si_3N_4 in H_2–H_2O and Ar–O_2 environments. *Journal of the American Ceramic Society* 73 (10): 3007–3014.

37 Kamino, T. and Saka, H. (1993). Newly developed high resolution hot stage and its applications to materials science microscopy. *Microanalanalysis and Miccrostructure* 4: 127–135.

38 Kishita, K., Sakai, H., Tanaka, H. et al. (2009). Development of an analytical environmental TEM system and its application. *Journal of Electron Microscopy* 58 (6): 331–339.

39 Zhao, Z., Wang, Z., Wang, D. et al. (2019). CFD modelling of gas flow characteristics for the gas-heating holder in environmental transmission electron microscope. *The Canadian Journal of Chemical Engineering* 97 (3): 777–784.

40 Tokunaga, T., Kawamoto, T., Tanaka, K. et al. (2012). Growth and structure analysis of tungsten oxide nanorods using environmental TEM. *Nanoscale Research Letters* 7 (1): 85.

41 Poppa, H. (2004). High resolution, high speed ultrahigh vacuum microscopy. *Journal of Vacuum Science and Technology A* 22 (5): 1931–1947.

42 Ross, F.M., Tersoff, J., and Tromp, R.M. (1998). Ostwald ripening of self-assembled Germanium Islands on silicon(100). *Microscopy and Microanalysis* 4 (3): 254–263.

43 Yeadon, M., Yang, J.C., Averback, R.S. et al. (1998). Sintering of silver and copper nanoparticles on (001) copper observed by in-situ ultrahigh vacuum transmission electron microscopy. *Nanostructured Materials* 10 (5): 731–739.

44 Kodambaka, S., Tersoff, J., Reuter, M.C., and Ross, F.M. (2006). Diameter-independent kinetics in the vapor–liquid–solid growth of Si nanowires. *Physical Review Letters* 96 (9): 096105.

45 Hansen Poul, L., Wagner Jakob, B., Helveg, S. et al. (2002). Atom-resolved imaging of dynamic shape changes in supported copper nanocrystals. *Science* 295 (5562): 2053–2055.

46 Helveg, S., Lopez-Cartes, C., Sehested, J. et al. (2004). Atomic-scale imaging of carbon nanofibre growth. *Nature* 427: 426.

47 Rose, A. (1948). Television pickup tubes and the problem of vision. In: *Advances in Electronics and Electron Physics* (ed. L. Marton), 131–166. Academic Press.

48 Akatay, M.C., Zvinevich, Y., Baumann, P. et al. (2014). Gas mixing system for imaging of nanomaterials under dynamic environments by environmental transmission electron microscopy. *Review of Scientific Instruments* 85 (3): 033704.

49 Yoshida, K., Tominaga, T., Hanatani, T. et al. (2013). Key factors for the dynamic ETEM observation of single atoms. *Microscopy* 62 (6): 571–582.

50 Koh, A.L., Gidcumb, E., Zhou, O. et al. (2016). Oxidation of carbon nanotubes in an ionizing environment. *Nano Letters* 16 (2): 856–863.

51 Garvie, L.A.J. and Buseck, P.R. (1999). Determination of Ce^{4+}/Ce^{3+} in electron-beam-damaged CeO_2 by electron energy-loss spectroscopy. *Journal of Physics and Chemistry of Solids* 60 (12): 1943–1947.

52 Johnston-Peck, A.C., Yang, W.-C.D., Winterstein, J.P. et al. (2018). In situ oxidation and reduction of cerium dioxide nanoparticles studied by scanning transmission electron microscopy. *Micron* 115: 54–63.

53 Sharma, R., Crozier, P.A., Kang, Z.C., and Eyring, L. (2004). Observation of dynamic nanostructural and nanochemical changes in ceria-based catalysts during in situ reduction. *Philosophical Magazine* 84: 2731–2747.

54 Zhang, D., Jin, C., Tian, H. et al. (2017). An in situ TEM study of the surface oxidation of palladium nanocrystals assisted by electron irradiation. *Nanoscale* 9 (19): 6327–6333.

55 Steinhauer, S., Wang, Z., Zhou, Z. et al. (2017). Probing electron beam effects with chemoresistive nanosensors during in situ environmental transmission electron microscopy. *Applied Physics Letters* 110 (9): 094103.

56 Hansen, T. and Wagner, J. (2012). Environmental transmission electron microscopy in an aberration-corrected environment. *Microscopy and Microanalysis* 18 (4): 684–690.

57 Wagner, J.B., Cavalca, F., Damsgaard, C.D. et al. (2012). Exploring the environmental transmission electron microscope. *Micron* 43 (11): 1169–1175.

58 Zhu, Y. and Browning, N.D. (2017). The role of gas in determining image quality and resolution during in situ scanning transmission electron microscopy experiments. *ChemCatChem* 9 (18): 3478–3485.

59 Vendelbo, S.B., Kooyman, P.J., Creemer, J.F. et al. (2013). Method for local temperature measurement in a nanoreactor for in situ high-resolution electron microscopy. *Ultramicroscopy* 133: 72–79.

60 Colby, R., Alsem, D.H., Liyu, A. et al. (2015). A method for measuring the local gas pressure within a gas-flow stage in situ in the transmission electron microscope. *Ultramicroscopy* 153: 55–60.

61 Wise, M.E., Biskos, G., Martin, S.T. et al. (2005). Phase transitions of single salt particles studied using a transmission electron microsocpe with an environmental cell. *Aerosol Scienec and Technology* 39: 849–856.

62 Mecklenburg, M., Hubbard, W.A., White, E. et al. (2015). Nanoscale temperature mapping in operating microelectronic devices. *Science* 347 (6222): 629–632.

63 Wang, C., W-CD, Y., Bruma, A. et al. (2021: p. . Under review). Endothermic reaction at room temperature enabled by deep-ultraviolet plasmons. *Nature Materials* 20: 346–352.

64 Mortensen, P.M., Hansen, T.W., Wagner, J.B. et al. (2015). Modeling of temperature profiles in an environmental transmission electron microscope using computational fluid dynamics. *Ultramicroscopy* 152: 1–9.

65 Vincent, J.L., Vance, J.W., Langdon, J.T. et al. (2020). Chemical kinetics for operando electron microscopy of catalysts: 3D modeling of gas and temperature distributions during catalytic reactions. *Ultramicroscopy* 218: 113080.

66 Li, M., Xie, D.-G., Zhang, X.-X. et al. (2021). Quantifying real-time sample temperature under the gas environment in the transmission electron microscope using a novel MEMS heater. *Microscopy and Microanalysis* 27 (4): 758–766.

67 Gai, P.L. and Boyes, E.D. (2003). *Electron Microscopy of Heterogeneous Catalysis. Series in Microscopy and Materials Science*. Bristol, PA: Institute of Physics Publishing.

68 Jinschek, J.R. (2014). Advances in the environmental transmission electron microscope (ETEM) for nanoscale in situ studies of gas–solid interactions. *Chemical Communications* 50 (21): 2696–2706.

69 Hansen, T.W., Wagner, J.B., and Dunin-Borkowski, R.E. (2010). Aberration corrected and monochromated environmental transmission electron microscopy: challenges and prospects for materials science. *Materials Science and Technology* 26 (11): 1338–1344.

70 Gai, P.L. and Calvino, J.J. (2005). Electronmicroscopy in the catalysis of alkane oxidation, environmental control, and alternative energy sources. *Annual Review of Materials Research* 35: 465–504.

71 Tao, F. and Crozier, P.A. (2016). Atomic-scale observations of catalyst structures under reaction conditions and during catalysis. *Chemical Reviews* 116 (6): 3487–3539.

72 Niu, Y. and Zhang, B. (2020). In situ investigation of nanocatalysts in gas atmosphere by transmission electron microscopy. *Current Opinion in Green and Sustainable Chemistry* 22: 22–28.

73 Uchiyama, T., Yoshida, H., Kuwauchi, Y. et al. (2011). Systematic morphology changes of gold nanoparticles supported on CeO_2 during CO oxidation. *Angewandte Chemie International Edition* 50 (43): 10157–10160.

74 Takeda, S. and Yoshida, H. (2013). Atomic-resolution environmental TEM for quantitative in-situ microscopy in materials science. *Microscopy* 62 (1): 193–203.

75 Vendelbo, S.B., Elkjær, C.F., Falsig, H. et al. (2014). Visualization of oscillatory behaviour of Pt nanoparticles catalysing CO oxidation. *Nature Materials* 13 (9): 884–890.

76 Yoshida, H., Omote, H., and Takeda, S. (2014). Oxidation and reduction processes of platinum nanoparticles observed at the atomic scale by environmental transmission electron microscopy. *Nanoscale* 6 (21): 13113–13118.

77 Vincent, J.L. and Crozier, P.A. (2021). Atomic level fluxional behavior and activity of CeO_2-supported Pt catalysts for CO oxidation. *Nature Communications* 12 (1): 5789.

78 Lawrence, E.L., Levin, B.D.A., Boland, T. et al. (2021). Atomic scale characterization of fluxional cation behavior on nanoparticle surfaces: probing oxygen vacancy creation/annihilation at surface sites. *ACS Nano* 15 (2): 2624–2634.

79 Chen, F.-R., Van Dyck, D., Kisielowski, C. et al. (2021). Probing atom dynamics of excited Co–Mo–S nanocrystals in 3D. *Nature Communications* 12 (1): 5007.

80 Yuan, W., Zhu, B., Fang, K. et al. (2021). In situ manipulation of the active Au-TiO_2 interface with atomic precision during CO oxidation. *Science* 371 (6528): 517–521.

81 Holse, C., Elkjær, C.F., Nierhoff, A. et al. (2015). Dynamic behavior of CuZn nanoparticles under oxidizing and reducing conditions. *The Journal of Physical Chemistry C* 119 (5): 2804–2812.

82 Jiang, Y., Li, H., Wu, Z. et al. (2016). In situ observation of hydrogen-induced surface faceting for palladium–copper nanocrystals at atmospheric pressure. *Angewandte Chemie International Edition* 55 (40): 12427–12430.

83 Baker, R.T.K., Harris, P.S., Thomas, R.B., and Waite, R.J. (1973). Formation of filamentous carbon from iron, cobalt and chromium catalyzed decomposition of acetylene. *Journal of Catalysis* 30: 86–95.

84 Yoshida, H., Seiji, T., Tetsuya, U. et al. (2008). Atomic-scale in-situ observation of carbon nanotube growth from solid state iron carbide nanoparticles. *Nano Letters* 9 (11): 3810–3815.

85 Hofmann, S., Sharma, R., Ducati, C. et al. (2007). In situ observations of catalyst dynamics during surface-bound carbon nanotube nucleation. *Nano Letters* 7 (3): 602–608.

86 Zhang, L., He, M., Hansen, T.W. et al. (2017). Growth termination and multiple nucleation of single-wall carbon nanotubes evidenced by in situ transmission electron microscopy. *ACS Nano* 11 (5): 4483–4493.

87 Picher, M., Lin, P.A., Gomez-Ballesteros, J.L. et al. (2014). Nucleation of graphene and its conversion to single-walled carbon nanotubes. *Nano Letters* 14 (11): 6104–6108.

88 Hofmann, S., Blume, R., Wirth, C.T. et al. (2009). State of transition metal catalysts during carbon nanotube growth. *Journal of Physical Chemistry C* 113 (5): 1648–1656.

89 Lin, P.A., Gomez-Ballesteros, J.L., Burgos, J.C. et al. (2017). Direct evidence of atomic-scale structural fluctuations in catalyst nanoparticles. *Journal of Catalysis* 349: 149–155.

90 Chao, H.-Y., Jiang, H., Ospina-Acevedo, F. et al. (2020). A structure and activity relationship for single-walled carbon nanotube growth confirmed by in situ observations and modeling. *Nanoscale* 12 (42): 21923–21931.

91 Yang, F., Wang, X., Zhang, D. et al. (2014). Chirality-specific growth of single-walled carbon nanotubes on solid alloy catalysts. *Nature* 510 (7506): 522–524.

92 Kodambaka, S., Tersoff, J., Reuter, M.C., and Ross, F.M. (2007). Germanium nanowire growth below the eutectic temperature. *Science* 316 (5825): 729–732.

93 Lang, C., Kodambaka, S., Ross, F.M., and Cockayne, D.J.H. (2006). Real time observation of GeSi/Si(001) island shrinkage due to surface alloying during Si capping. *Physical Review Letters* 97 (22): 4.

94 Kodambaka, S., Hannon, J.B., Tromp, R.M., and Ross, F.M. (2006). Control of Si nanowire growth by oxygen. *Nano Letters* 6 (6): 1292–1296.

95 van Dorp, W.F., van Someren, B., Hagen, C.W. et al. (2005). Approaching the resolution limit of nanometer-scale electron beam-induced deposition. *Nano Letters* 5 (7): 1303–1307.

96 Yang, J.C., Kolasa, B., Gibson, J.M., and Yeadon, M. (1998). Self-limiting oxidation of copper. *Applied Physics Letters* 73 (19): 2841–2843.

97 Cabrera, N. and Mott, N.F. (1949). Theory of the oxidation of metals. *Reports on Progress in Physics* 12 (1): 163–184.

98 Madras, P., Dailey, E., and Drucker, J. (2009). Kinetically induced kinking of vapor–liquid–solid grown epitaxial Si nanowires. *Nano Letters* 9 (11): 3826–3830.

99 Hofmann, S., Sharma, R., Wirth, C.T. et al. (2008). Ledge-flow controlled catalyst interface dynamics during Si nanowire growth. *Nature Materials* 7 (5): 372–375.

100 Kikkawa, J., Ohno, Y., and Takeda, S. (2005). Growth rate of silicon nanowires. *Applied Physics Letters* 86 (12): 123109.

101 Wen, C.Y., Reuter, M.C., Tersoff, J. et al. (2010). Structure, growth kinetics, and ledge flow during vapor–solid–solid growth of copper-catalyzed silicon nanowires. *Nano Letters* 10 (2): 514–519.

102 Diaz, R.E., Sharma, R., Jarvis, K., and Mahajan, S. (2012). Direct observation of nucleation and early stages of growth of GaN nanowires. *Journal of Crystal Growth* 341: 1–6.

103 Hillerich, K., Dick, K.A., Wen, C.-Y. et al. (2013). Strategies to control morphology in hybrid group III–V/group IV heterostructure nanowires. *Nano Letters* 13 (3): 903–908.

104 Hetherington, C., Jacobsson, D., Dick, K., and Wallenberg, L. (2020). In situ metal-organic chemical vapour deposition growth of III–V semiconductor nanowires in the Lund environmental transmission electron microscope. *Semiconductor Science and Technology* 35 (3): 034004.

105 Wen, C.-Y., Reuter, M.C., Bruley, J. et al. (2009). Formation of compositionally abrupt axial heterojunctions in silicon-germanium nanowires. *Science* 326 (5957): 1247–1250.

106 Rackauskas, S., Jiang, H., Wagner, J.B. et al. (2014). In situ study of noncatalytic metal oxide nanowire growth. *Nano Letters* 14 (10): 5810–5813.

107 Sun, X., Zhu, W., Wu, D. et al. (2020). Atomic-scale mechanism of unidirectional oxide growth. *Advanced Functional Materials* 30 (4): 1906504.

108 Ross, F.M. (2010). Controlling nanowire structures through real time growth studies. *Reports on Progress in Physics* 73 (11): 114501.

109 Mitsuishi, K., Shimojo, M., Han, M. et al. (2003). Electron-beam-induced deposition using a subnanometer-sized probe of high-energy electrons. *Applied Physics Letters* 83 (10): 2064–2066.

110 Crozier, P.A. (2007). Nanoscale oxide patterning with electron-solid-gas reactions. *Nano Letters* 7 (8): 2395–2398.

111 van Dorp, W.F., Hagen, C.W., Crozier, P.A. et al. (2006). One nanometer structure fabrication using electron beam induced deposition. *Microelectronic Engineering* 83: 1468–1470.

112 Liu, R.-J., Crozier, P.A., Smith, C.M. et al. (2005). Metal sintering mechanisms and regeneration of palladium/alumina hydrogenation catalyst. *Applied Catalysis* A282: 111.

113 Hwang, S., Chen, X., Zhou, G., and Su, D. (2020). In situ transmission electron microscopy on energy-related catalysis. *Advanced Energy Materials* 10 (11): 1902105.

114 Smigelskas, A.D. and Kirkendall, E.O. (1947). Zinc diffusion in alpha brass. *Transactions of AIME* 171: 131–142.

115 Chenna, S., Banerjee, R., and Crozier, P.A. (2011). Atomic-scale observation of the Ni activation process for partial oxidation of methane using in situ environmental TEM. *ChemCatChem* 3 (6): 1051–1059.

116 Wang, C.-M., Genc, A., Cheng, H. et al. (2014). In-situ TEM visualization of vacancy injection and chemical partition during oxidation of Ni–Cr nanoparticles. *Scientific Reports* 4 (1): 3683.

117 Jeangros, Q., Hansen, T.W., Wagner, J.B. et al. (2014). Oxidation mechanism of nickel particles studied in an environmental transmission electron microscope. *Acta Materialia* 67: 362–372.

118 Wang, C.-M., Schreiber, D.K., Olszta, M.J. et al. (2015). Direct in situ TEM observation of modification of oxidation by the injected vacancies for Ni–4Al alloy using a microfabricated nanopost. *ACS Applied Materials & Interfaces* 7 (31): 17272–17277.

119 Egerton, R.F. (2014). *Electron Energy-Loss Spectroscopy in the Electron Microscope*, 3e, 503. Boston, MA: Springer.

120 Loomer, D.D., Al T.A., W.L. et al. (2007). Manganese valence imaging in Mn minerals at the nanoscale using STEM-EELS. *The American Mineralogist* 92 (1): 72–79.

121 Cavé, L., Al, T., Loomer, D. et al. (2006). A STEM/EELS method for mapping iron valence ratios in oxide minerals. *Micron* 37 (4): 301–309.

122 Marris, H., Deboudt, K., Flament, P. et al. (2013). Fe and Mn oxidation states by TEM-EELS in fine-particle emissions from a Fe–Mn alloy making plant. *Environmental Science and Technology* 47 (19): 10832–10840.

123 Johnston-Peck, A.C., Winterstein, J.P., Roberts, A.D. et al. (2016). Oxidation-state sensitive imaging of cerium dioxide by atomic-resolution low-angle annular dark field scanning transmission electron microscopy. *Ultramicroscopy* 162: 52–60.

124 Xin, H.L., Pach, E.A., Diaz, R.E. et al. (2012). Revealing correlation of valence state with nanoporous structure in cobalt catalyst nanoparticles by in situ environmental TEM. *ACS Nano* 6 (5): 4241–4247.

125 Foucher, A.C., Marcella, N., Lee, J.D. et al. (2022). Dynamical change of valence states and structure in $NiCu_3$ nanoparticles during redox cycling. *The Journal of Physical Chemistry C* 126 (4): 1991–2002.

126 Frey, H., Beck, A., Huang, X. et al. (2022). Dynamic interplay between metal nanoparticles and oxide support under redox conditions. *Science* 376 (6596): 982–987.

127 Zou, L., Li, J., Zakharov, D. et al. (2017). In situ atomic-scale imaging of the metal/oxide interfacial transformation. *Nature Communications* 8 (1): 307.

128 Matsuda, J., Kawasaki, T., Futamura, S. et al. (2018). In situ transmission electron microscopic observations of redox cycling of a Ni–ScSZ cermet fuel cell anode. *Microscopy* 67 (5): 251–258.

129 Tavabi, A.H., Arai, S., Muto, S. et al. (2014). In situ transmission electron microscopy of ionic conductivity and reaction mechanisms in ultrathin solid oxide fuel cells. *Microscopy and Microanalysis* 20 (6): 1817–1825.

130 Yoshida, H., Kuwauchi, Y., Jinschek, J.R. et al. (2012). Visualizing gas molecules interacting with supported nanoparticulate catalysts at reaction conditions. *Science* 335 (6066): 317–319.

131 Yuan, W., Zhu, B., Li, X.-Y. et al. (2020). Visualizing H_2O molecules reacting at TiO_2 active sites with transmission electron microscopy. *Science* 367 (6476): 428–430.

132 Yang, W.-C.D., Wang, C., Fredin, L.A. et al. (2019). Site-selective CO disproportionation mediated by localized surface plasmon resonance excited by electron beam. *Nature Materials* 18 (6): 614–619.

133 Chenna, S. and Crozier, P.A. (2012). Operando transmission electron microscopy: a technique for detection of catalysis using electron energy-loss spectroscopy in the transmission electron microscope. *ACS Catalysis* 2 (11): 2395–2402.

134 Miller, B.K. and Crozier, P.A. (2014). Analysis of catalytic gas products using electron energy-loss spectroscopy and residual gas analysis for operando transmission electron microscopy. *Microscopy and Microanalysis* 20 (3): 815–824.

8

Multimodal and Correlative Microscopy

> *With the aid of these active experimental sciences man becomes an inventor of phenomena, a real foreman of creation; and under this head we cannot set limits to the power that he may gain over nature through future progress of the experimental sciences.*
>
> – Claude Bernard

Based on the research results, presented so far, we can safely say that in-situ TEM/STEM-based characterization techniques are not only desirable but essential for controlling nanomaterials design and properties. However, there are some inherent but significant limitations, such as effect of electron beam, the sampling size, and pressure gap, which may or may not result in an applicability gap. Fortunately, most of us are aware of these deficiencies, especially the electron beam effects are clearly visible and recommended steps to mitigate them are covered in almost every chapter of this book. Various multimodal and correlative techniques are also available to further substantiate outcomes of in-situ TEM experiments. In this chapter, we will describe multimodal and correlative methods currently being used. At first glance, these two appear to be the same, and we find that both terms are being used indiscriminately. Here, we plan to differentiate between the two using following qualifiers: Multimodal techniques are incorporated in the TEM column or the TEM sample holder such that multiple nano and micro-scale in-situ measurements are made under exactly the same conditions, whereas correlative methods verify in-situ TEM results using other characterization techniques outside the TEM column (even in a different lab or location) under similar experimental conditions. Alternatively, micro-scale in-situ characterization techniques, such as XRD, DSC, TGA, may be used to define experimental conditions to be used for in-situ TEM measurements.

First and foremost, we consider the alternative approaches that can address the materials size gap challenge, by associating nanoscale observations to the behavior of micro-scale sample. Figure 8.1 shows the relative volume, spatial and energy resolution for imaging techniques such as TEM/STEM, SEM, STXM, optical microscopes, and spectroscopies, such as EELS, EDS, XPS, extended X-ray absorption fine structure (EXAFS), and Raman. There are also differences in the

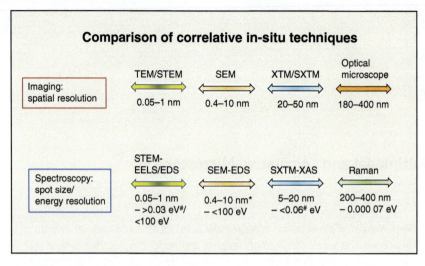

Figure 8.1 Block diagram showing the length scale, spatial and energy resolutions for various in-situ techniques currently employed. #Energy resolution is dependent on TEM configuration and is continuously improving. *spot size depends on the electron source and microscope configuration.

temporal resolution that may lead us to miss some intermediate steps. Therefore, alternative in-situ experiments using SEM and X-rays are commonly performed, although neither of these entirely addresses the sample damage probability, which is property of every high-energy source interacting with the sample.

> Extending the behavior of a few particles, observed during in-situ TEM observations to the expected dynamics of the entire sample is rarely a simple endeavor.
> – See Wee Chee, Fritz-Haber Institutes, Berlin

Correlative microscopy can also be extended to the experiments that apply multiple stimuli on the same sample under same experimental conditions to correlate their effects. For example, structural and compositional changes induced by mechanical strain and temperature, or biasing combined with heating or cooling, etc. These have been addressed in Chapters 1, 3, 4, 6, and 7, and are not included here. In this chapter, we will describe design and applications of multimodal and correlative approaches that can be used to address the applicability gap professed by in-situ TEM/STEM experiments.

8.1 Multimodal TEM

One of the best solutions to address both the electron beam effect and size gap is to be able to perform experiments with and without electrons at nano- and micro-scale under identical condition or concurrently. In past few years, multimodal approach for following reaction processes under the identical conditions using in-situ TEM observations has been reported. Main advantage is that the same sample is examined using a probe other than electrons.

However, such modifications are tedious, time-consuming, and expensive, as they need input from the TEM manufacturer, access to high-grade machine shop, personnel and financial resources. It is important to note that every TEM comes with its own physical specifications such as number and placements of the ports on the column, internal diameter of the port, objective pole-piece gap, etc. Moreover, you do not want to compromise the optics, spatial and energy resolutions of the parent instrument by your modifications. Therefore, it is important to work with your TEM manufacture's design team to either get the information or provide them with your design, including dimensions, for approval. It is a good idea to sign NDA with them such that intellectual properties of both parties are protected.

8.1.1 Parallel Ion Electron Spectrometry (PIES)

This is an unusual instrumental setup, and currently not commercially available, that combines secondary ion mass spectroscopy (SIMS) with TEM on the same instrument. As we know, high spatial and energy resolution for imaging and spectroscopy can be routinely obtained using TEM/STEM-based techniques. In-situ observations under various stimuli have revealed unprecedented information about the structure and reaction mechanisms. However, presence of trace amounts of elements in a matrix and/or to distinguish between the isotopes is beyond the detection limit of TEM/STEM imaging and spectroscopy techniques. On the other hand, SIMS is a powerful analytical technique with high sensitivity and ability to distinguish between all elements and isotopes in the periodic table. Moreover, it can generate a depth profile of all the elements present in a bulk sample, providing 3D distribution of all elements/isotopes present. But the lateral resolution is limited to few tens of nanometers even when using high brightness ion source, such as helium gun. Therefore, Yedra et al. have recently introduced a new platform that combines in-situ TEM and SIMS, called the Parallel Ion Electron Spectrometry (PIES) [1]. This setup utilizes TEM to find the interesting features at nanoscale that can then be analyzed using SIMS, alternatively SIMS can be used to locate the regions where trace elements or isotopes are present followed by TEM for high-resolution imaging of the hotspots. TEM and SIMS can also be performed iteratively and quickly without any need to transfer sample between two separate instruments and thereby avoid losing the area of interest and/or sample contamination by exposure to air. Moreover, any artifacts associated with SIMS, such as image distortion, can be corrected with the help of TEM imaging [1].

The twin pole pieces of a Tecnai F20 TEM are heavily modified to accommodate nozzles for the FIB source and SIMS detector, to sputter ions from the sample and collect them, respectively (Figure 8.2a). Both nozzles are at 68° angle with respect to the optic axis of the TEM (Figure 8.2b), and a FIB with a monoisotopic $^{69}Ga^+$ primary ion source and home-built compact high-performance double-focusing magnetic sector are mounted on the TEM column (Figure 8.2a). A TEM holder was also developed to obtain high-tilt capability required to select between TEM (Figure 8.2b) and SIMS (PIES) mode (Figure 8.2c) such that either the electron beam or SIMS nozzles are perpendicular to the sample. FIB nozzle is at 45° to the sample surface in this configuration (Figure 8.2c). Despite such heavy modification to the objective pole pieces and possibility of stray magnetic field from the mass spectrometer, the lattice resolution for TEM images was measured be 0.15 nm at 200 keV, and the image

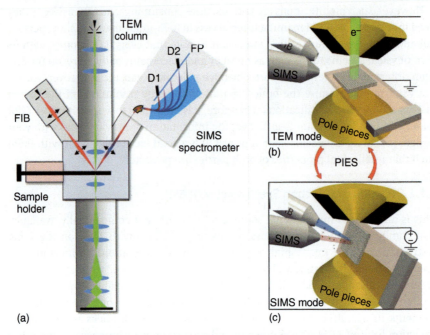

Figure 8.2 Schematic layout (a) of the PIES instrument. D1 and D2 indicate the two movable detectors along the Focal Plane (FP) of the magnet in the SIMS spectrometer. The instrument is a modified Tecnai F20 TEM equipped with a Ga FIB column and a dedicated magnetic-sector SIMS add-on system. In addition, the instrument is equipped with an EDS system. (b, c) Schematics of sample region showing location FIB and SIMS between the upper and the lower objective pole pieces during TEM (b) and SIMS (c) operating conditions. Source: Yedra et al. [1]/Springer Nature/CC BY 4.0.

resolution for SIMS was calculated to be below 60 nm. Also, this holder can be used to create a high-voltage (± 5 kV) electric field to improve the efficiency for collecting secondary ions.

There are limited applications of this instrument reported so far as it is not commercially available. It has been mostly used for biological samples, especially to locate the isotopes used for drug delivery in the cells [2]. Moreover, it has not been used for dynamic imaging, but the capability to obtain 3D elemental maps for trace elements or vacancy movements has potential applications for in-situ measurements in materials field.

8.1.2 Hybrid Microscope

Keeping in the spirit of the **multimodal approach**, various setups to focus the light on to a TEM sample to combine imaging, diffraction, electron spectroscopies with vibrational spectroscopies have been designed [3, 4]. Ohno et al. report introducing laser beam through an optical window attached to a side-entry port of the TEM and focusing it on the sample using a reflector with a 5 mm hole, placed in the column above the objective lens [3]. The signal is collected using an ellipsoidal mirror, placed

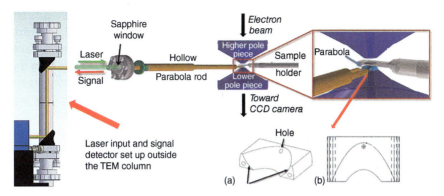

Figure 8.3 Schematic showing the external laser input and signal collection system located outside the TEM column, hollow parabola rod with a parabolic mirror attached to one end, the location of the parabola with respect to the sample within the pole-piece gap of a Titan 80 to 300 keV ETEM, magnified drawing showing the respective position of parabola and sample holder, and top view of the parabola showing the holes where the rod is attached and the hole to transmit the electron beam.

in the objective pole-piece gap. Applicability of this system (apparatus as the authors call it) was tested on CVD grown diamond samples [3].

Recently, Picher et al. have reported the design and application of a hybrid TEM, incorporating laser or light injection system, which can be focused on the sample using a parabola that also collects the spectroscopic signal generated by the impact of light or electrons on the sample and transmits it to a detector [4]. Both the laser/light source and the detector are mounted on a cart placed near the column but not attached to the microscope, and the light as well as spectroscopic signals travel through the same hollow rod, but in opposite direction, in free space beam path, i.e. without using optical fiber (Figure 8.3). Unlike a regular TEM/STEM, there is no port on the octagon of a Titan ETEM available to introduce any external stimuli. Therefore, the objective aperture rod was replaced by a hollow rod with a sapphire window on one end, and a parabola with a hole for the electron beam transmission on the other end such that it can be placed under the sample (Figure 8.3). There are multiple advantages for using the objective aperture rod; (i) the port is equipped with x, y positioning piezo drives, (ii) pole-piece touch signal is enacted for objective aperture rod that will keep the parabola from hitting the sample or the pole-piece, (iii) the parabola can be retracted easily when not in use. The parabola is the most important part, and its thickness, width, and curvature are defined by distance between the bottom of sample holder and top of the lower pole piece, and the diameter of the hole in the column port. The curvature of the parabola defines the focal length and signal collection efficiency. The numerical aperture (NA) of the parabola being used in the ESTEM at NIST is 0.8, and focal spot size is 10 μm [4].

> Due to the tight space between the lower pole piece, parabola, and the sample holder, a visual inspection using one of the ports on the octagon is required to make sure that there is enough clearance between them, i.e. the objective lens is protected.

The laser/light source and the optical components, needed to generate a parallel beam on the sapphire window of the hollow parabola rod, are mounted on a cart, located by the TEM column. The parallel beam travels though the hollow rod and hits the parabola that focuses the beam on the sample and collect response from the sample. This free space arrangement keeps the light source and detector decoupled from the TEM column and sample, reducing the possibility of introducing any vibrations. Also, signal losses from optical fibers due to internal reflection and/or joints are completely avoided, which is important given the low signal amount produced from the small sample size. Moreover, it allows us to collect both vibrational (Raman), optical (photoluminescence [PL]), and cathodoluminescence (CL) signals. This microscope was originally designed to follow in-situ growth of CNTs using Raman spectroscopy and HREM images concurrently such that micro- and nanoscale observations were made under identical synthesis conditions [4]. It has also been used to map the temperature variation on microfabricated heating chips and to map the CL signal from CdTe solar cells as a function of electron beam current, sample thickness, and grain-boundary chemistry [5].

8.1.3 Alternatives to Free Space Approach

Disadvantages of the free space design are that it requires extra space to mount the laser source, detector and the alignment of the free space optics can be time-consuming. Therefore, using optical fibers to transport light from a source, located outside the TEM column to the sample inside the column, is a preferred approach. However, this also requires careful design considerations as the some of the criteria, mentioned in the beginning of the chapter, such as availability of an empty port and diameter of its opening, pole-piece gap, remain relevant. Moreover, we also need to consider the properties of the optical fibers, such as (i) their wavelength dependence, (ii) diameter, (iii) bending radius without transmission loss, (iv) length, (v) cladding and protective material, etc. For example, the polymer protective cladding is not suitable for an optical fiber inside the high vacuum of the TEM column, a metal protection is therefore required. In principle, optical fibers can carry flux in the order of around 10^{10} W m^{-2}, which is enough for most experiments, in practice, the final flux is much reduced due to (i) diameter-dependent internal reflection, (ii) bending radius, and (iii) number of connections, etc. Therefore, optical fibers and the method to bring them as near as possible to the sample inside a TEM must be chosen carefully. Still, optical fibers provide an attractive alternative to the free space optics described above.

Therefore, another design using a lens attached to an optical fiber to focus laser and collect signal has been incorporated on a TEM column [6]. Here, custom-built lensed optical fiber was used as an excitation probe as well as signal collector (Figure 8.4a) and is mounted on an x–y translations stage on one of the available vacuum ports on the TEM column at an angle of 45° to the port. The specimen holder is also modified such that a side cut-out in the tip region provides direct access to the laser probe (Figure 8.4b). The signal cone, shown as pink shaded area in Figure 8.4a, subtends an angle of 60° with probe tip located at a distance

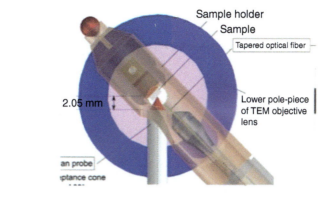
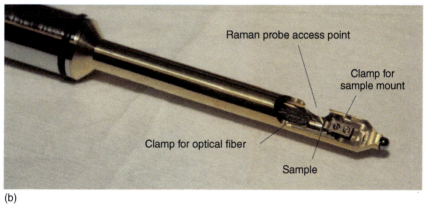

Figure 8.4 (a) Layout of in-situ Raman assembly in the sample chamber of the TEM. (b) Photo showing the tip of the sample holder. Source: Allen et al. [6]/from ELSEVIER.

of ≈ 2 mm from the sample. The spot size and NA of the probe are ≈ 2 μm and 0.2, respectively. The system was tested to compare EELS and Raman signals from a MoS_2 flake and laser melting was used to observe transformation of amorphous Si into a single crystal [6, 7]. Advantage of this design is its provision for near field optics, and the disadvantage is low SNR that requires long collection times.

Although neither of these systems has been used extensively, interest in developing techniques to combine TEM/STEM with vibrational and optical spectroscopy continues as evidenced by recent report of a modified TEM specimen holder [8]. Our current experience indicates that incorporating external stimuli in a holder is more popular than modifying a TEM column. The reasons can be attributed to the fact that holders can be modified without taking the TEM out of service during the development period, and the basic design can be replicated for other TEMs with minimal effort. Modified holder also employs optical fiber to direct light onto the sample and collect vibrational and optical signals, excited by light or electrons, from the sample.

The design of the specimen holder, shown in Figure 8.5a, consists of modified holder body to insert bundle of optical fibers, a 4f demagnified system (using

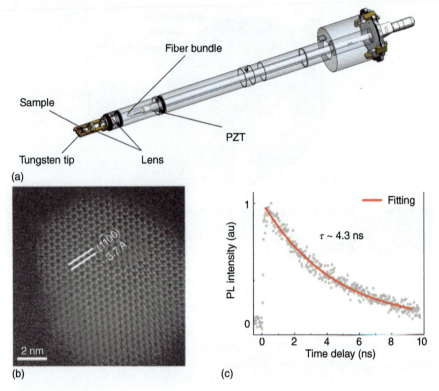

Figure 8.5 (a) The schematic design of the specimen holder. (b, c) In-situ time-resolved PL of CdSe/ZnS quantum dots in AC-TEM: (b) the STEM image of the CdSe/ZnS quantum dot and (c) in-situ measurement of PL lifetime of CdSe/ZnS quantum dots by using time correlating single photon counting (TCSPC). Source: Liu et al. [8]/from AIP Publishing LLC.

Fourier optics with two lenses), translation stages, and motion control system (not shown) [8]. The holder body is made of a hollow aluminum tube with tungsten tip, located in the specimen holder tip region on which the sample is loaded. The tungsten tip of desired length is prepared by electrochemical etching. The 4f demagnifying system, consisting of two lenses, is fixed near the tip and is used to deliver pulse beam onto the sample from a confocal system located outside the column. Commercially available fiber bundle, for single mode or a few mode cores, were inserted through the central hollow core of the TEM holder body along with several electrical feed-throughs. A piezo-driven translation stage (PZT) is located at the distal end of the fiber bundle to shift the laser spot within 20 nm, resulting in focal spot of 10 nm due to 4f demagnified system. Outside the holder, the fiber bundle is mounted on a three-dimensional translation stage to optimize the laser coupling and scan the focal spot in the x–y plane. The optical system for femtosecond pulse generation, collimation, and focusing it onto fiber bundle on the exit side of the sample holder, though important, is outside the scope of this book but is explained well in Ref. [8]. The application of this system to image (Figure 8.5b) and measure PL lifetime of CdSe/ZnS quantum dots (Figure 8.5c) is an important step toward measuring Raman or CL signals with high spatial and

8.1.4 Introducing Light for Other Applications

Apart from introducing the laser to excite vibrational modes (Raman) or collect photon (PL or CL), described above, it can also be used for heating or to activate reactions by photon energy instead or thermal energy (photocatalysis). Latter is to mimic the natural photosynthesis phenomenon in plants using solar light. Photocatalysis is an important field of research as the idea is to exploit photon (solar) energy to activate chemical reactions instead of thermal energy is very attractive. Water splitting and CO_2 reduction are two of the main catalytic reactions where photons have been successfully used. Recently, photons have also been used to excite localized surface plasmon resonances (LSPR) in certain metals that in turn can replace heat to initiate chemical reactions at RT [9–13]. Therefore, it is not surprising that several research groups have used one of the available ports with access to the sample chamber or modified sample holders to shine light on the sample under reactive environment. While a focused light source is needed for local heating or for photon-induced spectroscopy, flooding the entire sample with photon is acceptable for photon-induced chemical processes.

> Caution: Electron beam can behave like white light, therefore will initiate the chemical reaction just like UV light would! Therefore, observations must be made with electron beam off when the sample is exposed to light in reactive environment and/or observed in the absence of reactive environment.

Currently, it is achieved in following two ways:

8.1.4.1 Through Sample Chamber Port

Introducing light through one of the available ports in TEM using optical fiber is one of the approaches similar to the one used by Allen et al. [6] described above (Section 8.1.3). Main difference here is that the light may or may not be focused on a small part of the sample. For example, Yoshida et al. report introducing an optical fiber (Figure 8.6a) connected to a light source outside the TEM column [14]. The light intensity carried by the fiber was measured to be $10\,\mathrm{m\,W\,cm^{-2}}$ ($\lambda = 360\,\mathrm{nm}$) on the sample plane and wavelength of the light can be changed by changing the source. This setup was used to observe decomposition of organic film deposited on TiO_2 layer [16]. It is important to remember that UV lasers can heat the sample, and they found that the heating also caused a drift of $\approx 1\,\mathrm{mm\,h^{-1}}$ in the sample.

In another example, similar arrangement was followed to introduce UV light in the sample area of an ETEM (Figure 8.6b) to follow the photocatalytic water splitting reaction over TiO_2 (anatase) nanocrystals [15]. It is comparatively simple design that overcomes the constraints posed by the diameter of the port opening and objective pole-piece gap (Figure 8.6b). Here a laser-driven broadband light source (200 to 800 nm) was used as it closely mimics the solar spectrum. The light was

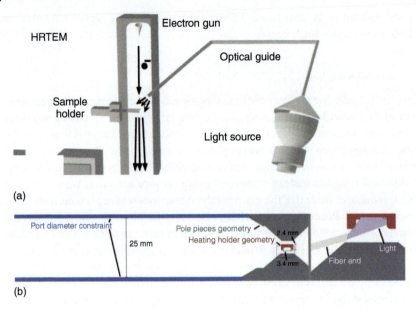

Figure 8.6 (a) Schematic diagram of an in-situ TEM observation system for photochemical reaction. Source: Adapted from Yoshida et al. [14]. Ultraviolet light is introduced into a 200 kV TEM through an optical guide. (b) The fiber tip in its working position, in a different design, is shown in the right side of the figure, with the light beam striking the sample. Source: Miller and Crozier [15]/Cambridge University Press.

introduced using optical fibers after being focused and collimated using the mirrors through one of the free ports on the octagon of the Tecnai ETEM [15]. The optical fiber used inside the TEM column is specially designed with core diameter of 600 μm protected by aluminum instead of polymer-based protective buffer and jacket. The limited space available in the sample chamber, the pole-piece gap, and restricting bending radius of the optical fibers is overcome by cutting the end of the fiber, facing the sample, by 30° such that a cone of light illuminates the sample (Figure 8.6b). It was used to observe atomic-scale changes in the surface layer of TiO_2 nanocrystals when exposed to water vapor and light or electron irradiation. They report the formation of a self-limiting amorphous layer containing Ti^{3+}, as determined by using EELS and XPS. This layer was formed only during water splitting reaction, and its thickness did not increase with prolonged exposure to light or electrons in water-vapor-rich environment [17]. On the other hand, Lu et al. have used a liquid cell holder containing TiO_2 nanoparticles immersed in water to follow the photocatalytic reaction and report the formation of self-hydrogenated shell around nanoparticles and hydrogen bubbles when irradiated with UV light [17, 18].

8.1.4.2 Through Sample Holder

As mentioned above, modification of a TEM holder to focus light on the sample is advantageous and simpler compared to modifying a TEM column. Therefore, incorporation of light carrying optical fiber in the sample holder has also been achieved

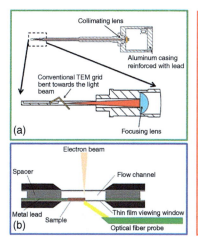
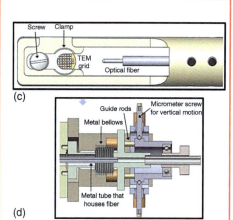

Figure 8.7 (a) Schematic cross-sectional view of the lens-based specimen holder. (Top) The feed-through on the left hosts a laser diode, which is connected through a mini-DIN connector installed in the lateral port. Two lenses (shown in blue), selected for the chosen laser wavelength, collimate, and focus the light onto the sample. (Bottom) A close-up cross-sectional side view of the tip. The sample (shown in yellow) is bent in order to allow it to be exposed to both light and electron beams. Source: Adapted from Cavalca et al. [20]. (b) Schematic of assembled liquid cell with integrated optical fiber for photo illumination for a Hummingbird Scientific liquid optical TEM holder. Source: Karki et al. [21]. (c) Top view of the tip section of a holder with another design for bringing light through optical fiber to the sample and the TEM grid clamp. (d) Motion control design; micrometer stage at the rear handle of the holder. Source: Martis et al. [19]/Cambridge University Press.

to either heat selective areas of sample, instead of entire sample, or to interrogate atomic level changes during photocatalytic reactions [8, 19, 20]. Here we describe a few holder designs that range from simple to more complicated with some similar fundamental elements, such as shining or focusing the laser light on the sample using optical fiber inserted through the sample rod, and/or using piezo drive for translating the laser spot on the sample.

One such holder design, shown in Figure 8.7a, was constructed by drilling hole through the specimen holder to feed the optical fibers enclosed in a lead cylinder [20]. The lead casing acts as a shield for the X-rays, and the TEM vacuum extends up to the optical fiber feed through. The tip of the specimen holder was also modified by axially drilling up to the sample location such that the sample is fixed in place using a single clamp. A bundle of fibers, compatible with different laser wavelengths, is used. A laser diode casing is located on axis in the feed through near the first of the two lenses used to focus the light on the sample to minimize the power losses. A five-pin mini-DIN connector is used to supply power to the diode through three contacts while the other two can be used for sample heating or biasing. The sample grid is tilted toward the light beam during the operation (Figure 8.7a). This holder was successfully employed to follow the photo-induced degradation of cuprous oxide and reduction of Pt loaded on GaN-ZnO support using a combination of HREM imaging, electron diffraction, and EELS [20]. Figure 8.7b shows schematic of another modified liquid cell holder for photocatalytic measurements of battery materials [21] that is now commercially available (https://hummingbirdscientific.com/products/optical-bulk-liquid).

An alternative design, also using optical fibers enclosed in a metal tube through a hollow sample rod, has been reported for spectrally selective photo-excitation in the TEM [19]. Here, a micro-lens, with a built-in reflector, fused with optical fiber, is placed at a working distance of ≈ 1 mm to the sample to reduce the optical flux loss (Figure 8.7c). The reflector helps to mount the fiber in required position with respect to the sample. A spot size of 20 and 100 µm was achieved using single-mode and a 50 µm multi-mode fiber, respectively. The tube is vacuum-sealed using Viton seals such that the fibers can be easily changed. The spot can be centered at the desired location on the TEM grid with help of a three-axis external micrometer stage (Figure 8.7d), located at the end specimen holder and attached to the metal tube carrying the fiber-lens assembly. Three screws and the attached guide rod to the micrometer stage enable in-situ translation and focus of the laser spot at a desired location without noticeable drift or backlash [19]. This holder is being used for photo-absorption microscopy using electron analysis on TEM platform.

Murakami et al. have developed a TEM holder with two probes, one of them could be replaced by an optical fiber to shine laser light on the sample and other can be used to measure photocurrent [22] or modify electric or magnetic property of the sample [23]. Note that the holder is designed for JEOL microscopes, where the wider tip of the sample holder has enough space to accommodate two probes. Introduction of the two probes in the sample area is through the hollow specimen holder as described above for the other holders and shown in Figure 8.8a. Both probes are mounted on two arms, arm 1 and arm 2 (Figure 8.8b) that can be independently manipulated in three dimensions, using micrometers and/or piezoelectric elements housed at the other end of the holder, that is, outside the TEM column (Figure 8.8a). Figure 8.8b shows photographs of the tip area with location of the sample, probe, and arms marked for clarity. The ends of both probes are located in the tip area of the holder near the sample and can be moved using manual micrometers (coarse adjustment) and piezo drives (fine adjustments) in X, Y, and Z directions to bring them to a desired position with respect to the sample. The micrometers drive the probes over ± 1.0 mm along the X direction and ± 0.5 mm along the Y and Z directions and the motion by piezo elements is limited to ± 3 µm in all three directions. In one of the applications, an optical fiber was placed as one the probes to focus the light on the sample and commercial Pt–Ir needles, similar to the one used in a STM, were used to measure the current. Both probes were brought in contact with two different particles on the same sample to measure the I–V curves. They found that the resistance between two Ag particles that appear to be connected in a TEM image was too large and concluded that they were actually not connected [22].

In another experiment, using this multifunctional specimen holder, keeping one probe to introduce laser, the other replaced by a glass rod (Figure 8.8c) to generate electrostatic charge by rubbing the glass probe against commercial organic photoconductors (used in electrophotography), with or without laser radiation, the charge of the sample was measured using electron holography [23]. A Mo plate was placed

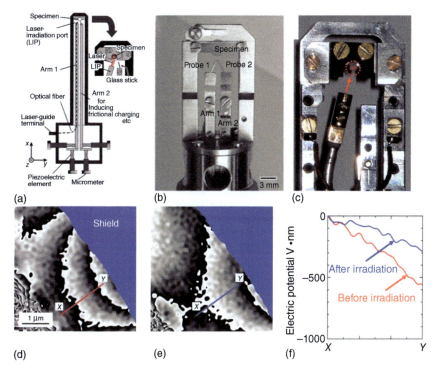

Figure 8.8 (a) Schematic illustration of a specimen holder equipped with a piezo driving probe and a laser irradiation port. An enlargement of the tip section is inset on the right-hand side [23]. (b) Optical mage of the top portion of the specimen holder showing the locations of sample, probes, and two arms. Source: Murakami et al. [22]/from AIP Publishing LLC. (c) An optical image of slightly modified design where probe 1 is replaced by an optical fiber to focus laser light and probe 2 is a glass rod [23]. (d–e) Reconstructed phase images showing the electric potential distribution around the organic photoconductor (d) after frictional charging (before laser irradiation) and (e) after laser irradiation, the phase is amplified by 3. (f) Change in the projected electric potential distribution, due to the laser irradiation. The red and blue curves represent the electric potential observed along the red and blue lines in (d) and (e). Source: Shindo et al. [23]/from Oxford University Press.

as electric shield on the organic photoconductor under observation in order to avoid emission of electron-induced secondary electrons. Formation of negative charges after rubbing the sample with glass rod was obtained by recording holograms, and the electric potential distribution around the photoconductor could be seen in the reconstructed phase image (Figure 8.8d). Time-resolved experiments confirmed that the friction generated charge did not die out for a period of three minutes, giving enough time to make controlled measurement. The change in the electric potential in the sample upon 30 seconds exposure to laser light is visible in the reconstructed phase images acquired without (Figure 8.8d) and with laser illumination (Figure 8.8e). The difference was further confirmed by measuring electrical potential in x–y directions (Figure 8.8f) along the red and blue lines marked on phase images [23]. These two experiments indicate that this multifunctional holder can be used to observe light-induced phenomenon other than photocatalysis or Raman.

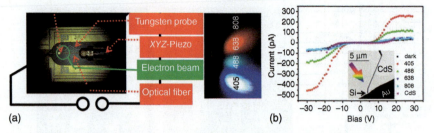

Figure 8.9 (a) Optical holder schematics: four laser diodes with 405, 488, 638, and 808 nm wavelengths were utilized for in-situ TEM optoelectronic measurements on CdS-p-Si heterojunctions. (b) A photocurrent response of an individual CdS/p-Si axial nanowire junction under a dark condition and during illumination with the light of various wavelengths with fixed intensity. The inset is the low-magnification TEM image of the CdS nanowire structure under testing. The colored arrow shows the incoming light incidence; the black arrow marks a short Si-branch segment. Source: Zhang et al. [24]/from IOP Publishing Limited.

Zhang et al. modified a piezo-driven optical holder from Nanofactory Instruments (this company is no longer in business) to introduce laser light with different wavelength to measure wavelength-dependent photo response of nanostructures (Figure 8.9a). A multimode optical fiber, that can be connected to four laser diodes with 405 nm, 488 nm, 638 nm, and 808 nm wavelengths and tunable operating power and temperature [24], was used. In this experiment, p-Si nanowires were transferred to TEM after CdS nanowires to make a junction between them. A tungsten probe was used for in-situ manipulation and making a junction between the two (inset in Figure 8.9b) using a special electron beam soldering glue. HRTEM images were used to confirm a clean junction formation between the two nanowires [24]. Optical wavelength-dependent currents, measured as function of bias, with and without light exposure for the CdS-p-Si nanowires, are shown in Figure 8.9b. The photo response for 405 nm (3.06 eV) laser is higher than for 488 nm (2.54 eV) but of the same order for the other two wavelengths. The higher response for the CdS-p-Si junction than the band gap of individual CdS and Si nanowires (2.4 and 1.5 eV, respectively) is attributed to a complex band diagram for the heterostructure that results in more light absorption at higher energy and hence more photo-induced carriers.

A modified holder, developed in collaboration with Gatan, has also been used to compare laser and electron-induced LSPR for PdH_x formation reaction [12].

Above examples provide an overview of ways to focus light on the sample to combine imaging, diffraction, and electron spectroscopies with vibrational and optical spectroscopies. These modified TEM columns or holder can also be used to follow the mechanisms of photon induced chemical reactions. A comparison between the modified column and holder is given in Table 8.1. Overall, modification to TEM column does not require any special consideration for sample geometry or choice of holders with other stimuli such as heating or cooling. On the other hand, modified holder can be used in any TEM from the same vendor or for any microscope with minimal modifications.

Table 8.1 Modifications reported to focus light on the sample for in-situ TEM experiments.

Modification type	Method	Advantages	Limitations	Reported applications	Reference
Modified TEM/Free space	Light introduced through available port	High SNR, sample holders with other stimuli can be used	Complicated alignment procedure, available port needed	Raman, CL	[4]
Modified TEM/ Optical fiber	Light introduced through available port	Simple alignment, sample holders with other stimuli can be used	Low SNR, available port needed	Raman, laser heating, photo-catalysis	[6, 7, 13]
Modified holder/ Optical fiber	Light introduced through sample holder	Simple alignment	Low SNR, other stimuli must be incorporated in the same holder	PL, photo-catalysis	[8]
Modified holder/ Optical fiber/ Multifunc-tional	Multiple optical and physical stimuli incorporated in the holder	Simple alignment, possible to incorporate other probes	Low SNR, experiment specific design	Optoelec-tronics mea-surement	[22–24]

8.1.5 Laser Alignment

Laser alignment can be divided in two parts: first is to focus it as a point source at the entrance of the TEM column (window or at the fiber bundle), the second is to focus it on the sample such that it coincides with the electron beam focus. The first part requires a set of optics, mirrors, or reflectors that will make the beam from laser source parallel before focusing it at the window or fiber bundle. This is relatively an easy step as laser path can be visually followed using appropriate eyeglasses and focus spot can be visualized using special paper. However, the second part is difficult as unlike STM probes, laser is not visible on the sample area in the TEM. The best way is to find the region that has been irreversible modified by the laser irradiation, such as burning a hole in a thin carbon film, or an irreversible transformation of the sample, find the area using stage translation, and then use the piezo drives to make the two focus spots concentric. The location of the laser spot size can then be noted for future experiments.

8.2 Correlative Approaches

Multimodal approaches, described in Section 8.1, do not entirely address the applicability of experimental challenge arising from the materials size and electron beam effects. Moreover, for gas environment experiments, pressure gap is also a limitation and liquid experiments lack stirring often needed for large-scale applications, is an issue. Alternative solution is to make experimental observations outside the TEM column, using an alternate probe with larger sampling size, possibly under identical conditions. Correlative liquid electron and optical microscopy is now a developing area of research for biological materials. While SEM is an alternative to address the size gap, X-ray diffraction, imaging, spectroscopies, and optical spectroscopies can address most of the applicability gap.

Therefore, performing experiments using any of these techniques and confirming TEM results, or vice versa, are highly desirable and have started to get traction. There are two ways to perform experiments in gas and/or liquid environments using different platforms: (i) build similar experimental setup as used for in-situ TEM experiments or (ii) use the same microfabricated sample holder for all experiments. However, the number of reports using multimodal or complementary correlative techniques for a project by the same group is still limited as most of the instrumentation capabilities are not universally available.

8.2.1 TEM and SEM

SEM provides a natural extension to the in-situ TEM/STEM experiments by narrowing the materials size gap by an order of magnitude if not more. Some of the concurrent applications of SEM imaging include EBSD, EDS, and imaging using either secondary or backscattered electrons. Unlike TEM, bulk samples can be used, and the images provide information about the topography of the sample.

One of the typical examples reported is the role of Cu_2O catalyst nanoparticles during electrochemical (EC) reduction of CO_2 into valuable commodities. The reaction ($CO_2 + H_2O$) results in producing CO, H_2, CH_4, and/or C_2H_4, based on the nature of the catalyst and applied voltage during the EC reaction [25]. The motivation was to relate electrical voltage, nanoparticle size, and morphology, to the product formed such that reaction conditions for selective production of C^{2+} can be identified. A combination of in-situ STEM imaging in liquid electrochemical cell, GC-MS to measure the products in a bench-top setup under identical conditions and ex-situ imaging using STEM and SEM was used (Figure 8.10). Three different size of Cu_2O cubes, 390, 170, and 80 nm, in CO_2 saturated 0.1 M $KHCO_3$ solution, were loaded in EC-TEM holder for three independent experiments. Time-resolved morphological changes under different electrochemical conditions, i.e. in open circuit and after applying $-1.1 V_{RHE}$, were followed by recording videos in STEM mode. Bubbles were observed to form after nine minutes of observations. It is more likely that the bubbles are the gaseous reaction products (i.e. C_2H_4, CH_4, CO, and H_2) here, rather due to electron beam interaction with water, as has been reported previously [26]. The cubes were also observed to become porous and shrink in size with smaller particles leaching out, but the bubble formation posed a problem for following the same particle for longer period as the particles were swept away by the bubbles. Therefore, $-0.9 V_{RHE}$ was applied for the subsequent experiments that increased the observation time to about 60 minutes. Images extracted from the videos after 45 minutes of applied voltage show that large cubes, 390 and 170 nm, are relatively more stable than 80 nm particles, which fragmented during the same period and under identical reaction conditions (Figure 8.10a). The fragmentation

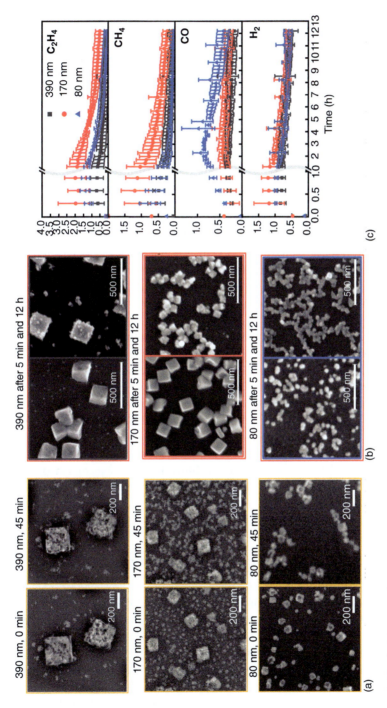

Figure 8.10 (a) Morphological evolution of Cu_2O cubes and re-deposited NPs captured under operando conditions after 45 minutes of CO_2RR in electrochemical cell-TEM. Images, shown at 0, and 45 minutes, were extracted from the videos of tracking 390, 170, and 80 nm cubes after applying a potential of $-0.9\,V_{RHE}$. Each image is an average of 10 frames. The electron flux used is $3.5\,e^-\,\text{Å}^{-2}\,s^{-1}$. (b) SEM images of Cu cubes with three different average size distributions measured ex situ after 5 minutes and 12 hours of CO_2RR over Cu_2O cubes deposited on glassy carbon. (c) Partial current densities for H_2, CO, CH_4, and C_2H_4, obtained from time-resolved gas chromatography data acquired over 12 hours. The currents are normalized by the exposed surface area of the as-synthesized Cu_2O cubes. Note that this normalization does not consider the dynamic morphological changes observed under reaction conditions. Source: Grosse et al. [25]/from Springer Nature, CC BY 4.0.

leading to an increased number of nanoparticles in the samples with 80 nm cubes was further confirmed by ex-situ SEM images, acquired after longer reaction time than in-situ experiments (Figure 8.10b).

Ex-situ TEM analysis of the cubes after reaction confirmed the dynamic morphological change observed can be related to the partial reduction of Cu_2O to Cu resulting in the fragmentation and re-deposition of smaller nanoparticles. Distribution of various products, namely, CO, H_2, CH_4 and/or C_2H_4, as measured using a GC system attached the benchtop reactor, show that the production of CH_4 and C_2H_4, decreased over 12 hours period at $-1.1\,V_{RHE}$ (Figure 8.10c). Furthermore, the production of these two species was observed to be highest for the sample with 170 nm particles and could be related to higher number of catalytic particles as observed in the ex-situ SEM image after 12 hours period (Figure 8.10b). These correlated measurements elucidate that C^{2+} production is controlled by the initial cube size and surface coverage. The formation of porous catalyst particles and re-deposition of leached material can sustain the selectivity to produce hydrocarbons. It is clear from these results that the dynamics of morphological changes in catalyst particles, leading to its activity and selectivity, is revealed using in-situ TEM images, but the observation period was restricted due to electron beam effects, a practical limitation to experiments that require lengthy observation period. However, these changes can be related to ex-situ measurements of reaction products and SEM imaging over a longer time and length scale, respectively.

Liu et al. have reported a detailed investigation of structural evolutions in K403 superalloys, using XRD, SAED, EBSD, SEM, and STEM imaging, and spectroscopy (EDS), during heating up to 950 °C [27]. K403 is a Ni–Cr-based superalloy and used for high-temperature applications such as turbine rotor blades below 900 °C. Therefore, it is important to understand the structural evolution with temperature to predict the failure mechanism and thereby enhance the reliable lifetime for the parts made from K403 alloy. The formation of different strengthening phases, dislocation, and particle coarsening were followed using in-situ XRD and compared using SEM, EBSD, TEM, HAADF-STEM images, electron diffraction, and elemental maps [27]. This work is an excellent example of combining various imaging, diffraction, and spectroscopies to obtain detailed information about the structural and morphological changes as a function of temperature at different length scale.

8.2.2 Electron and X-ray Microscopies and Spectroscopies

Electrons, photons, and X-rays are three main sources used to probe the morphology, crystal structure, and/or chemistry of materials. Photons used for optical microscopy have limited applications for nanomaterials due to spatial resolution limit. X-rays, on the other hand, are ideally suited to fill the materials size gap limitations of the electron microscopies. X-ray diffraction is an ensemble method, routinely used to check the overall quality of as synthesized material with respect to its crystallinity, structure, and particle size. With the development and availability of synchrotron facilities, the applications of X-rays have extended to imaging and spectroscopies with reasonable spatial resolutions [28, 29]. Although, these two techniques are complementary to each other to establish synthesis, structure–property relationship at nano- and micro-scale [30], examples of correlative study using them by the same group are limited. On the other hand, different groups have exploited one or the other technique for the same material system as access to both techniques at the same place is limited.

One of the recent advancements is the development of transmission or scanning transmission X-ray microscopy (TXM/STXM) at synchrotron facilities, which can be considered comparable to TEM/STEM, with lower spatial but high energy resolution for spectroscopy. Although, a detailed description of STXM is beyond the scope of this book, a simple functioning of the setup is provided to understand the similarities and differences between the two. Figure 8.11a shows a setup for STXM that was used for in-situ observations of gas–solid reactions using a modified gas cell. A conventional high-energy-resolution synchrotron beam line provides high brightness illumination at a Fresnel zone plate that also acts as condenser.

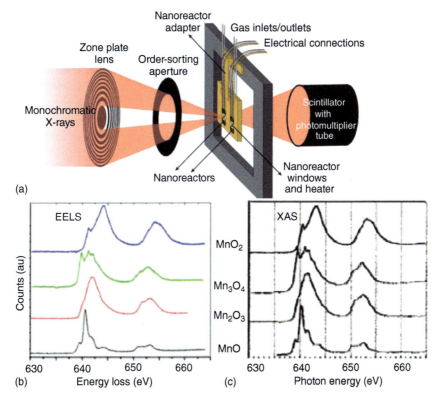

Figure 8.11 (a) Diagram of the in-situ STXM technique. Soft X-ray light is focused on the sample using a Fresnel-type zone plate lens. An order-sorting aperture filters out higher-order diffraction peaks. The nanoreactor containing the sample is placed in an adaptor that holds up to two nanoreactors at the same time. The adaptor can be translated with nanometer precision by an interferometrically controlled (x, y, z) piezoelectric stage, allowing the acquisition of raster scans. Source: Smit et al. [31]/Springer Nature. (b) Fine structure at the Mn $L_{2,3}$ edges of Mn oxides ELNES (EELS) at 0.2 eV resolution and (c) XANES (XAS) with 0.1 eV resolution show excellent agreement.

The sample (nanoreactor in Figure 8.11a) is placed at the focus of the zone plate and transmitted signal is detected using either single photon counting or current detector [28, 31, 32]. Modern STXM setup includes a mechanical interferometry-based control system to align various X-ray absorption spectra (XAS), which includes EXAFS and X-ray absorption near-edge structure (XANES), for core edges of different elements to provide large ranges for microanalysis. Compared with TEM/STEM, it is better suited for chemical analysis of relatively large sample areas with higher spectral (energy) resolution and lower radiation damage [33]. Both XANES and EXAFS provide detailed information about the chemical state and coordination environment, as the former represents the dipole-induced electronic transitions near absorption edges, and latter reflects the oscillations observed after core edge in the extended spectrum that arises from near neighbor interactions [30]. However, the spatial resolution is much lower than STEM, around 20 nm, depending on the energy of the X-rays used and is limited by our ability to fabricate outer zone with a width of less than 10 nm. The spatial resolution limit, imposed by X-ray optics, can be overcome by using diffraction-based methods (e.g. ptychography), and chemical maps with 5 nm resolution have also been reported [34].

Therefore, STEM-EELS and STXM-XAS offer a comparable platform for in-situ investigation of the chemistry of nanoparticles under reaction conditions. Whereas using EELS we measure the energy lost by an electron as it travels through the sample, XAS is used measure the energy absorbed the sample to determine the chemistry. In both cases, energy lost (by electrons) or absorbed (for X-ray) is used to promote an atomic electron from its initial state to an unoccupied state that is not only element-specific but also depends on the bonding environment and oxidation state. Whereas the energy loss near-edge structure (ELNES) of an element represent its oxidation state, similar information is obtained from XANES (Figure 8.11b,c). However, it is important to note that number of peaks in the two spectra is different due to different energy resolution achievable in either case. Whereas the EELS resolution can be improved by using a monochromated electron source, which can make the energy resolution for EELS and XAS comparable, the accessibility to a larger length scale even with lower spatial resolution for XAS is an advantage.

Based on the criteria explained above, it is conceivable that STEM and STXM should become popular for in-situ investigations, especially for gas–solid reaction, using portable nano or microreactors that are modeled after microfabricated gas cell for TEM/STEM applications or building an independent system, as described below. The application of STXM is limited mainly due to a limited number of synchrotron facilities available around the world.

8.2.2.1 Portable Reactor for Various Platforms

Gas-cell reactors, modified using TEM holder designs, have also been used for SEM and X-ray platforms. One such reactor, using the microfabricated chips from DENSsolutions, was used to perform complementary (to TEM) X-ray ptychography, and X-ray fluorescence measurements of catalysts, and other functional materials using synchrotron radiation (Figure 8.12 [35]. The reactor setup, as used at the

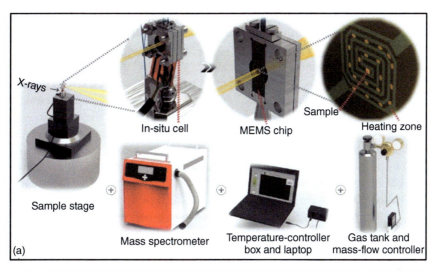

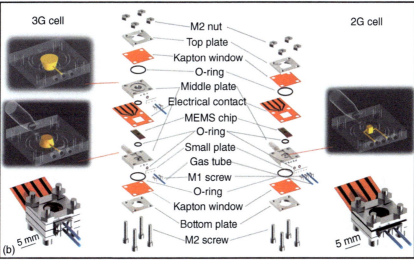

Figure 8.12 (a) Illustration of the in-situ setup for 2D and 3D ptychography at the P06 nanoprobe end station of PETRA III. Zooming into the in-situ cell (top), and gas and temperature control infrastructure and product analytical components (bottom).
(b) Illustration of the disassembled 2G (right) and 3G (left) cell with their internal gas-flow volumes (highlighted in yellow). The fully assembled setup is shown at the bottom. Source: Fam et al. [35]/International Union of Crystallography/CC BY-4.0.

synchrotron facility, PETRA III (Hamburg, Germany) consists of sample stage, which holds the in-situ cell with heating chip, connected to gas cylinder through gas inlet tube, and the gas outlet is connected to a mass spectrometer (Figure 8.12a). The electrical connections are used to heat the chip by Joule heating. The sample can be heated up to 1300 °C in 100 kPa pressure with 23 nm spatial resolution in hard X-ray ptychography images. A limited range of ±35° tilt for ptychographic tomography is achievable within the reactor and ±65° on sample loaded on chip only. The gas flow, pressure, and temperature are controlled using a computer-based program that also measures the temperature of the sample and gas composition at the outlet [35].

A detailed construction of the nanoreactor is shown in Figure 8.12b, where each component and their assembly process for the two types of cells are clearly marked. The main difference between the "2G" and "3G" cell is the additional plate used in the assembly of the latter that increases the internal volume of the cell from ≈ 2.6 to 24.5 ml. Also, 2G cell is suited for 2D imaging, diffraction, and XRF, while 3G cell allows limited tilt angle (±35°) for ptychographic X-ray computed tomography (PXCT). The 2G nanoreactors were successfully used to acquire X-ray ptychography and TEM images using the same microfabricated chip, to follow the morphology of Au nanoparticles in He gas environments during heating form RT to 650 °C and annealing for 30 minutes. The two complementary experiments performed under similar experimental conditions demonstrated the ability to probe the sample at multiple length scales. The application of PXCT, within limited angular range, is exhibited using zeolite under in-situ and reaction conditions as an example of catalyst and functional material. It is feasible to extend the application of these reactors to compare elemental compositions using XRF and EDS, determine crystal structures from SAED and XRD and oxidation states from XAS and EELS data.

Zhao et al. have also reported a microreactor, based on Hummingbird Scientific TEM holder design, compatible with various probes other than electrons (TEM), e.g. X-ray and optical (IR), to observe and measure gas–solid reactions under similar conditions (Figure 8.13a). This reactor is similar to the gas-cell TEM holder chip but modified to make it compatible to be used in synchrotron source at Brookhaven

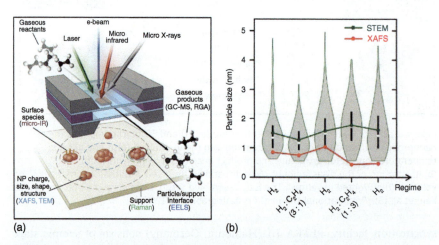

Figure 8.13 (a) Schematic of the microcell: the catalyst is confined between two silicon nitride windows, and the reaction gases flow through the system to interact with the confined catalyst at atmospheric pressure. The four different beams (electron, X-ray, IR, and laser) can probe different parts of the catalytic system. Source: Zhao et al. [36]/John Wiley & Sons. (b) The experimental results from XAFS and STEM: Combined EXAFS and STEM results are displayed under varying gas compositions. Source: Li et al. [37]/Springer Nature/CC BY 4.0.

National Laboratory (BNL, USA) for X-ray and IR micro-spectroscopies and is connected to a GC-MS system to measure the reaction products [36]. Total sample loading, gas pressure and temperature conditions can be kept identical while nano to micro-scale characterization is performed by changing the probed area. This reactor was used to follow the Pt nanoparticles during catalytical hydrogenation of ethylene using EXAFS data at synchrotron radiation and STEM imaging [37]. They followed the particle size evolution as the feed gas composition was varied from $H_2 \rightarrow (3:1)\, H_2{:}C_2H_4 \rightarrow H_2, \rightarrow (1:3)\, H_2{:}C_2H_4$, and found that the particle size measured by EXAFS and STEM imaging was not the same (Figure 8.13b). The observed discrepancy could be explained by recognizing the fact that while average Pt–Pt coordination number in sample area is used for EXAFS measurements, only clusters of Pt atoms above a certain size, visible in STEM images due to reduced spatial resolution when using windowed gas cell, are used to obtain the particle size distribution. In other words, the two methods may not be directly comparable due to complex structural dynamics of supported metal catalysts and special analytical approaches, such as weighted average, is needed to reconcile the results [37]. Ethylene hydrogenation reaction with Pd as catalyst was also followed using similar protocol where a metal hydride phase was observed to form in hydrogen-rich environment. The hydride phase transformed to carbide phase in olefin-rich environment via an intermediary metallic phase. The metal particles acted as a catalyst to decompose olefin to produce C atoms that were used for metal carbide formation [38].

Another example includes a combination of SEM imaging, EDS mapping, ETEM-STEM imaging, EELS, and STXM ptychography to understand the stability of Cu-based core ($CuO/ZnO/Al_2O_3$) with zeolite (H-ZSM-5) shell catalyst during dimethyl ether synthesis in reducing environment [39]. Baier et al. report using a microfabricated modified TEM gas cell based on Protochips, Inc. design for in-situ X-ray ptychography experiments at the PETRA III (Hamburg, Germany) synchrotron facility [40]. STEM images of as prepared 100 nm thick Cu-containing (core)-zeolite (shell) sample, recorded at RT in H_2, show the core–shell structure as bright and dark regions, respectively (Figure 8.14a, top row). The contrast in STEM images is sensitive to the distribution Cu rich particles within the core, whereas a uniform outer gray region represents zeolite shell as marked on the images (Figure 8.14). Disintegration of copper core into smaller bright particles in 1.1 mbar H_2 at 250 °C is accompanied by a reduction in core volume and intensities of Cu $L_{2,3}$ peaks (Figure 8.14b, top row). Their observations indicate the reduction of CuO to Cu was reversible as both the image and EELS data upon oxidation in 3.2 mbar of O_2 is similar as of the starting material (Figure 8.14c, top row). However, the shell remained intact during the redox reaction [39].

For in-situ ptychography experiments, 300 to 400 nm lamella were prepared, transferred to the microfabricated chips, and loaded in the modified gas cell. Thicker samples can be used due to the higher penetration power of X-rays that addresses the materials size gap of the TEM observations. However, phase-contrast images obtained by X-ray ptychography remained unchanged at length scales up to 1 μm in both reducing and oxidizing environments. It is possible that the nanoscale changes observed in ETEM images at 250 °C could not be resolved by in-situ ptychography.

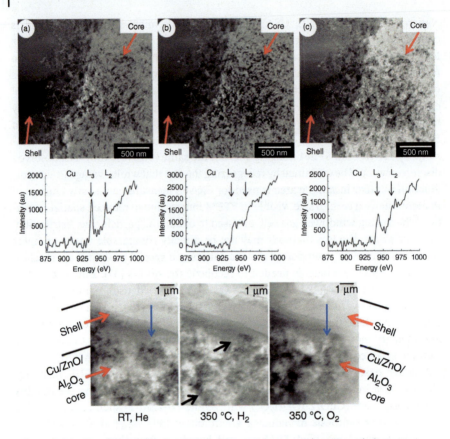

Figure 8.14 Top: Scanning transmission electron microscopy images and electron energy-loss spectra: (a) as-prepared catalyst studied at room temperature (RT) in H_2 by environmental transmission electron microscopy; (b) catalyst under reducing conditions at 250 °C in 1.1 mbar of H_2; (c) catalyst under oxidizing conditions at 250 °C in 3.2 mbar of O_2. Bottom: phase contrast images obtained by in-situ X-ray ptychography under different gas atmospheres and at different temperatures. Source: Baier et al. [39] /from Microscopy Society of America.

However, changes in the core structure became pronounced at 350 °C in reducing environments (marked by black arrows in the center image in the bottom row of Figure 8.14). X-ray ptychography images in oxidizing environment revealed further changes in the core structure, evidenced by an increase dark area, instead of returning to the original form. Authors believe it could be arising from the ptychography reconstruction artifacts, but their cumulative results emphasize that TEM images to be most suitable to observe nanoscale changes in the material.

8.2.2.2 Independent Correlative Measurements

Alternative approach is to make independent measurements using electron and X-ray sources to address the materials size gap. For example, Tsoukalou et al. have investigated the structural transformation of In_2O_3 catalyst nanoparticles during CO_2 hydrogenation to methanol as function of time on stream (TOS) and reaction

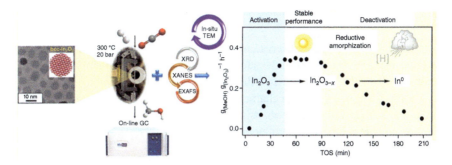

Figure 8.15 Summary chart showing the TEM image of synthesized particles and atomic arrangement, in-situ gas cell connected to a GC, correlative techniques (TEM, XRD, XANES, EXAFS) used for in-situ characterization and results summary graph showing the relationship between the catalyst structure and reaction rate. Source: Tsoukalou et al. [41]/from American Chemical Society.

temperature [41]. They combined in-situ XRD, ETEM imaging, and diffraction with XAS (XANES and EXAFS) and GC to relate the methanol production to the structural changes to determine the active component. Figure 8.15 summarizes the catalyst ensemble, reactor setup at synchrotron beam line, investigative probes, and methanol production as a function of TOS of the reactive gases. Structural transformation with time in $H_2:CO_2$ (3:1) mixture (20 bar) was followed using XRD and HREM imaging (800 mbar) at 300 °C. Despite pressure difference and electron beam effects, crystalline nanoparticles were observed to become amorphous, due to the reduction of In_2O_3 to metallic indium that melts below 200 °C, in both experiments.

X-ray experiments were also used to follow the chemical changes in the catalyst particles using XANES and EXAFS as well as measure the methanol production as a function of TOS using GC (Figure 8.15). The change in the population In—In and In—O bonds with time was used to measure the In/In_2O_3 ratio in the sample and compared it with the methanol production rate. Combined results show four reaction regimes, incubation, linear growth in product, plateau, and decline in product formation. Time-resolved XRD, TEM, and XAS results reveal a negligible change in catalyst structure in the beginning (incubation period), with start of reduction of In_2O_3 resulted an increase in vacancy formation (In_2O_{3-x}) (reaction rate growth) and finally converting to amorphous material (decline in product formation), indicating complete reduction of oxide to metal (Figure 8.15). The comparison of structural change with methanol production indicates that partially reduced In_2O_3, i.e. with vacancies, improves the reactivity of the catalyst [41]. The converging results highlight the complimentary nature of the two techniques that addressed both the materials and pressure gap for applicability of the results to large scale processes.

Recently, in-situ liquid cell observation has also been combined with ex-situ time-resolved XRD, STEM imaging, EXAFS, and atom probe tomography to understand the growth mechanism and conditions for cubic and hexagonal phases of sodium yttrium fluoride ($NaYF_4$) [42].

8.3 Take Home Messages

Comparative experimental observations are required to address the materials size gap and electron beam effects to address the limitations of in-situ TEM/STEM results. Multimodal experiments on the TEM platform can be used to address electron beam effects using another characterization probe, such as ions or photons, but designing correlative experiments using TEM and X-rays at synchrotron facilities is more suitable to address the materials size gap and provide comparable results, especially for microanalysis [30, 32].

Ion and light sources can be introduced using an available port on the TEM but requires knowledge of the internal stage design and dimensions. Therefore, making multimodal TEM holders is more popular.

A number of groups have modified TEM holders to irradiate sample with white light or laser to follow photon-induced reaction by high-resolution imaging and spectroscopy. Basic steps for building these holders are quite similar, i.e. (i) a hollow rod for the TEM side-entry holder is used to introduce optical fiber and electrical wiring into the column; (ii) optical fiber or bundle is selected based on the length, diameter, protective jacket, etc. for optimum performance; (iii) the number of joints between optical fiber and lens/reflector is kept to minimum; (iv) combination of micrometer (coarse) and piezo drive (fine), located at the outer end of the holder, is used to align the light on the sample; (v) light source (laser diode) is mounted on a separate platform and focused on the optical fiber using a combination of lenses.

Currently, spatial resolution for STEM imaging and EELS is much higher (near atomic level) compared with STXM and XAS, while the reverse is true for energy resolution. However, the energy resolution for STEM-EELS is improving (even for core-loss edges) by using monochromated electron sources. Similarly, the diffraction limit for spatial resolution of STXM imaging and XAS is being overcome by using ptychography.

References

1 Yedra, L., Eswara, S., Dowsett, D., and Wirtz, T. (2016). In-situ isotopic analysis at nanoscale using parallel ion electron spectrometry: a powerful new paradigm for correlative microscopy. *Scientific Reports* 6 (1): 28705.

2 Wirtz, T., Philipp, P., Audinot, J.-N. et al. (2015). High-resolution high-sensitivity elemental imaging by secondary ion mass spectrometry: from traditional 2D and 3D imaging to correlative microscopy. *Nanotechnology* 26 (43): 434001.

3 Ohno, Y. and Takeda, S. (1995). A new apparatus for in situ photoluminescence spectroscopy in a transmission electron microscope. *Review of Scientific Instruments* 66 (10): 4866–4869.

4 Picher, M., Mazzucco, S., Blankenship, S., and Sharma, R. (2015). Vibrational and optical spectroscopies integrated with environmental transmission electron microscopy. *Ultramicroscopy* 150: 10–15.

5 Yoon, Y., W-CD, Y., Ha, D. et al. (2019). Unveiling defect-mediated charge-carrier recombination at the nanometer scale in polycrystalline solar cells. *ACS Applied Materials & Interfaces* 11 (50): 47037–47046.
6 Allen, F.I., Kim, E., Andresen, N.C. et al. (2017). In situ TEM Raman spectroscopy and laser-based materials modification. *Ultramicroscopy* 178: 33–37.
7 Xiang, B., Hwang, D.J., In, J.B. et al. (2012). In situ TEM near-field optical probing of nanoscale silicon crystallization. *Nano Letters* 12 (5): 2524–2529.
8 Liu, C., Ma, C., Xu, J. et al. (2021). Development of in situ optical spectroscopy with high temporal resolution in an aberration-corrected transmission electron microscope. *Review of Scientific Instruments* 92 (1): 013704–013704.
9 Narayan, T.C., Hayee, F., Baldi, A. et al. (2017). Direct visualization of hydrogen absorption dynamics in individual palladium nanoparticles. *Nature Communications* 8: 14020.
10 Yang, W.-C.D., Wang, C., Fredin, L.A. et al. (2019). Site-selective CO disproportionation mediated by localized surface plasmon resonance excited by electron beam. *Nature Materials* 18 (6): 614–619.
11 Wang, C., W-CD, Y., Bruma, A. et al. (2021: p. . Under review). Endothermic reaction at room temperature enabled by deep-ultraviolet plasmons. *Nature Materials* 20 (3): 346–352.
12 Sytwu, K., Vadai, M., Hayee, F. et al. (2021). Driving energetically unfavorable dehydrogenation dynamics with plasmonics. *Science* 371 (6526): 280–283.
13 Swearer, D.F., Bourgeois, B.B., Angell, D.K. et al. (2021). Advancing plasmon-induced selectivity in chemical transformations with optically coupled transmission electron microscopy. *Accounts of Chemical Research* 54 (19): 3632–3642.
14 Yoshida, K., Nozaki, T., Hirayama, T. et al. (2007). In situ high-resolution transmission electron microscopy of photocatalytic reactions by excited electrons in ionic liquid. *Journal of Electron Microscopy* 56 (5): 177–180.
15 Miller, B.K. and Crozier, P.A. (2013). System for in situ UV–visible illumination of environmental transmission electron microscopy samples. *Microscopy and Microanalysis* 19 (2): 461–469.
16 Yoshida, K., Yamasaki, J., and Tanaka, N. (2004). In situ high-resolution transmission electron microscopy observation of photodecomposition process of poly-hydrocarbons on catalytic TiO_2 films. *Applied Physics Letters* 84 (14): 2542–2544.
17 Zhang, L., Miller, B.K., and Crozier, P.A. (2013). Atomic level in situ observation of surface amorphization in anatase nanocrystals during light irradiation in water vapor. *Nano Letters* 13 (2): 679–684.
18 Lu, Y., Yin, W.-J., Peng, K.-L. et al. (2018). Self-hydrogenated shell promoting photocatalytic H_2 evolution on anatase TiO_2. *Nature Communications* 9 (1): 2752.
19 Martis, J., Zhang, Z., Li, H.-K. et al. (2021). Design and construction of an optical TEM specimen holder. *Microscopy Today* 29 (5): 40–44.
20 Cavalca, F., Laursen, A.B., Kardynal, B.E. et al. (2012). In situ transmission electron microscopy of light-induced photocatalytic reactions. *Nanotechnology* 23 (7): 075705.

21 Karki, K., Kumar, P., Verret, A. et al. (2020). In situ/operando study of photoelectrochemistry using optical liquid cell microscopy. *Microscopy and Microanalysis* 26 (S2): 2446–2447.

22 Murakami, Y., Kawamoto, N., Shindo, D. et al. (2006). Simultaneous measurements of conductivity and magnetism by using microprobes and electron holography. *Applied Physics Letters* 88 (22): 223103.

23 Shindo, D., Takahashi, K., Murakami, Y. et al. (2009). Development of a multifunctional TEM specimen holder equipped with a piezodriving probe and a laser irradiation port. *Journal of Electron Microscopy* 58 (4): 245–249.

24 Zhang, C., Xu, Z., Tian, W. et al. (2015). In situ fabrication and optoelectronic analysis of axial CdS/p-Si nanowire heterojunctions in a high-resolution transmission electron microscope. *Nanotechnology* 26 (15): 154001.

25 Grosse, P., Yoon, A., Rettenmaier, C. et al. (2021). Dynamic transformation of cubic copper catalysts during CO_2 electroreduction and its impact on catalytic selectivity. *Nature Communications* 12 (1): 6736.

26 Schneider, N.M., Norton, M.M., Mendel, B.J. et al. (2014). Electron–water interactions and implications for liquid cell electron microscopy. *The Journal of Physical Chemistry C* 118 (38): 22373–22382.

27 Liu, J., Li, J., Hage, F.S. et al. (2017). Correlative characterization on microstructure evolution of Ni-based K403 alloy during thermal exposure. *Acta Materialia* 131: 169–186.

28 Howells, M., Jacobsen, C., Warwick, T. et al. (2007). Principles and applications of zone plate X-ray microscopes. In: *Science of Microscopy* (ed. P.W. Hawkes and J.C.H. Spence), 835–926. New York, NY: Springer New York.

29 Kirz, J. and Rarback, H. (1985). Soft X-ray microscopes. *Review of Scientific Instruments* 56 (1): 1–13.

30 Alayoglu, S. and Somorjai, G.A. (2015). Nanocatalysis II: in situ surface probes of nano-catalysts and correlative structure–reactivity studies. *Catalysis Letters* 145 (1): 249–271.

31 de Smit, E., Swart, I., Creemer, J.F. et al. (2008). Nanoscale chemical imaging of a working catalyst by scanning transmission X-ray microscopy. *Nature* 456 (7219): 222–225.

32 Goode, A.E., Porter, A.E., Ryan, M.P. et al. (2015). Correlative electron and X-ray microscopy: probing chemistry and bonding with high spatial resolution. *Nanoscale* 7 (5): 1534–1548.

33 Hitchcock, A.P., Dynes, J.J., Johansson, G. et al. (2008). Comparison of NEXAFS microscopy and TEM-EELS for studies of soft matter. *Micron* 39 (3): 311–319.

34 Acremann, Y., Strachan, J.P., Chembrolu, V. et al. (2006). Time-resolved imaging of spin transfer switching: beyond the macrospin concept. *Physical Review Letters* 96 (21): 217202.

35 Fam, Y., Sheppard, T.L., Becher, J. et al. (2019). A versatile nanoreactor for complementary in situ X-ray and electron microscopy studies in catalysis and materials science. *Journal of Synchrotron Radiation* 26 (5): 1769–1781.

36 Zhao, S., Li, Y., Stavitski, E. et al. (2015). Operando characterization of catalysts through use of a portable microreactor. *ChemCatChem* 7 (22): 3683–3691.

37 Li, Y., Zakharov, D., Zhao, S. et al. (2015). Complex structural dynamics of nanocatalysts revealed in Operando conditions by correlated imaging and spectroscopy probes. *Nature Communications* 6 (1): 7583.

38 Zhao, S., Li, Y., Liu, D. et al. (2017). Multimodal study of the speciations and activities of supported Pd catalysts during the hydrogenation of ethylene. *The Journal of Physical Chemistry C* 121 (34): 18962–18972.

39 Baier, S., Damsgaard, C.D., Klumpp, M. et al. (2017). Stability of a bifunctional Cu-based core@zeolite shell catalyst for dimethyl ether synthesis under redox conditions studied by environmental transmission electron microscopy and in situ X-ray ptychography. *Microscopy and Microanalysis* 23 (3): 501–512.

40 Baier, S., Damsgaard, C.D., Scholz, M. et al. (2016). In situ ptychography of heterogeneous catalysts using hard X-rays: high resolution imaging at ambient pressure and elevated temperature. *Microscopy and Microanalysis* 22 (1): 178–188.

41 Tsoukalou, A., Abdala, P.M., Stoian, D. et al. (2019). Structural evolution and dynamics of an In_2O_3 catalyst for CO_2 hydrogenation to methanol: an operando XAS-XRD and in situ TEM study. *Journal of the American Chemical Society* 141 (34): 13497–13505.

42 Bard, A.B., Zhou, X., Xia, X. et al. (2020). A mechanistic understanding of nonclassical crystal growth in hydrothermally synthesized sodium yttrium fluoride nanowires. *Chemistry of Materials* 32 (7): 2753–2763.

9

Data Processing and Machine Learning

> *Without big data, you are blind and deaf and in the middle of a freeway.*
> – Geoffrey Moore

Data processing has always been a part of transmission electron microscopy (TEM) image and spectroscopy analysis. In the past, we had selected one image or spectroscopy data and extracted the information using manual analysis methods available. However, manual analysis of selective data is not applicable for in-situ measurements as we must analyze each and every image and/or spectrum recorded to decipher the transformation mechanisms as a function of applied stimulus and time. The situation has become worse as our quest for combining high spatial and temporal resolution has started to become reality. Now we are collecting thousands of images in 1 second by using direct electron (DE) detectors that translates into data size in the order of GBs s^{-1}. Therefore, we must start to take advantage of machine learning (ML) to tackle the so called "big data" problem. We need to transfer, save, and analyze the data collected in a smart and effective way. In this chapter, we will revisit the historical aspect of image/data processing, understand the basic concepts of ML, open-source tools available, select applications, future needs, and limitations. Note that ML is a part of artificial intelligence (AI) with vast number of possible applications that fall outside the scope of this book. The objective of this chapter is to describe the basic principles, open sources available and provide references that can guide you to find appropriate methods suitable to your needs.

9.1 History of Image Simulation and Processing

Image simulations and processing form an integral part of our ability to decipher TEM/scanning transmission electron microscopy (STEM) data with confidence. Atomic scale images and/or chemical maps require unambiguous interpretation of the data to obtain conclusive evidence about the nature of dynamic changes and their relationship to the properties. Our goal here is to understand the fundamentals behind the image processing and simulations as they are now being accomplished using ML applications (see Section 9.4).

9.1.1 Image Simulations

Image matching has been fundamental requirement of image interpretation process since the contrast in TEM images is a function of the imaging conditions, such as crystal alignment, beam tilt, defocus, astigmatism, etc. After the acquisition of first structural resolution images that appeared to have one-to-one resemblance to the atomic structure, as determined by X-ray diffraction methods, it became obvious that the validity of this interpretation must be verified by some independent means. Therefore, image simulation, using dynamical multi-slice image formulation, suggested by Cowley and Moodie's [1], was developed and used to confirm the image interpretation [2]. Simulated images, using the available crystal structural information, and matching it to the experimental TEM image, can be considered a rudimentary step of image recognition. However, the first simulation method was quite slow as it required to include multiple beam interactions that traveled through the sample. Later, Ishizuka developed a different scheme using fast Fourier method that reduced simulation time [3]. Further reduction for simulation time was achieved by using real space calculation [4] based on theory proposed by van Dyke [5]. However, for in-situ experiments, the full potential of computer simulations is relevant only if an immediate comparison between the computed and experimental image can be performed "live" during an experiment, requiring accurate and rapid algorithms. Pierre Stadelmann from EPFL has been working on developing platforms to reduce the simulation time to seconds for both TEM and STEM images, using Bloch waves [6, 7]. It is interesting to note that some of these simulation programs use the ML algorithms developed by Cooley and Turkey in 1965 [8]. However, the intensities in the two images were found to differ by a factor of 3 to 4 that has been known as Stobbs factor (named after one of the authors) [9]. It has been argued that thermal diffuse scattering, present in the experimental images, not being included in the simulations, kept us from making a quantitative comparison between the two images [10]. There has been continued effort to find methods, such as incorporating an Einstein frozen-phonon model of thermal diffuse scattering to decrease or mitigate the Stobbs factor over the years, without major success [11, 12]. We should keep this in mind as image matching is important step when using ML-based networks or algorithms. On the other hand, Stobbs factor is not as critical for annular dark field (ADF) or STEM images, and further improvements to the simulation algorithms have been made as explained in Section 9.2.

9.1.2 Image Processing

Image processing algorithms were developed to improve the contrast and thereby confidence in interpreting TEM images during 1970s. Most widely used program during 1980s, Semper, based on Fortran (commonly used programming language then), was developed by Saxton et al. from Cambridge University [13]. The goal of the image processing routines was to improve the image contrast such that atomic position or structural information became distinguishable from the background noise. Fundamental idea behind semper was to divide various processing steps into small sub-routines that can be run on smaller machines, thus avoiding the need of large computer systems. It was a flexible and modular program and had various application areas besides TEM image processing. Users could define their own processing steps, drift correction routines, choose filters, imaging conditions, etc. from a set of commands defined in Semper to obtain images with reduced noise [14]. However, with the advent of simpler and widely available programing platforms, which included some of these processing steps, such as Digital Micrograph (DM) from Gatan and ImageJ from NIH (publicly available free program), it slowly lost its utility for image processing.

Both DM and ImageJ, among others, provide from simple to more complicated steps, for improving image contrast by applying filters, water-shedding, particle size measurements, image drift correction, etc. Several filters such as Gaussian Filter (based on Gaussian distribution of image intensities), Mean filter (using simple arithmetic average of neighboring pixel intensities), Top Hat filter (shape with flat top) are available to choose from. ImageJ is also an open-source program and allows users to define their own filters and small routines to process large set of data. Also regression analysis can be used to find the noise function and remove it from the images [15].

Another image processing method involves calculating the FFT (fast Fourier transform or power spectrum) of the image and mask the diffuse intensities around the central spot, then calculate reverse FFT to obtain an image of the noise, which can be subtracted from the original experimental image to remove the noise. Alternately, a donut-shaped mask can be placed around each bright spot (equivalent to Bragg spot) in the FFT, and the inverse FFT forms an image using only the selected spots, thus filtering out the noise.

This method can also be used to extract structure of individual layers, stacked in the electron beam direction. Sidorov et al. have reported using it to determine a complicated structure in which three types of Hg lattice were found to be intercalated in the TiS_2 layered structure, resulting in a long range order along c-axis (electron beam direction) [16]. They observed the presence of a long-range modulation in the images (Figure 9.1a), giving rise to satellite spots in diffraction as well as in the FFT from thick section of the image (Figure 9.1b). After mask operations (Figure 9.1c,d), the long-range periodicity became clearly visible (Figure 9.1e). Thus, three stacking variants of Hg sub-lattice could be identified in the inverse transform images formed by selecting the appropriate Bragg reflections in the FFT (Figure 9.1g,h). Note that this simple operation can also be used to extract structure of individual layers in stacked 2D materials, find the angle between them or identify the composition of multicomponent stacked heterostructures. Furthermore, it can assist in following intercalation and deintercalation mechanisms relevant to battery materials.

Geometric phase analysis (GPA) technique, also based on using inverse FFT, has been used to isolate phase images of two domains present in a $PbTiO_3$ crystal and quantitatively measure the strain generated at the domain wall [17]. Zhu et al. have tested the accuracy of this method for aberration corrected STEM images and report that with a properly chosen Fourier mask, size, and strain profile direction/width, the lattice displacements can be measured with 1 pm accuracy [18].

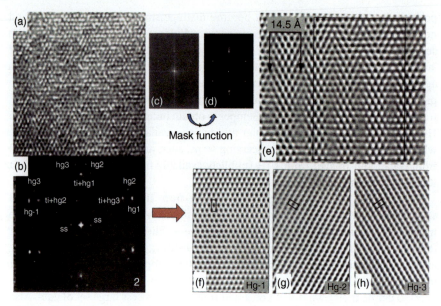

Figure 9.1 Separation of the host and Hg sublattices by Fourier image processing incorporating only the Bragg peaks associated with these sublattices. (a) A noisy experimental HRTEM image, recorded during in-situ deintercalation of Hg, showing long-range modulation. (b) FFT of the image showing satellite Bragg spots belonging to different location of Hg in TiS_2 layers. The main hexagonal set of peaks represents the periodic component of the image. (c and d) The mask operation. (e) Fourier-filtering helped to clearly show the long-range modulation of the lattice. (f–h) The three orientational variants of the Hg sublattice extracted from the FFT of the image in (a). Intralayer sublattice unit cells are outlined in images (f–h). Source: Sidorov et al. [16]/from American Chemical Society.

Before the advent of nanoprobe forming electron sources, FFT was routinely used to identify the structure of nanoparticles and their transformation during nucleation and growth [19]. The method is similar to the one used to obtain structural information using SAEDP and requires existence of two-dimensional lattice vector and knowledge of chemical elements present. The d-spacing and the measured angle between the two lattice vectors is then used to identify the structure. It can be a time-consuming process as multiple compounds containing same elements with similar d-spacing are often possible. For example, there exist a number of Fe oxides (Fe, FeO, Fe_2O_3, Fe_2O_5, etc.) and carbides (Fe_2C, Fe_3C, Fe_2C_5, etc.), with similar d-spacing but different composition and structure. Therefore, algorithms were developed that find a match between the measured lattice data (d-spacing and angles) and a structure reported in International Centre for diffraction Data (ICDD or similar library). Figure 9.2 shows the schematic of an algorithm developed at NIST. It requires two-d lattice fringes in the image to form an FFT, measure the d-spacing and angle between them using routines available in DM and/or ImageJ, and X-ray diffraction data of possible structures from a library, such as ICDD or similar source. The measured data and reference files are used to find matching d-spacing, and angle between lattice vectors within the measurement errors (generally between 2%

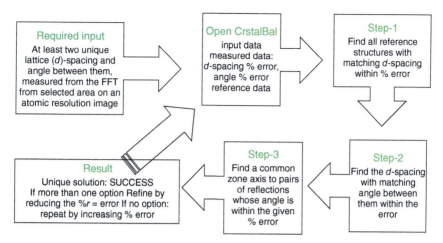

Figure 9.2 Schematic description of the Crystal Ball Plus program to unequivocally identify the structure from HRTEM images using FFT (https://notabug.org/stefano-m/crystalballplus).

and 10%) defined by the user. Finally, a common zone axis for matching planes is identified to provide a unique solution. This method has been used to follow the structural transformation occurring in catalyst nanoparticles during carbon nanotube growth and has helped in revealing the structures of active and inactive catalyst [20, 21].

9.2 Current Status

As we will see in Sections 9.2.1 and 9.2.2, both image processing and image simulations have been revolutionized during past couple of decades due to availability of fast computers and simple programming platforms such as Python and MATLAB. However, the processing opportunities available in ImageJ and DM may still be enough for simple processing or could be combined with other faster routines. Detailed aspects of data processing and handling can be found in a recent book by Park and Ding [15]. Other image simulation programs (apart from JEMS [Stadelmann] and xHRTEM [Ishizuka] described above) using the multislice approach have also become available in recent years, including Dr. Probe [22], QSTEM [23, 24], MuSTEM [25, 26], abTEM [27, 28], and others. Many of these modern software packages are graphic processing unit (GPU)-accelerated and some of them, making use of distributed high-performance computing resources, have vastly improved simulation scales and quality.

9.2.1 Progress for Image Simulations

The reduction in simulation time to fractions of seconds is an important development for applying simulated images as training data for automation and/or autonomous experiments in future (see Section 10.4.3). On the other hand, the discrepancy in contrast between simulated and experimental ADF images, has been reduced by including a physically reasonable source size in simulations that enables us to obtain quantitative data [29]. Furthermore, quantitative analysis of STEM-electron energy loss spectroscopy (EELS) maps has been achieved by using a relativistic multi-pole core-loss theory proposed by Dwyer [30, 31]. A recently developed image simulation program, named PRISM (or Prismatic), combines Bloch wave and multislice methods for atomic-resolution STEM images at realistic dimensions [32, 33]. This program, using a Fourier interpolation factor f, which has typical values between 4 and 20 for atomic resolution simulations, can provide an acceleration that scales with f^4 compared to multislice simulations, with a negligible loss of accuracy [32]. Recently, an algorithm, compatible with PRISM, to partition the STEM probe into "beamlets," given by a natural neighbor interpolation of the parent beams, has been developed for 4-D STEM diffraction simulations [34]. In future, this progress may enable the comparison of simulated and experimental images on-site during "live" in-situ experiments.

9.2.2 Progress in Data (Image) Processing

As we have found methods to increase temporal resolution, the dataset to be analyzed has increased exponentially (in the order of few $GB\,s^{-1}$), requiring us to find ways to automate data analysis. Therefore, data processing has shifted from manual analysis of a few select frames (extracted from videos) to automated platforms that process all of the image frames recorded. As an example, an automated image processing was developed to meet the challenge of deciphering the fluctuations between two structures, present in a 5 nm catalyst particle, occurring during single walled carbon nanotube (SWCNT) growth (see also Section 7.5.2) [35]. We faced two primary challenges: (i) FFTs from the two regions could not be generated as they were too small and continuously changing in size, (ii) we needed to determine the structures of the two regions from at least 1000 frames to determine the structural fluctuation rate. The crystal structure determination method described in Figure 9.2 is applicable only if the entire nanoparticle under observation has the same structure and thus could not be used. Similar fluctuations have also been reported during nanoparticle growth [36]. Hussaini et al. reported an automated image analysis scheme, by combining ImageJ and MATLAB routines [37]. It employs peak-finding routine to determine the atomic positions in one of the images and applies it to all images after drift correction, measures the nearest neighbor distances to identify the structure evolution in thousands of images extracted from a video sequence. This method has been successfully used not only to identify the structural fluctuation but also measure the fluctuation rate [35, 38]. Other methods to find atomic positions are also available. Peak-finding approaches have been developed in Python, such as the open-source *Atomap* package [28], which can identify crystal symmetry and measure various lattice distortions. Another possibility is to use a non-rigid registration to determine the atomic positions with picometer precisions after averaging a series of Z-contrast STEM images [39].

With the increasing complexity and amount of data generated by new detectors, data processing faces new challenges. Our ultimate goal is to perform data analysis on-site during "live" experiments as described in the workshop reports, identified in Section 10.3.1, but this requires better solutions than described above. Therefore, we need to design new methods that align with high-throughput automation and data-intensive workflows for rapid experimental feedback from significantly large volume rather than a limited subset. With the emergence of ML for image recognition and data analysis, attention has shifted to take advantage of neural network technologies for image processing and interpretation as described more in Section 9.4.

> Data is a precious thing and will last longer than the systems themselves.
> – Tim Berners-Lee

9.3 Data Management

One of the other major challenges facing in-situ TEM/STEM measurements is to save, transfer, and store large datasets at the rate they are generated. The rate of data acquired with high spatial and temporal resolution, as mentioned earlier, is in the order of few GB s^{-1}. For example, an acquisition rate of 50 frame per second for image size of 166 MB will generate 1 TB data in 30 minutes [40]. Most of the computers associated with the microscope and image acquisition systems do not have memory and/or hard drive large enough to store the dataset that will be accumulated in 1 hour of experimental session. Therefore, for in-situ experiments, acquisition and moving the dataset at a high throughput rate is a formidable task. One of the solutions will be to develop procedures to reduce the data size before transferring to improve the rate and reduce storage space needed. Datta et al. have reported an efficient and flexible data reduction and compression scheme (ReCoDe) that preserves the individual events by retaining the spatial and temporal resolution [40]. The methodology developed relies on the fact that all useful information in a single raw image is contained within "secondary electron puddles," which are digitized as the high-energy electrons pass through the detector's sensor. Therefore, first the data are reduced to keep the information that only identifies electron puddles, which are then further compressed. Their program constitutes of four reduction levels and several types of lossless compression algorithms that can be applied before storing the data onto attached network drives. Using on-the-fly ReCoDe compression, a continuous data transfer rate reduction of 3 GB s^{-1} raw data, input stream from DE-16 detector, to 200 MB s^{-1} could be achieved [41]. The data stored are 100× smaller than the raw data and can be re-processed to remove data artifacts and retain the electron puddle shape and intensity information needed for image analysis [40].

Another collective need is to store metadata that includes all the experimental and imaging conditions with universal identifiers such that it can be accessed and understood in the future. Without this information, electron microscopy data cannot adhere to the *FAIR data principles*, i.e. scientific data should be Findable, Accessible, Interoperable, and Reusable (FAIR) [42]. To this end, NIST has developed a laboratory wide system, NexusLIMS [43, 44]. The motivation behind the development of the system is to (i) centralize the data storage, (ii) make it easy to collect and store metadata for future use, and (iii) mitigate the problem of forgotten nomenclature used for data collected in the past or by collaborators. Users are asked to input identifiers for their sample and basic experimental metadata, while experimental conditions and instrument settings are automatically extracted from the EM data files. This information is stored in a central location that the user can access to evaluate and plan their experimental sessions without extensive notetaking. Moreover, the data can be accessed and analyzed from any location connected to the network. A web-based portal provides access to all of the research records, allowing users to search for prior experiments by date, user name, instrument, sample, or any other metadata parameter [43].

Keeping exact records of experimental parameters, such as temperature, nature of gas and pressure, flow rate, property of liquid, data collection rate, magnification, lens apertures, dose rate, date, start and end time of the experiments, etc. is important for the reproducibility of experimental results and to make conclusive remarks from the in-situ observations. It is a herculean task and keeping manual records while operating the TEM is next to impossible. Therefore, Protochips, Inc., has taken the initiative to build a module based platform, called Axon, which can be synchronized with the TEM and sample holder controls such as to keep the sample locked in place for continuous observations and for recording experimental conditions (https://www.protochips.com/products/axon). Gatan's DM software also has the possibility to record certain reaction conditions along with the images or spectroscopy data.

For in-situ gas cell experiments, correlating observed events with measured gas composition and temperature can be difficult, especially as mentioned in Chapter 2, the gases take time to equilibrate [45]. Therefore, Zhang et al. have developed algorithms and scripts for automatic data synchronization in operando gas and heating during real-time experiments and/or post experiments performed using nanoreactor [46].

9.4 Data Processing and Machine Learning (ML)

In the last several years, AI and ML have taken central stage in the development of science and engineering applications, from automatic control systems to data processing and image recognition. They are routinely used by web-based search engines and for data processing by almost all big companies. These developments are of interest to the in-situ TEM community, as they can help us in controlling in-situ experimental parameters, automated image/spectroscopy processing and simulations to match experimental results with simulated data (see Section 10.4.3). The main goal of this section is to make you aware about the power and applications of ML to decipher in-situ data using relevant examples and references. Therefore, consult the references to obtain detailed information. Here we will first go over some of the basic building blocks of ML in simple terms, then provide an overview of the current status, challenges, and wish list.

9.4.1 What Is Machine Learning?

ML is considered a part of AI, which utilizes computer-based networks, designed to mimic human brain to solve complex problems, with or without external input. The main attraction, especially for in-situ TEM community, is that these methods can handle large datasets and process vastly larger amounts of experimental data than could be done manually. Therefore, ML, either unsupervised or supervised, is an attractive approach to deal with the "big data" challenge as explained below.

9.4.1.1 Unsupervised ML

Unsupervised ML is a collection of multivariate statistical approaches where the large dataset is analyzed without human interaction. Unsupervised algorithms are used to extract information from experimental data without requiring a prior knowledge or previously labeled data. It has been extensively used in analyzing spectroscopy data for phase mapping, deconvolving overlapping signals, etc. [47]. Hence, the use of statistical unsupervised learning, decorrelation, and visualization techniques can assist us to harness the signals that may not be visually obvious. Common examples of unsupervised learning are principal component analysis (PCA), independent component analysis (ICA), non-negative matrix factorization (NMF), and clustering, where ML algorithms are used to find statistical correlations within the experimental data, which can be used to denoise and/or identify unique and potentially interesting signals within the data [48].

⇒ **Principal component analysis (PCA)** is an example of unsupervised analysis that has been used for spectroscopy as an easy way to visualize a multidimensional (spectrum imaging) dataset. The technique performs a matrix decomposition to identify "principal components" that are ranked in the order of their information content, and can be used for denoising data, or identifying areas that have the highest variance. The results include eigenvectors (i.e. representative spectra, when applied to spectrum imaging data) and their respective loading (weight). Due to a lack of constraints, PCA typically results in non-physical components (eigenvectors) that cannot be directly interpreted and is better suited for exploratory analysis or denoising [49, 50].

⇒ **Independent component analysis (ICA)** is designed to decompose a multi-dimensional dataset while attempting to maximize the statistical independence between the resulting spectra and loading maps. This constraint usually results in more physically interpretable results, but it is important to remember that not all samples consist of independent areas (e.g. solid solutions), and care should be taken when interpreting results [51].

⇒ **Non-negative matrix factorization (NMF)** is another multidimensional signal separation technique that performs a decomposition while imposing a constraint that both the eigenvectors (components) and loadings must be positive.

This constraint conforms with the physics of spectral acquisitions; the Poissonian nature of the data collection implies data can only be non-negative. NMF has been widely used to deconvolve difficult signal overlaps in TEM- and SEM-based spectroscopies [52].

⇒ **Clustering** is used to divide data points into a number of groups with similar values, characteristics, or other properties (e.g. identifying different "types" of spectra present in a spectrum image). Various algorithms to cluster the data are available such as K-means, where the data are divided based on the mean value within a cluster [53]. It typically classifies the data into a predetermined number of clusters and often provides a "prototypical" signal for each cluster. Clustering is a widely used statistical technique in many scientific disciplines and has been successfully used to analyze STEM-EELS data [48].

9.4.1.2 Supervised ML

In supervised ML, a data-based training or learning system is used to analyze and draw conclusions from experimental dataset with minimum human involvement after an initial "training" process. For example (Section 9.1.2), whereas a simple algorithm was used to find a match for measured FFT within select set of reference files (Figure 9.2) and the final decision was made by the user, ML can be used to find a match within thousands of files, without human input and much faster (Figure 9.6) [55]. Deep learning (DL), a subset of ML that utilizes three or more layers of neural networks, is often used to learn from a large training dataset such that they can apply the experience learnt to classify and recognize experimental data/images [56]. Here complex algorithms are employed to build various types of neural networks that progressively extract higher-level features from the raw input images. It has also materialized as an effective tool for "big data" analysis, allowing for automated image classification, segmentation, anomaly detection, etc. [57, 58].

Although the physical and mathematical basis of the computational design may differ based on the properties of the experimental dataset and the knowledge to be gained, fundamental requirements are the same. Mainly, we need two types of datasets: one is called training dataset and the other is the input dataset for supervised learning. The training dataset could be obtained from previously unequivocally identified experimental datasets or simulation results obtained by using known parameters. The training process is the most important step, computationally expansive and slow, but once working, they are fast and effective for analyzing large dataset and thereby useful during "live" in-situ experiments. Therefore, DL techniques require large amounts of training data that are sufficiently accurate, and the learned models can be biased by the type of training data provided. For example, if your goal is to recognize a certain type of atomic defect, but if this defect is not included in your training data (or appears very rarely), the resulting model may never work for the intended purpose. Thus, careful consideration must be given to the types of data chosen for training purposes.

Most of the available algorithms and/or software platforms are based on a few fundamental design strategies. The neural network generated using the training dataset is then used to analyze and decipher the experimental dataset, decreasing number of possible matches after each node, without any human input. The algorithms for building neural networks, where number of nodes are defined for each iteration, are commonly written in Python or MATLAB but any other computer language is equally viable. **As explained in previous paragraph, the depth of training dataset controls the success of the operation.** Fundamental details about ML can be found in following two books: Data Analytics: Models and Algorithms for Intelligent Data Analysis [59] and The Elements of Statistical Learning: Data Mining, Inference and Prediction [60].

> Data analysis techniques for TEM and related modalities are developing as fast (and perhaps faster!) than hardware advances in recent years. Certainly not all microscopists will be experts in these emerging areas, which is why it is critical to build collaborations and work alongside data and computer scientists to be able to effectively and accurately extract information from experimental data.
> – Joshua Taillon, National Institute of Standards and Technology (NIST)

As mentioned above, there are multiple ways to create network structures for DL to derive information, functionalities, and predictive behavior of materials from big datasets collected during TEM/STEM experiments. A detailed discussion and reviews of these processes can be found in Refs. [48, 57, 61–65], and a brief description for some of commonly used networks is given below.

⇒ **Artificial neural network (ANN)** constitutes algorithms that simulate learning for a specific task and a class of functions. It is employed to estimate unknown functions that may have a very large number of inputs.
⇒ **Convolutional neural network (CNN)** is widely used for image classification, i.e. object class recognition within an image; object recognition, i.e. classification and object detection within an image and generate the bounding box around the object; for semantic segmentation. Here pixel to pixel classification of an image; and regression model where continuous values rather than probabilities of discrete functions are collected such that the neural network gathers continuous values rather than probabilities of discrete classes.
⇒ **Fully convolutional networks (FCNs)** are used to train end-to-end (pixels to-pixels), result in state of the art in semantic segmentation and are used for pixelwise prediction from supervised pre-training datasets [66].
⇒ **Deep kernel learning (DKL)** is effective for Gaussian processes by combining the representational power of neural networks with the reliable uncertainty estimates [67].

Each ML workflow is dependent on the knowledge to be extracted from the experimental dataset but follows some basic steps, such as preparing a labeled training dataset, selecting a suitable neural network, sematic segmentation (labeling information in each pixel as independent entity [discrete class]), analysis, and confirmation sets (human involvement), tuning the training parameters until ideal results are achieved. Currently, scalable workflows are being designed to handle

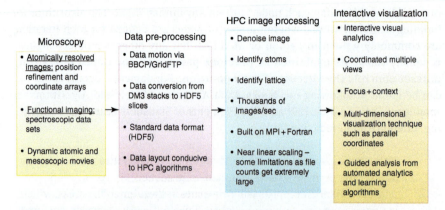

Figure 9.3 High-performance computation microscopy workflow. Life cycle of near-real-time analysis of large data from local imaging and spectroscopy experimental measurements. Source: Belianinov et al. [48]/Springer Nature/CC BY 4.0.

big data, assuming the data have been transferred and stored [48]. An example of a typical workflow using high-performance computing (HPC) algorithms is shown Figure 9.3. We collect digital data from the TEM/STEM interface during the progression of experiments (Figure 9.3, panel 1). After completion of the experiment, the workflow begins with data transfer to a high-performance storage, where it is pre-processed (Figure 9.3, panel 2). After the data are converted to a suitable HPC based format, such that it can be analyzed using appropriate algorithms or networks build by the researcher or from publicly available domains and is ready for further applications (Figure 9.3, panels 3 and 4) [48].

> Do not hesitate to reach out to commercial companies, involved in developing ML-based architecture to meet the ever-growing demands in public sector. They are often willing to share their knowledge or collaborate.

9.4.2 Motivation

As mentioned earlier, our quest to achieve high spatial and temporal resolution has resulted in big data problem and ML can help us to address it. The incorporation of the big-, deep-, and smart-data approaches in imaging coupled with computational-based simulations could potentially enable breakthroughs in the rate and quality of material discoveries [68]. As we will see in the examples described in Section 9.5, the imaging data now offer considerably more than a simple picture; in fact, it contains quantitative structural and functional information. This information is spatially distributed and often has a complex multidimensional nature. Additionally, new opportunities in materials design, enabled by the availability of big data in imaging and data analytics approaches, are waiting to be realized. For example, DL architectures have been applied to atomic resolution images for extracting physical and chemical parameters to establish structure–property relationships [69, 70].

Figure 9.4 shows a framework for translating unknown material structures into "quantifiable, physically meaningful descriptors and model representations" [71]. As described in Chapters 1, 3, 4, 6, and 7, there are several technical entities, such as time-resolved (in-situ) imaging, diffraction, and spectroscopy within the TEM/STEM platform, which we employ to decipher unknown parameters and their relationship. These entities provide us information about microstructure, overall structure, dynamics, and chemistry of the material under observation. The process from Figure 9.4 can be encompassed into three overlapping categories: experimental design, feature extraction, and knowledge discovery. So far, we have been working within this broad framework and succeeded in revealing a number of unknown entities, such as structures of defects, grain boundaries, and established relationships between structure and properties, functioning and properties of nanostructures in controlled environments, etc. We have been analyzing the data to obtain the information identified in Figure 9.4 manually, which is labor-intensive,

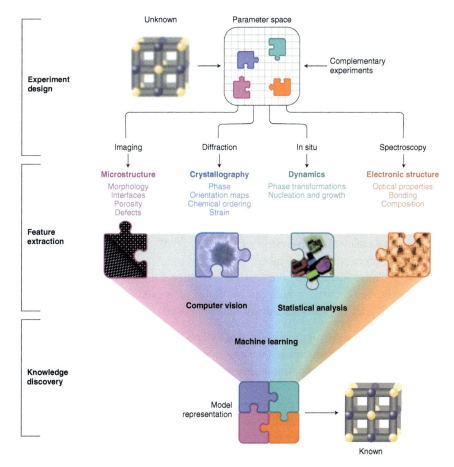

Figure 9.4 The electron microscopy framework. The framework for translating unknown structures or processes into quantifiable, physically meaningful descriptors and model representations. Source: Spurgeon et al. [71]/from Springer Nature.

time-consuming, and affected by human bias. With the exponential increase in the data collected with high spatial and temporal resolution, we need faster and impartial analytical platforms.

Establishing a pipeline between microscope and experiment specific data to materials and functionalities with specific descriptors will enable us to convert the large experimental dataset into all-inclusive unbiased knowledge much faster. Currently, we face four interconnected challenges in processing atomically resolved images: (i) transferring and managing the data flow from (potentially multiple) detectors; (ii) utilizing knowledge base to obtain material specific information; (iii) generate structure–property (materials behavior) or chemistry–property relationship from the knowledge gained; and (iv) predicting materials' functionalities from the knowledge gained or using it as feedback to tweak microscope operation or experimental parameters. Our aim is to extract the knowledge contained in the "big data" to the fullest extent, using DL capabilities and convert it to "smart data" that can then relate imaging or spectroscopy data to materials' functionalities and predictive behavior [68]. Our ambitious goal is to use the knowledge gained to control the microscope and/or experimental parameters for automation and/or autonomous operations as explained in Section 10.4.3 [72].

9.4.3 Current Status

A common ground between TEM/STEM data analysis and the image recognition algorithms developed during last couple of decades for broad spectrum of applications (e.g. search engines, robots, autonomous vehicles, etc.) can be imagined. Several research groups have designed DL or other ML-based software and neural networks to solve their specific problem and have made them publicly available. Recent advances are community-driven collaborative efforts with intent to share the outcome on an open platform. As a result, growth of community-developed open-source software has greatly accelerated the advances in EM data processing and democratized access to state-of-the-art analysis techniques (typically for zero cost). Some of the open-source tools and their applications, data sources, and libraries, which can be used as is or modified to suit a particular system (with relevant reference), are outlined below:

- **HyperSpy:** HyperSpy is an open-source Python library that facilitates interactive data analysis of multidimensional datasets, with a focus on EM data processing. It provides easy access to common unsupervised ML techniques (PCA, ICA, NMF, and others) and serves a base platform for many extension packages that provide additional functionality (https://hyperspy.org) [73].
- **Py4DSTEM:** py4DSTEM is an open-source set of Python tools for processing and analysis of four-dimensional scanning transmission electron microscopy (4D-STEM) data (https://github.com/py4dstem/py4DSTEM) [74].
- **Pyxem:** An open-source project enabling analysis of data from pixelated electron detectors (including 4D-STEM), providing functions for orientation mapping of crystals, strain mapping, virtual dark field imaging, differential phase contrast,

structural characterization of amorphous materials, and more (https://github.com/pyxem/pyxem) [75, 76].
- **LiberTEM:** LiberTEM is an open-source platform designed to match the growing performance requirements of EM data processing (https://libertem.github.io/LiberTEM) [77, 78].
- **PyTorch:** A general purpose (not specific to EM) open-source ML framework that accelerates the path from research prototyping to production deployment (https://pytorch.org).
- **TensorFlow:** A general-purpose (not specific to EM) open-source end-to-end ML platform to find solutions to accelerate ML tasks at every stage of the workflow and to create production-grade models (https://www.tensorflow.org) [79].
- **scikit-learn:** A general-purpose open-source ML Python library featuring various classifications, regressions, and clustering algorithms. The scikit-learn library is interoperable with the Python numerical and scientific libraries NumPy and SciPy, and its algorithms are used extensively in the scientific Python ecosystem (including by HyperSpy) (https://scikit-learn.org) [80].
- **Deep kernel learning:** In a deep kernel learning software, data are explored layer by layer to obtain knowledge and have been successfully applied for autonomous experimentation in 4D STEM (https://github.com/kevinroccapriore/AE-DKL-4DSTEM) [69].
- **DefectSegNet:** It is a convolutional neural network (CNN) architecture that performs semantic segmentation of common crystallographic defects in structural alloys as applied to analyze dislocation lines, precipitates, and voids (https://github.com/rajatsainju/DefectSegNet) [81].
- **DefectTrack:** A DL-based multi-object tracking (MOT) algorithm for quantitative defect analysis of in-situ TEM videos in real-time. It is not a simple detection model, instead, it is a tracking model (https://figshare.com/s/9e7f6c0870e828dbc1a2) [82].
- **ATOMAI:** AtomAI is a Pytorch-based package for deep and ML analysis of microscopy data and doesn't require any advanced knowledge of Python or ML. The intended users are domain scientists with a basic understanding of how to use NumPy and Matplotlib. (https://github.com/pycroscopy/atomai) [83].
- **Citrination platform:** It is an open, cloud-based, AI platform service for materials and chemicals located at http://citrination.com and developed by Citrine Informatics, Inc. It is open to academics, researchers, and other non-commercial users, or can be used in an enterprise environment with a license agreement (https://citrination.com/faq).
- **Nion Swift:** Nion Swift is an open-source scientific image processing software integrating hardware control, data acquisition, visualization, processing, and analysis using Python, published by the Nion company, and can be easily extended. It runs on Windows, Linux, and macOS (https://github.com/nion-software/nionswift) [84].
- **ImageNet:** It is a dataset library of over 15 million labeled high-resolution images belonging to roughly 22 000 categories. The images were collected from the web and labeled by human labelers using Amazon's Mechanical Turk crowd-sourcing

tool. It was started in 2010, as part of the Pascal Visual Object Challenge, an annual competition called the ImageNet Large-Scale Visual Recognition Challenge (ILSVRC) (https://doi.org/10.1145/3065386) [85].

○ **Matminer:** It is a general Python-based software package, designed to facilitate the development, reuse, and reproducibility of data pipelines for materials' informatics applications and to read data from existing library (https://github.com/hackingmaterials/matminer) [86].

A NIST-developed repository, Joint Automated Repository for Various Integrated Simulations (**JARVIS**), is designed to automate materials discovery and optimization, using classical force-field, density functional theory, ML calculations and experiments. Access to the database is a free resource with registration (https://jarvis.nist.gov) [87].

As mentioned earlier, building an appropriate training dataset is the most important step for designing a neural network as it is used then to extract information from the experimental data. The training dataset could be obtained from a set of experimental data with well-defined parameters or from libraries such as Matminer [86], JARVIS [87], or Imagenet [85]. Again, a training set for supervised learning uses the structure models/image or dataset based on prior knowledge, either simulated or experimental.

> It is a capital mistake to theorize before one has data.
> – Sherlock Holmes

9.5 Select Applications

During the last decade or so, various architectures and models, as described in Section 9.4.2, have been developed and applied to extract information from the large datasets. One of the goals have been to revolutionize the way image/data processing and interpretation has been done in the past such that same results are achieved without significant human bias and much faster. DL architectures are also employed to extract physical and chemical parameters from atomic resolution images to establish structure–property relationships [69, 70, 88]. These approaches tend to resemble the logic of a human brain, such that the mode of analyzing raw experimental data and extracting information can be significantly scaled up.

9.5.1 Noise Reduction

Apart from large dataset, low SNR images/spectra are also a product of high spatial and temporal resolution. For example, low-dose STEM images acquired with a dwell time of 10^{-7} seconds with less than 10 electrons per pixel can be very noisy. The situation is worse for phase-contrast TEM images recorded during in-situ experiments using a DE camera with temporal resolution of ms. Using such noisy images as training or experimental dataset for any image matching algorithms is a recipe for failure. Therefore, denoising is frequently the first step of pre-processing the phase-contrast

TEM images before they can be used as training or experimental set. As explained in Section 9.1.2, manual filtering was commonly performed for TEM images, but lately DL has been employed to denoise the images.

Sadre et al. have developed a general and simple DL routine based on U-Net architecture that produces more accurate and robust results than the conventional algorithm [89]. They compare this routine with the conventional Bragg filtering methods (see Section 9.1.2) by applying both to perform the segmentation of complex features in phase-contrast high-resolution transmission electron microscopy (HRTEM) images of monolayer graphene. They report that the U-Net filter outperformed the conventional Bragg method in every performance metrics tested and accurately revealed the structurally important regions. On the other hand, Vincent et al. have developed a CNN-based architecture to denoise images of Pt/CeO_2 catalyst, acquired using DE camera during in-situ experiments. They employed large set of simulated images as training set and compared the performance of their network against other denoising methods and show that it outperformed them all [90]. Pulse coupled and wavelet neural networks are some of the other methods used for denoising, and they all give satisfactory results for specific parameters used. However, the parameters need to be experimentally measured and each method can reduce only one type of noise, and methods for reduction of all types of noise still need to be worked out.

Noise2Atom is another DL-based unsupervised denoising method developed for STEM images [91]. The model assumes consecutive structural similarity (CSS) for image quality assessment, i.e. the structure remains the same between consecutive frames acquired within small scan intervals. This model does not require paired training images, prior signal, or noise model estimation, and all information is obtained from the experimental datasets. Another CNN model treats the image denoising as a plain discriminative learning problem, i.e. noise is separated from a noisy image by feed-forward denoising convolutional neural networks (DnCNN). They show that residual learning and batch normalization boosts the denoising performance and can speed up the training [92].

A general overview of using DL for image denoising, not specific to TEM data, can be found in Ref. [93].

9.5.2 Structure Determination

Extracting the crystal or defect structure from TEM/STEM images is one of the main analytical functions but can be quite complicated due to electron-optical functions involved in image formation. Electron diffraction can be used to obtain 2-D information, but it is not optimum to determine 3D structure of an unknown material. As explained in Section 9.1, we have used prior knowledge about the materials, such as chemistry, to derive structural information in various ways. The entire process is not only time-consuming but requires sufficient knowledge about the materials and results will be biased due to the human factor. Therefore, DL architectures, models, and/or platforms have been developed and applied to extract unbiased information from the atomic resolution images and/or diffraction patterns (DPs) at a faster rate. Some such examples are given below.

9.5.2.1 Diffraction Pattern Analysis

A method that extends to find matching structure for experimental diffraction patterns or images (using FFTs) within all crystal structures reported, without any knowledge about the chemistry of the material, was reported recently [55]. Aguiar et al. take advantage of the ML approach and generate 571 340 individual crystal structures, divided among seven families, 32 genera, and 230 space groups as a training set for the network (Figure 9.5). Unlike the Crystal Ball Plus algorithm (Section 9.1.2; Figure 9.2), no prior knowledge about the structure is needed here. This DL approach was used to find crystal structures of alloys and two-dimensional materials and found that the network narrows down the space group within 70% confidence in the worst case and up to 95% in the common cases. They tested several models before deciding that CNNs are most suitable for the analysis with highest accuracy [55]. This method has also been applied for real-time analysis of the HRTEM images of metal nanoparticles on oxide support, single layer graphene, and is expected to identify chemical species and height of atomic columns [94].

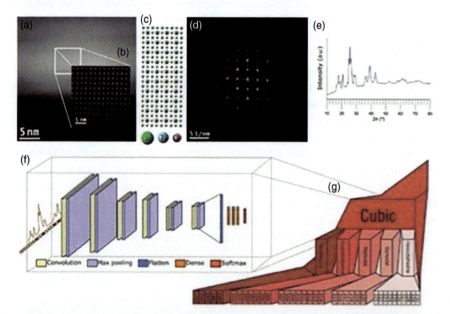

Figure 9.5 Neural network data architecture and workflow for crystal space group determination from experimental high-resolution atomic images and diffraction profiles. (a) Any material in a STEM, in this case crystalline strontium titanate (STO) islands distributed on a rock salt MgO substrate, can be simultaneously imaged with (b) high-resolution atomic mass contrast STEM imaging and (c) decoupled with a selective area. (d) FFT to reveal the material's structural details. (e) On the basis of either electron diffraction or FFT of an atomic image, a two-dimensional azimuthal integration translates this information into a relevant one-dimensional diffraction intensity profile from which the relative peak positions in reciprocal space can be indexed. (au, arbitrary units.) (f) Seeding the prediction of crystallography is a hierarchical classification using a one-dimensional convolution neural network model replicated at (g), each layer from family to space group forming a nested architecture. Source: Aguiar et al. [55]/from American Association for the Advancement of Science - AAAS, CC BY-NC 4.0.

A deep CNN has also been reported to automatically analyze the position averaged convergent beam electron diffractions (CBEDs) by calibrating the zero-order size, center position, and rotation without pretreating the data [95]. Additional networks are then used to measure sample thickness and tilt from the large dataset of CBED patterns at a processing rate of 0.1 second per pattern that is suitable for automatically process large 4D-STEM data. The source code is available at https://github.com/subangstrom/DeepDiffraction.

9.5.2.2 Image Analysis

An example of a DL workflow, used by Ghosh et al., to apply the knowledge gained from microscopic images to identify atomic features in graphene, is shown in Figure 9.6 [54]. Their focus of this project was to employ CNN to identify atomic features (type and position), use them for DFT simulations to unearth the optimized geometry that can be further used to study the dynamics of the ad-atoms and defect formation upon heating. They start (stage 1) with STEM images and use DL to label each pixel to generate input 1 [54]. At stage 2, an atomistic scale model is generated for simulations to reveal the co-ordination and bond information and combined information is used to obtain structural information from experimental images [54]. This kind of workflow can be applied to other datasets with some modification.

Application of DL approaches have also been demonstrated to extract information about the location of the atomic species and type of defects from atomically resolved STEM images. The mechanism and chemistry of defects formed in monolayer graphene were explored using a "weakly supervised" approach that employed the

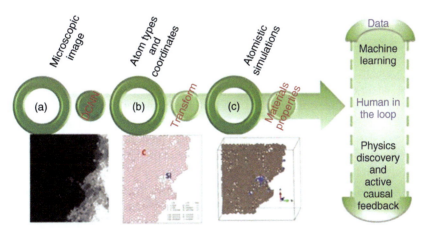

Figure 9.6 Schematic of the workflow, from images to evaluating material properties. Figure shows three primary stages of the framework. Stage 1 (a) consists of how deep learning models take microscopic images (STEM image of graphene) as inputs and identify features such as atoms, defects, and respective positions. In Stage 2 as shown in (b), the coordinates are put together to build a simulation object (c) followed by performing atomistic simulations in Stage 3 to study physical phenomena. The results from such simulations are used later to better build and guide physics informed experiments. Source: Ghosh et al. [54]/from Springer Nature, CC BY 4.0.

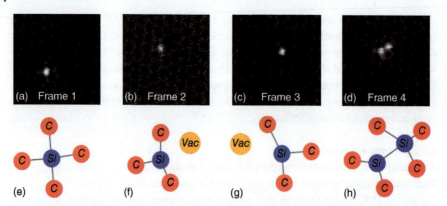

Figure 9.7 Tracking complex defect transformations on the surface. (a–d) STEM-based tracking of defect transformations on graphene surface. (e–h) Corresponding single defect structures extracted by applying FCN, LoG, and graph representations to data in (a–d). Imaging area in (a–d) is 2 nm × 2 nm. Source: Ziatdinov et al. [96]/from American Chemical Society.

atomic coordinates in the image, extracted via a FCN model [96]. This training data were then used to identify various defects, such as missing or substitutional Si atoms present in the STEM images. Si adatoms could be easily identified in STEM images from the contrast difference (Figure 9.7a) and agree with the atomic model with Si in fourfold coordination (Figure 9.7e). During the atomic manipulation, using the STEM probe, a vacancy was observed to form that resulted in the coordination of Si atom to change from fourfold to threefold, termed as "rotor" (Figure 9.7b,f) that rotated by 60° in the next frame (Figure 9.7c,g). Another Si atom was added during the experiment and the distance between Si atoms in the corners of a triangle increases by ~30% in the Si dimers compared to the threefold Si-rotor structure (Figure 9.7d,h). This change indicates that the bonds between carbon and Si atoms in the rotor structure were broken to accommodate another Si atom. On the hand, average length of the Si—Si bonds was similar to that for the carbon lattice. The knowledge about the coordinates in the building blocks of the individual defect can be used to obtain the information about their expected behavior and evolution with time. Based on the details provided in this paper, the authors "envision a development of a 'self-driving' microscope for atomically resolved imaging in the not too distant future" [96].

This group at Oak Ridge National lab has developed multiple architectures and approaches for defect analysis. For example, a complete and reliable information for atomic positions may not be available in the images recorded during a chemical transformation. They suggest and implement bottom-up description of systems to explore the dynamic changes and show the feasibility of this approach by applying it to electron-beam induced changes to Si-doped graphene layers [97]. Another approach to detect defect formations is based on unsupervised ML using a one-class support vector machine (OCSVM). Here, two schemes, based on Patterson function of each segments as input, are successfully employed to resolve point as well as line defects [98].

Building upon the success of TEM/STEM imaging modes to detect defect structure with utmost clarity, Roberts et al. have built a CNN architecture, called DefectSegNet (see Section 9.4.2) to perform semantic segmentation of three types of crystallographic defects: dislocations, precipitates and voids, commonly observed in structural alloys [81]. They employed this method to retrieve information from high-quality images of steel samples. A high pixel-level accuracy was achieved using supervised training subset: for dislocations ($91.60 \pm 1.77\%$), precipitates ($93.39 \pm 1.00\%$), and voids ($98.85 \pm 0.56\%$), obtained from a direct comparison between the DefectSegNet prediction and ground truth results. They also discuss the uncertainties in CNN prediction and the training data but find that on an average defect quantification using DefectSegNet prediction outperforms human experts. Other ML pipelines for segmentation and defect identification from HRTEM data have also been reported [99].

9.5.2.3 Atomic Column Heights (3-D Structure)

Most of the reported applications described above are for STEM images, where the image intensity can be directly related to the relative number (thickness) or nature of atoms. However, the intensities in a HRTEM image depend on many factors other than the number of atoms in Z direction or the nature (atomic number) of the atoms. Therefore, Rogane et al. have developed a framework based on CNN to estimate the atomic column heights in Au nanoparticles from TEM images. A structure model obtained from Wulff construction of a 4 nm Au particle was used to simulate images along different orientation for training dataset and a model-based regression scheme was applied. This regression-based CNN approach was then used to obtain atomic height from the atomic resolution phase-contrast TEM image of 2.8 nm (Figure 9.8a), 3.9 nm (Figure 9.8b), and 4 nm (Figure 9.8c) Au particles. In Figure 9.8, the experimental image with a detailed view of one particle is shown in left, the predictive model in the center and plotted atomic column heights on the right. As expected, the central column of a particle has the most number of atoms for all three particles and decreases moving away from the center. The maximum number of atoms identified was 6, 10, and 11 for 2.8, 3.9, and 4 nm particles, respectively. The minimum was predicted to be 1 or 2, but a sharp decrease in the number atoms was observed for 4 nm particle where the majority of column had 8 to 10 atoms [100].

Although the number of atoms and their spread within the columns are qualitative, as a larger training dataset is needed for quantitative CNN predictions, a trend can be established and used for accelerated or "on-the-fly" relative analysis of nanoparticles. This framework may also be developed further to extend DL models to other broad applications.

9.5.2.4 Other Applications

The power of ML-based networks and architectures to find solutions to challenges we are facing for fast, unbiased, and reliable analysis of large datasets, generated during in-situ experiments and 4D-STEM imaging, is being realized in many aspects. It is an ever-expanding field and covering all of the examples is beyond the scope of this book. However, some of the other relevant reports are summarized below.

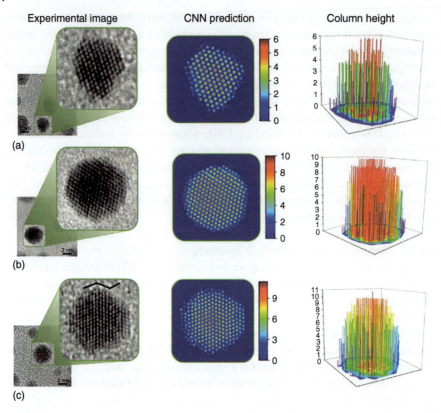

Figure 9.8 CNN predictions of the atomic column heights in the HRTEM experimental image with the nanoparticle size of 2.8 nm (a), 3.9 nm (b), and 4 nm (c). Source: Ragone et al. [100]/from ELSEVIER.

The most relevant example for in-situ TEM experiments is *DefectTrack*, reported by Sainju et al. [82] (also see Section 9.4.2) as it is developed to track the defect formation and their dynamic behavior, i.e. life-time, from real-time videos. It is a DL-based one-shot MOT model that can track a cascade of defect clusters as they are being formed. This approach takes advantage of the inherent property of MOT, which is to predict the trajectories of the objects of interest in videos or image sequences. The training dataset was established using an experimental in-situ TEM video by applying the standard tracking annotation protocol. They employed this architecture to track the number defects formed, their location, and lifetime during irradiation of pure Ni at 700 °C with 1 MeV Kr^{2+} ions and found an MOT accuracy of 66.43%. Pre-processing images to reduce noise was found to help improve the detection accuracy. Their statistical evaluations on the defect lifetime distribution, tested against four manual analyses, suggest that the DefectTrack outperformed human experts in accuracy and speed [82].

As mentioned earlier, 4D-STEM images also generate large dataset and a number of DL architectures and models have been developed to obtain physics and chemistry from such datasets. Moreover, automated experiments have been implemented to apply 4D-STEM to probe local structures, symmetry-breaking distortions, internal electric and magnetic fields in complex materials [69]. A DKL approach was also employed to discover the behavior of twisted bilayer graphene and mesoscale ordered patterns in $MnPS_3$, a 2D structure [88].

A CNN model has also been developed to automatize the determination of chiral indices of SWCNT by using large number of atomic-model-based simulated images, representing different chiral angles and diameters, as training dataset [101]. An excellent agreement between manual and auto-assigned classification was found most of the time, with some exceptions. Further development of this much faster approach will help in making meaningful statistical analysis of SWCNT images as well in automation of in-situ TEM synthesis of SWCNTs (see Section 7.5.2).

Some of the other relevant applications include, but are not limited to, a model predicting electronic quantum transmission coefficients as a function of electron energy for one-dimensional channels [102], ML methods to explore the self-assembly behavior of colloids and nanoparticles for building soft materials, such as crystal structure prediction, phase diagram calculations, enhanced sampling techniques, and reverse-engineering methods [61, 103], possibility to employ the Gaussian process (GP) methods for analysis and reconstruction of EELS datasets in STEM [104], identification and correction of temporal and spatial distortions in STEM [105], image registration of low signal-to-noise cryo-STEM data [106], reducing discovery time for materials and molecular modeling, imaging, informatics, and integration [107], NMF, to solve problems associated with the materials analysis of STEM, electron energy-loss spectroscopy and energy-dispersive X-ray spectroscopy data [65], The application of Cascade Mask-RCNN to identify and measure particles in TEM images (ParticlesNN) [108].

> Torture the data, and it will confess to anything
> – Ronald Coase

9.6 Future Needs

We have just started to take advantage of ML/DL to extract previously undetectable information in atomic resolution images and spectroscopy data. As is evident from the examples cited in this chapter, our main goal has been to process and retrieve unbiased or almost unbiased structural and chemical information at faster speed to meet the challenge of analyzing large datasets. However, the images also contain information about the physical and chemical properties, such as spin states and bonding environment, etc. which can be used to extract their nanoscale properties. As we have realized, there will always be specific needs for specific application or project, but there are also some general needs that can be considered universal as described below:

1. **Data transfer rate:** We need fast transfer rate from the experimental station (TEM/STEM) instrument to the HPC environments to start data processing. Lossless data reduction algorithms, during or after data acquisition could be helpful.

2. **Fast one-off experiments:** Establish networks to run fast experiments to check the feasibility. It will reduce number of failed experiments, increase efficiency on costly instrumentation time, increase scalability, microscope optimization, and increase productivity, as a result.
3. **Information recovery:** There is a need for big data analytics for unsupervised learning, dimensionality reduction, and clustering techniques. Also, a unified cyber infrastructure designed for materials' electron microscopy must be created to accelerate integration of data science approaches.
4. **Materials science in the AI age:** high-throughput library generation, ML, and a pathway from correlations to the underpinning physics and chemistry [69].
5. **Image and DP data base:** we need to create a publicly available data base that can be used as a part of training set similar to existing databases, such as Inorganic Crystal Structure Database (ICSD) [94]; Materials Project database of first-principles calculations [95]; the Royal Society of Chemistry's ChemSpider [54], proposed depository FAIR from M313 program at KAIST (Reducing Time to Discovery: Materials and Molecular Modeling, Imaging, Informatics, and Integration) [92].
6. **High-performance computing facility is a fundamental requirement**.

Our final goal is to apply "on the fly" image/data recognition to set up automated or autonomous experiments (see Section 10.6). Researchers have recognized this need and its importance, and there are a few reports already published [48, 63, 69, 70, 107]. For example, an automated experiment is set up to manipulate

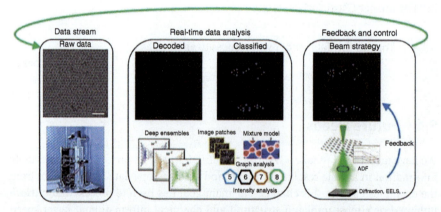

Figure 9.9 Atomic manipulation workflow. Image is acquired and passed to trained ensemble of neural networks, where raw image is decoded, and atomic coordinates are extracted. Different colors in neural networks represent different initialized weights. Image patches centered at each coordinate are generated and used for analysis of local atomic environment, where coordinates are classified using mixture models, in addition to graph analysis, intensity comparison, or a combination of all, depending on material system and possible defect evolution. According to chosen strategy, electron beam is directed to specific coordinates based on determined positions of classified atoms and defects, with possibility of using detector feedback to guide beam conditions such as dwell time. A new image is acquired once the selected beam irradiation strategy has completed, and the scheme repeats.

electron-beam-induced defect creation using STEM and identify them using DL algorithms (Figure 9.9). The program is called "ensemble learning and iterative training" (ELIT) and is available on GitHub (https://github.com/kevinroccapriore/ELIT-live) [109], but we have just scratched the tip of the iceberg and more needs to be done. Fortunately, it is a fast-growing field, and I am optimistic for the future.

9.7 Limitations

Apart from the future needs, outlined above, there are some fundamental limitations to applicability of the ML-based networks for in-situ TEM applications. First and foremost is the availability of training sets that must be either experimentally acquired or simulated. Moreover, the success of the network depends on the size, breadth, and appropriateness of the training dataset. As we can imagine, it is difficult task as we will need to run multiple experiments with the same results to be fed as training set. Similarly, simulating images of unknown structures is not possible. However, with some assumptions, it can help in saving time to analyze a large number of images.

It is not possible to keep "human bias" completely out of the data analysis as the training set must include the defect or structure information that we are looking for in our experimental data.

Also, most of the networks are being developed for STEM images, where image contrast is easily interpretable to find atomic positions. On the other hand, finding precise atomic positions in phase contrast TEM images, especially the ones acquired using DE with low SNR, is not easy.

ML networks are sensitive to artifacts and slight change in image or data acquisition condition may result as catastrophic failure of the model.

Last but not the least is the difficulty for setting up a workflow for automatic control on in-situ parameters as there are too many of them. Unlike, controlling the electron beam position, we need to control, temperature, image drift, liquid, or gas flow, etc.

However, these limitations should not stop us from using ML-based networks where they can be helpful.

9.8 Take Home Messages

ML-based networks and algorithms have replaced the image simulation methods we were using not too long ago. These techniques can be used to handle large datasets generated by high-speed detectors combined with high spatial resolution.

Moreover, the methods can be used for precise location of atoms, deduce bond distance and angles that can be used to reveal the physical and chemical behavior at nanoscale. There are number of networks, algorithms, and data sources publicly available. It is a fast growing field of research, and we expect it be an integral and essential part of designing and controlling in-situ TEM experiments in near future.

Last but not the least, a note from **Sergei Kalinin, ONRL** (private communication) for ML applications to TEM data: "The first step in applying ML in STEM is the application of supervised learning methods in post-acquisition data analysis, when the trained network essentially performs human-level tasks but faster and in more scalable fashion. Human can label several images, and ML can analyze hundreds. The transition of thus trained ML methods to the in-situ data analysis is remarkably complex, since the microscope settings can vary from experiment to experiment, and different objects can look similar under different microscope settings. This is where the ensemble networks and technique such invariant risk minimizations and networks with strong inferential biases become useful. The next step is the adoption of machine learning tools that allow automated experiment, e.g. dynamic sampling to reduce beam damage, choice of locations for EELS or 4D STEM. For these, if the objects of interest or decision-making logic is known in advance, the workflows can be deployed once the engineering controls are in place. The land of opportunity here are the algorithms where the decision making (e.g. policy in ML speak) is made by ML algorithm and modified in line with the experiment. Simples example of such beyond human workflows are deep kernel learning experiments when ML algorithms learns which microstructural elements gives rise to required physical signatures in EELS and 4D STEM, but this is very clearly but the beginning."

References

1 Cowley, J.M. and Moodie, A.F. (1957). The scattering of electrons by atoms and crystals. I. A new theoretical approach. *Acta Crystallographica* 10 (10): 609–619.
2 O'Keefe, M.A., Buseck, P.R., and Iijima, S. (1978). Computed crystal structure images for high resolution electron microscopy. *Nature* 274 (5669): 322-324.
3 Ishizuka, K. and Uyeda, N. (1977). A new theoretical and practical approach to the multislice method. *Acta Crystallographica Section A* 33 (5): 740-749.
4 Kilaas, R. and Gronsky, R. (1983). Real space image simulation in high resolution electron microscopy. *Ultramicroscopy* 11 (4): 289-298.
5 Dyck, D.V. (1980). Fast computational procedures for the simulation of structure images in complex or disordered crystals: a new approach. *Journal of Microscopy* 119 (1): 141-152.
6 Stadelmann, P. (2003). Image analysis and simulation software in transmission electron microscopy. *Microscopy and Microanalysis* 9 (S03): 60-61.
7 Stadelmann, P. (2020). JEMS. http://www.jems-saas.ch.
8 Cooley, J.W. and Tukey, J.W. (1965). An algorithm for the machine calculation of complex Fourier series. *Mathematics of Computation* 19 (90): 297-301.
9 Hÿtch, M.J. and Stobbs, W.M. (1994). Quantitative comparison of high resolution TEM images with image simulations. *Ultramicroscopy* 53 (3): 191-203.
10 Forbes, B.D., D'Alfonso, A.J., Findlay, S.D. et al. (2011). Thermal diffuse scattering in transmission electron microscopy. *Ultramicroscopy* 111 (12): 1670-1680.
11 Krause, F.F., Müller, K., Zillmann, D. et al. (2013). Comparison of intensity and absolute contrast of simulated and experimental high-resolution transmission

electron microscopy images for different multislice simulation methods. *Ultramicroscopy* 134: 94–101.

12 Loane, R.F., Xu, P., and Silcox, J. (1991). Thermal vibrations in convergent-beam electron diffraction. *Acta Crystallographica Section A* 47 (3): 267–278.

13 Saxton, W.O., Pitt, T.J., and Horner, M. (1979). Digital image processing: the semper system. *Ultramicroscopy* 4 (3): 343–353.

14 Saxton, W.O. and Koch, T.L. (1982). Interactive image processing with an off-line minicomputer: organization, performance and applications. *Journal of Microscopy* 127 (1): 69–83.

15 Chiwoo, P. and Ding, Y. (2021). Data science for nano image analysis. In: *International Series in Operations Research & Management Science*, vol. 308 (ed. C.C. Price), 376. Switzerland: Springer.

16 Sidorov, M., McKelvy, M., Sharma, R. et al. (1995). Structural investigation of mercury-intercalated titanium disulfide. 2. HRTEM of Hg_xTiS_2. *Chemistry of Materials* 7 (6): 1140–1152.

17 Hÿtch, M.J., Snoeck, E., and Kilaas, R. (1998). Quantitative measurement of displacement and strain fields from HREM micrographs. *Ultramicroscopy* 74 (3): 131–146.

18 Zhu, Y., Ophus, C., Ciston, J., and Wang, H. (2013). Interface lattice displacement measurement to 1 pm by geometric phase analysis on aberration-corrected HAADF STEM images. *Acta Materialia* 61 (15): 5646–5663.

19 de Ruijter, W.J., Sharma, R., McCartney, M.R., and Smith, D.J. (1995). Measurement of lattice-fringe vectors from digital HREM images: experimental precision. *Ultramicroscopy* 57 (4): 409–422.

20 Mazzucco, S., Wang, Y., Tanase, M. et al. (2014). Direct evidence of active and inactive phases of Fe catalyst nanoparticles for carbon nanotube formation. *Journal of Catalysis* 319: 54–60.

21 Chao, H.-Y., Jiang, H., Ospina-Acevedo, F. et al. (2020). A structure and activity relationship for single-walled carbon nanotube growth confirmed by in situ observations and modeling. *Nanoscale* 12 (42): 21923–21931.

22 Barthel, J. (2018). Dr. Probe: a software for high-resolution STEM image simulation. *Ultramicroscopy* 193: 1–11.

23 Sarahan, M. (2017). QSTEM: Quantitative TEM/STEM Simulations. www.physik.hu-berlin.de/en/sem/software/software_qstem.

24 Koch, C.T. (2002). Determination of core structure periodicity and point defect density along dislocations. Physics and Astronomy. ProQuest Dissertations and Theses. Arizona State University.

25 Allen, L.J., D'Alfonso, A.J., and Findlay, S.D. (2015). Modelling the inelastic scattering of fast electrons. *Ultramicroscopy* 151: 11–22.

26 Brown, H.G. (2020). MuSTEM multislice electron microscopy simulation code. https://github.com/HamishGBrown/MuSTEM.

27 Madsen, J. and Susi, T. (2021). The abTEM code: transmission electron microscopy from first principles. Open Res Europe.

28 Nord, M., Vullum, P.E., MacLaren, I. et al. (2017). Atomap: a new software tool for the automated analysis of atomic resolution images using two-dimensional Gaussian fitting. *Advanced Structural and Chemical Imaging* 3 (1): 9.

29 LeBeau, J.M., Findlay, S.D., Allen, L.J., and Stemmer, S. (2008). Quantitative atomic resolution scanning transmission electron microscopy. *Physical Review Letters* 100 (20): 206101.

30 Dwyer, C. (2005). Multislice theory of fast electron scattering incorporating atomic inner-shell ionization. *Ultramicroscopy* 104 (2): 141–151.

31 Xin, H.L., Dwyer, C., and Muller, D.A. (2014). Is there a Stobbs factor in atomic-resolution STEM-EELS mapping? *Ultramicroscopy* 139: 38–46.

32 Ophus, C. (2017). A fast image simulation algorithm for scanning transmission electron microscopy. *Advanced Structural and Chemical Imaging* 3 (1): 13.

33 DaCosta, L.R., Brown, H.G., Pelz, P.M. et al. (2021). Prismatic 2.0 – simulation software for scanning and high resolution transmission electron microscopy (STEM and HRTEM). *Micron* 151: 103141.

34 Pelz, P.M., Rakowski, A., DaCosta, L.R. et al. (2021). A fast algorithm for scanning transmission electron microscopy imaging and 4D-STEM diffraction simulations. *Microscopy and Microanalysis* 27 (4): 835–848.

35 Lin, P.A., Gomez-Ballesteros, J.L., Burgos, J.C. et al. (2017). Direct evidence of atomic-scale structural fluctuations in catalyst nanoparticles. *Journal of Catalysis* 349: 149–155.

36 Xin, H.L. and Zheng, H. (2012). In situ observation of oscillatory growth of bismuth nanoparticles. *Nano Letters* 12 (3): 1470–1474.

37 Hussaini, Z., Lin, P.A., Natarajan, B. et al. (2018). Determination of atomic positions from time resolved high resolution transmission electron microscopy images. *Ultramicroscopy* 186: 139–145.

38 Zhu, W., Winterstein, J.P., Yang, W.-C.D. et al. (2017). In situ atomic-scale probing of the reduction dynamics of two-dimensional Fe_2O_3 nanostructures. *ACS Nano* 11 (1): 656–664.

39 Yankovich, A.B., Berkels, B., Dahmen, W. et al. (2014). Picometre-precision analysis of scanning transmission electron microscopy images of platinum nanocatalysts. *Nature Communications* 5 (1): 4155.

40 Datta, A., Ng, K.F., Balakrishnan, D. et al. (2021). A data reduction and compression description for high throughput time-resolved electron microscopy. *Nature Communications* 12 (1): 664.

41 pyReCoDe (2021). NDLOHGRP. https://github.com/NDLOHGRP/pyReCoDe.

42 Wilkinson, M.D., Dumontier, M., Aalbersberg, I.J. et al. (2016). The FAIR guiding principles for scientific data management and stewardship. *Scientific Data* 3 (1): 160018.

43 Taillon, J.A., Bina, T.F., Plante, R.L. et al. (2021). NexusLIMS: a laboratory information management system for shared-use electron microscopy facilities. *Microscopy and Microanalysis* 27 (3): 511–527.

44 Gibbon, G.A. (1996). A brief history of LIMS. *Laboratory Automation and Information Management* 32 (1): 1–5.

45 Crozier, P.A. and Chenna, S. (2011). In situ analysis of gas composition by electron energy-loss spectroscopy for environmental transmission electron microscopy. *Ultramicroscopy* 111 (3): 177–185.

46 Zhang, F., Pen, M., Spruit, R. et al. (2022). Data synchronization in operando gas and heating TEM. *Ultramicroscopy* 238: 113549.

47 Keenan, M.R. (2007). Multivariate analysis of spectral images composed of count data. In: *Techniques and Applications of Hyperspectral Image Analysis* (ed. H.F. Grahn and P. Geladi), 89–126. Wiley Online Library.

48 Belianinov, A., Vasudevan, R., Strelcov, E. et al. (2015). Big data and deep data in scanning and electron microscopies: deriving functionality from multidimensional data sets. *Advanced Structural and Chemical Imaging* 1 (1): 6.

49 Trebbia, P. and Bonnet, N. (1990). EELS elemental mapping with unconventional methods I. Theoretical basis: image analysis with multivariate statistics and entropy concepts. *Ultramicroscopy* 34 (3): 165–178.

50 Bosman, M., Watanabe, M., Alexander, D.T.L., and Keast, V.J. (2006). Mapping chemical and bonding information using multivariate analysis of electron energy-loss spectrum images. *Ultramicroscopy* 106 (11): 1024–1032.

51 de la Peña, F., Berger, M.H., Hochepied, J.F. et al. (2011). Mapping titanium and tin oxide phases using EELS: an application of independent component analysis. *Ultramicroscopy* 111 (2): 169–176.

52 Nicoletti, O., de la Peña, F., Leary, R.K. et al. (2013). Three-dimensional imaging of localized surface plasmon resonances of metal nanoparticles. *Nature* 502 (7469): 80–84.

53 Hartigan, J.A. and Wong, M.A. (1979). Algorithm AS 136: a K-means clustering algorithm. *Journal of the Royal Statistical Society: Series C: Applied Statistics* 28 (1): 100–108.

54 Ghosh, A., Ziatdinov, M., Dyck, O. et al. (2022). Bridging microscopy with molecular dynamics and quantum simulations: an atomAI based pipeline. *npj Computational Materials* 8 (1): 74.

55 Aguiar, J.A., Gong, M.L., Unocic, R.R. et al. (2019). Decoding crystallography from high-resolution electron imaging and diffraction datasets with deep learning. *Science Advances* 5 (10): eaaw1949.

56 Ede, J.M. (2021). Deep learning in electron microscopy. *Machine Learning: Science and Technology* 2 (1): 011004.

57 LeCun, Y., Bengio, Y., and Hinton, G. (2015). Deep learning. *Nature* 521 (7553): 436–444.

58 Garcia-Garcia, A., Orts-Escolano, S., Oprea, S. et al. (2017). A review on deep learning techniques applied to semantic segmentation. *arXiv:1704.06857*.

59 Runkler, T.A. (2012). *Data Analytics: Models and Algorithms for Intelligent Data Analysis*. Vieweg + Teubner Verlag.

60 Hastie, T., Tibshirani, R., and Friedman, J. (2009). *The Elements of Statistical Learning: Data Mining, Inference and Prediction*, 2e. Springer.

61 Dijkstra, M. and Luijten, E. (2021). From predictive modelling to machine learning and reverse engineering of colloidal self-assembly. *Nature Materials* 20 (6): 762–773.

62 Kalinin, S.V., Lupini, A.R., Dyck, O. et al. (2019). Lab on a beam—big data and artificial intelligence in scanning transmission electron microscopy. *MRS Bulletin* 44 (7): 565–575.

63 Vasudevan, R.K., Choudhary, K., Mehta, A. et al. (2019). Materials science in the artificial intelligence age: high-throughput library generation, machine learning, and a pathway from correlations to the underpinning physics. *MRS Communications* 9 (3): 821–838.

64 Vasudevan, R.K., Ziatdinov, M., Vlcek, L., and Kalinin, S.V. (2021). Off-the-shelf deep learning is not enough, and requires parsimony, Bayesianity, and causality. *npj Computational Materials* 7 (1): 16.

65 Muto, S. and Shiga, M. (2019). Application of machine learning techniques to electron microscopic/spectroscopic image data analysis. *Microscopy* 69 (2): 110–122.

66 Shelhamer, E., Long, J., and Darrell, T. (2017). Fully convolutional networks for semantic segmentation. *IEEE Transactions on Pattern Analysis and Machine Intelligence* 39 (4): 640–651.

67 Ober, S.W., Rasmussen, C.E., and van der Wilk, M. (2021). The promises and pitfalls of deep kernel learning. *arXiv:2102.12108*.

68 Kalinin, S.V., Sumpter, B.G., and Archibald, R.K. (2015). Big–deep–smart data in imaging for guiding materials design. *Nature Materials* 14 (10): 973–980.

69 Roccapriore, K.M., Dyck, O., Oxley, M.P. et al. (2022). Automated experiment in 4D-STEM: exploring emergent physics and structural behaviors. *ACS Nano* 16 (5): 7605–7614.

70 Deng, H.D., Zhao, H., Jin, N. et al. (2022). Correlative image learning of chemo-mechanics in phase-transforming solids. *Nature Materials* 21 (5): 547–554.

71 Spurgeon, S.R., Ophus, C., Jones, L. et al. (2021). Towards data-driven next-generation transmission electron microscopy. *Nature Materials* 20 (3): 274–279.

72 Stach, E., DeCost, B., Kusne, A.G. et al. (2021). Autonomous experimentation systems for materials development: a community perspective. *Matter* 4 (9): 2702–2726.

73 de la Pena, F., Ostasevicius, T., Fauske, V.T. et al. (2017). Electron microscopy (big and small) data analysis with the open source software package HyperSpy. *Microscopy and Microanalysis* 23 (S1): 214–215.

74 Savitzky, B., Zeltmann, S., Hughes, L. et al. (2021). py4DSTEM: a software package for four-dimensional scanning transmission electron microscopy data analysis. *Microscopy and Microanalysis* 27 (4): 712–743.

75 Johnstone, D., Crout, P., Nord, M. et al. (2022). pyxem/pyxem: pyxem 0.14.2. https://zenodo.org/record/6645923#.Y1w7PILMI-A.

76 Cautaerts, N., Crout, P., Ånes, H.W. et al. (2022). Free, flexible and fast: orientation mapping using the multi-core and GPU-accelerated template matching capabilities in the Python-based open source 4D-STEM analysis toolbox Pyxem. *Ultramicroscopy* 237: 113517.

77 Clausen, A., Weber, D., Ruzaeva, K. et al. (2020). LiberTEM: software platform for scalable multidimensional data processing in transmission electron microscopy. *Journal of Open Source Software* 5 (50): 2006. https://doi.org/10.21105/joss.02006.

78 LiberTEM (2020). LiberTEM – open pixelated STEM platform. https://libertem.github.io/LiberTEM/.

79 Abadi, M., Agarwal, A., Barham, P. et al. (2016). Tensorflow: large-scale machine learning on heterogeneous distributed systems. *arXiv:1603.04467*.

80 Pedregosa, F., Varoquaux, G., Gramfort, A. et al. (2012). Scikit-learn: machine learning in Python. *Journal of Machine Learning Research* 12: 2825–2830.

81 Roberts, G., Haile, S.Y., Sainju, R. et al. (2019). Deep learning for semantic segmentation of defects in advanced STEM images of steels. *Scientific Reports* 9 (1): 12744.

82 Sainju, R., Chen, W.-Y., Schaefer, S. et al. (2022). DefectTrack: a deep learning-based multi-object tracking algorithm for quantitative defect analysis of in-situ TEM videos in real-time. *Scientific Reports* 12 (1): 15705.

83 Ziatdinov, M., Ghosh, A., Wong, T., and Kalinin, S.V. (2022). AtomAI: a deep learning framework for analysis of image and spectroscopy data in (scanning) transmission electron microscopy and beyond. *arXiv:2105.07485*.

84 Nion Swift (2022). Nion Swift. https://github.com/nion-software/nionswift.

85 Krizhevsky, A., Sutskever, I., and Hinton, G.E. (2017). ImageNet classification with deep convolutional neural networks. *Communications of the ACM* 60 (6): 84–90.

86 Ward, L., Dunn, A., Faghaninia, A. et al. (2018). Matminer: an open source toolkit for materials data mining. *Computational Materials Science* 152: 60–69.

87 Choudhary, K., Garrity, K.F., Reid, A.C.E. et al. (2020). The joint automated repository for various integrated simulations (JARVIS) for data-driven materials design. *npj Computational Materials* 6 (1): 173.

88 Roccapriore, K.M., Kalinin, S.V., and Ziatdinov, M. (2021). Physics discovery in nanoplasmonic systems via autonomous experiments in scanning transmission electron microscopy. *arXiv:2108.03290*.

89 Sadre, R., Ophus, C., Butko, A., and Weber, G.H. (2021). Deep learning segmentation of complex features in atomic-resolution phase-contrast transmission electron microscopy images. *Microscopy and Microanalysis* 27 (4): 804–814.

90 Vincent, J.L., Manzorro, R., Mohan, S. et al. (2021). Developing and evaluating deep neural network-based denoising for nanoparticle TEM images with ultra-low signal-to-noise. *Microscopy and Microanalysis* 27 (6): 1431–1447.

91 Wang, F., Henninen, T.R., Keller, D., and Erni, R. (2020). Noise2Atom: unsupervised denoising for scanning transmission electron microscopy images. *Applied Microscopy* 50 (1): 23.

92 Zhang, K., Zuo, W., Chen, Y. et al. (2017). Beyond a Gaussian denoiser: residual learning of deep CNN for image denoising. *IEEE Transactions on Image Processing* 26 (7): 3142–3155.

93 Liu, B. and Liu, J. (2019). Overview of image denoising based on deep learning. *Journal of Physics: Conference Series* 1176: 022010.

94 Madsen, J., Liu, P., Kling, J. et al. (2018). A deep learning approach to identify local structures in atomic-resolution transmission electron microscopy images. *Advanced Theory and Simulations* 1 (8): 1800037.

95 Xu, W. and LeBeau, J.M. (2018). A deep convolutional neural network to analyze position averaged convergent beam electron diffraction patterns. *Ultramicroscopy* 188: 59–69.

96 Ziatdinov, M., Dyck, O., Maksov, A. et al. (2017). Deep learning of atomically resolved scanning transmission electron microscopy images: chemical identification and tracking local transformations. *ACS Nano* 11 (12): 12742–12752.

97 Kalinin, S.V., Dyck, O., Jesse, S., and Ziatdinov, M. (2021). Exploring order parameters and dynamic processes in disordered systems via variational autoencoders. *Science Advances* 7 (17): eabd5084.

98 Guo, Y., Kalinin, S.V., Cai, H. et al. (2021). Defect detection in atomic-resolution images via unsupervised learning with translational invariance. *npj Computational Materials* 7 (1): 180.

99 Groschner, C.K., Choi, C., and Scott, M.C. (2021). Machine learning pipeline for segmentation and defect identification from high-resolution transmission electron microscopy data. *Microscopy and Microanalysis* 27 (3): 549–556.

100 Ragone, M., Yurkiv, V., Song, B. et al. (2020). Atomic column heights detection in metallic nanoparticles using deep convolutional learning. *Computational Materials Science* 180: 109722.

101 Förster, G.D., Castan, A., Loiseau, A. et al. (2020). A deep learning approach for determining the chiral indices of carbon nanotubes from high-resolution transmission electron microscopy images. *Carbon* 169: 465–474.

102 Lopez-Bezanilla, A. and von Lilienfeld, O.A. (2014). Modeling electronic quantum transport with machine learning. *Physical Review B* 89: 235411.

103 Jackson, N.E., Webb, M.A., and de Pablo, J.J. (2019). Recent advances in machine learning towards multiscale soft materials design. *Current Opinion in Chemical Engineering* 23: 106–114.

104 Kalinin, S.V., Lupini, A.R., Vasudevan, R.K., and Ziatdinov, M. (2021). Gaussian process analysis of electron energy loss spectroscopy data: multivariate reconstruction and kernel control. *npj Computational Materials* 7 (1): 154.

105 Roccapriore, K.M., Creange, N., Ziatdinov, M., and Kalinin, S.V. (2021). Identification and correction of temporal and spatial distortions in scanning transmission electron microscopy. *Ultramicroscopy* 229: 113337.

106 Savitzky, B.H., El Baggari, I., Clement, C.B. et al. (2018). Image registration of low signal-to-noise cryo-STEM data. *Ultramicroscopy* 191: 56–65.

107 Hong, S., Liow, C.H., Yuk, J.M. et al. (2021). Reducing time to discovery: materials and molecular modeling, imaging, informatics, and integration. *ACS Nano* 15 (3): 3971–3995.

108 Nartova, A.V., Mashukov, M.Y., Astakhov, R.R. et al. (2022). Particle recognition on transmission electron microscopy images using computer vision and deep learning for catalytic applications. *Catalysts* 12 (2): 135.

109 Roccapriore, K.M., Boebinger, M.G., Dyck, O. et al. (2022). Probing electron beam induced transformations on a single defect level via automated scanning transmission electron microscopy. *arXiv:2207.12882*.

10

Future Vision

> *With the aid of these active experimental sciences man becomes an inventor of phenomena, a real foreman of creation; and under this head we cannot set limits to the power that he may gain over nature through future progress of the experimental sciences*
>
> – Claude Bernard

We can now conclude that in-situ transmission electron microscope (TEM)/scanning transmission electron microscope (STEM) observations/measurements are commonly used to understand and establish relationships between synthesis–structure–properties–functioning of nanomaterials. The shrinking of the nanoworld around us challenges us to understand the reaction mechanism not only during synthesis but also in the working environment for the nanomaterials. Technical developments in terms of TEM/STEM operation to improve spatial, temporal, and energy resolution have tried to match the pace of nanotechnology developments. Aberration-corrected electron optics and monochromated electron source have helped us to achieve atomic-level spatial resolution and up to 5 meV energy resolution, respectively. On the other hand, UEM and DTEM, stroboscopic TEM, and direct electron detectors have brought the temporal resolution down to femto second level. However, we still have not been able to achieve atomic-scale spatial and 5 meV energy resolution with femto second time resolution concurrently. At the same time, image processing techniques are being developed to improve spatial resolution affected by low signal-to-noise resolution (SNR) in the images acquired by high temporal resolution. Machine learning (ML) and artificial intelligence (AI) techniques are being developed to reduce the human bias and time for collecting and analyzing large datasets produced during in-situ experiments (see Chapter 9). In this chapter, we will review the historical aspect of recent progress, current status, ideas under consideration for future developments, including application of machine learning for experimental design control. The motivation of this chapter is to encourage readers to think outside the box to generate new ideas, develop new technologies, and/or implement the ideas presented here.

In-Situ Transmission Electron Microscopy Experiments: Design and Practice, First Edition. Renu Sharma.
© 2023 WILEY-VCH GmbH. Published 2023 by WILEY-VCH GmbH.

10.1 Historical Aspect

The history of in-situ TEM is almost as old as TEM itself, since the potential of observing the effect of external stimuli on defect and dislocation formation in materials has been quite tempting. Development of sample holders for heating in vacuum or gaseous environments was one of the earlier approaches (Chapters 3 and 7). Moreover, the interest to observe biological and/or geological materials in their native state led to the development of window liquid cell holder during this period (Chapter 6). However, during the last couple of decades, the availability of TEM holders for indentation, biasing, magnetization, etc. has led to an explosion in other in-situ TEM applications such as measuring the material strength, elasticity, current, etc. at nanoscale and/or of nanomaterials. Some of these holders are equipped with multiple stimuli such as indentation with heating or combined heat and bias. The liquid cell holders with heating and biasing are also available that permit detailed mechanistic understanding of charge/discharge processes in battery electrode materials and chemical reactions at electrode–electrolyte interfaces (Chapter 6). Similarly, gas cell holders with heating are available to follow catalytic reactions down to the atomic scale (Chapter 7).

Modification to TEM column, such as ESTEM, UEM, and DTEM, has also led to new capabilities for in-situ TEM experiments. On the other hand, development of algorithms based on ML/AI has transformed the field of data acquisition and analysis (Chapter 9).

10.2 Current Status

As mentioned above, currently we have access to TEM/STEM instruments equipped with monochromated electron source, aberration correctors, energy-dispersive X-ray spectroscopy (EDS), high-resolution electron energy loss spectroscopy (EELS), high-speed direct electron cameras, and pixelated detectors. Similarly, modified sample holders equipped with external stimuli such as heating, cooling, biasing, indentation, and liquid and gas cell are now commercially available.

Apart from the base TEM/STEM instruments, we know that the TEM column has been modified to incorporate external stimuli such as sample environment, laser, ion radiation, etc. Following is a short summary of reported instruments.

10.2.1 ETEM

Sample region modified to enable gas pressures up to 20 mbar in the sample chamber while keeping the rest of the TEM under high vacuum conditions. Applications include observation of gas–solid interactions and synthesis of nanomaterials that may be combined with other stimuli such as heating, biasing, straining, magnetization, etc. using specialty TEM holders (Section 7.2.2).

10.2.2 UEM and DTEM

Both the sample region and gun chamber are modified to include pulsed laser source to (i) extract electrons and (ii) to heat the sample such that image acquisition rate depends on the laser pulse rate and is independent of detector speed [1–3]. The time resolution, the fourth dimension for in-situ observation is improved at the cost of poor SNR that reduces the effective spatial resolution. Whereas UEM operates in stroboscopic mode for electron emission, DTEM employs a "pump and probe" strategy to observe reversible processes such as phase transformation and melting (Section 1.6.3).

10.2.3 Stroboscopic TEM

It is another modification of UEM and DTEM where the laser excitation for electron pulse generation is eliminated by placing RF cavity combs as tunable electron pulser between electron source and condenser lenses and has achieved picosecond resolution by generating electron pulses at GHz rate (also see Section 1.6.3) [4].

10.2.4 PIES

A new platform that combines in-situ TEM and secondary ion mass spectrometer or spectrometry (SIMS), called the Parallel Ion Electron Spectrometry (PIES), has been recently developed. It is a multimodal approach for correlative measurements that decrease the elemental analysis limit to ppm level (using SIMS). This is an unusual instrumental setup and currently not commercially available (also see Section 8.1.1) [5].

> I like the word "nanotechnology." I like it because the prefix "nano" guarantees it will be fundamental science for decades; the "technology" says it is engineering, something you're involved in not just because you're interested in how nature works but because it will produce something that has a broad impact.
>
> – Richard E. Smalley

10.3 Technical Challenges

We have discussed the limitations of in-situ TEM in Chapters 1, 3, 4, 6, and 7, and now we should have a good idea about making the best of what is available to us. However, we also know that the functionalities of materials, especially nanomaterials, are constantly evolving at a rapid rate: therefore, we need to continuously find ways to overcome the challenges of today as we did for the challenges of yesterday. In this section, we will report how various workshops, sponsored mostly by government agencies, have put forward roadmaps for future research direction and technique developments in last 20 years. We will find that researchers have been dreaming of addressing the limitations and improving in-situ capabilities using TEM-based platforms, and the outcome of these workshops have helped in finding new solutions. The main motivation to describe these workshops and their outcome is to excite new generation to devise ways to meet the current challenges. We will also provide some examples of the developments that have been made to meet the challenges identified by discussion groups followed by proposed visions.

10.3.1 List of Major Workshops

Following is list of some of the major workshops initiated by funding agencies, challenges identified, and progress made to meet them within last two decades:

1. **"Dynamic in-situ electron microscopy as a tool to meet the challenges of the nanoworld" 2006, Sponsored by NSF, Organized by Renu Sharma, Peter Crozier, M.M.J Treacy, January 3–6, 2006** [6].
   Following major challenges were identified by the workshop participants:
   o Expand in-situ TEM into a Molecular Observatory for the synthesis and observation of active nanostructures.
   o Develop (ultra)fast image detectors to enable rapid observation of processes in active nanostructures at the atomic length scales.
   Relevant progress to date
   A noteworthy progress has been made to address these challenges, for example, development of MEMS-based liquid and gas holders, heating chips for higher heating rate, and cooling chips for continuous temperature control have enabled us to observe synthesis and functioning of materials at nanoscale (see Chapters 3–7). Also, both UEM and DTEM [3, 7], which were at the developmental stages at the time, have progressed to the point where 0.2 nm and 10 fs spatial and temporal resolution, respectively, is now achievable by using a stroboscopic cavity approach [4].

2. **"Current Status and Future of in-situ TEM," 2012, Sponsored by NIST and others, Organized by Mitra Taheri, Eric Stach and Renu Sharma, 2013** [8].
   The challenges identified during this workshop were divided into four categories (i) general for all in-situ measurements, such as drift correction, local temperature measurement, improved temporal resolution, data acquisition, and processing, concurrent imaging and spectroscopy, simultaneous acquisition of diffraction and imaging and specific challenges in the field of (ii) catalysis, (iii) electrochemistry and (iv) phase transformation.
   Relevant progress to date
   Some of the notable **progresses** made to achieve these goals are software development for drift correction and data processing (as discussed in detail in Chapter 9) [9].
   Development of 4D-STEM, i.e. recording 2D convergent beam electron diffraction (CBED) or nano beam electron diffraction (NBED) diffraction pattern at

each probe position as it is scanned across the sample in two dimensions using STEM mode, generating a 4D data cube that can be analyzed further [10]. Gammer et al. show that diffraction patterns acquired at each beam position in STEM mode using a probe with small convergence angle or a tilt series of darkfield images can be used to reconstruct diffraction pattern, bright-, dark-, or annular dark-field images using virtual apertures [11].

Development of active pixel detector (APS) or direct electron detectors (DEDs) has helped in improving temporal resolution for HREM imaging from 30 to 1600 fps (0.033 to 0.000 62 seconds) [12, 13]. Hybrid pixelated array detectors (PADs), which have enabled 4D-STEM technique [14], are now commercially available.

Methods to measure temperature at nanoscale under reaction condition have also been reported [15–22]. Similarly, an alternate approach to retrieve image information from diffraction patterns has been proposed [11, 23].

3. **"Future of electron scattering and diffraction" 2014, Sponsored by DOE-BES Ernest Hall, Organized by Susanne Stemmer, Haimei Zheng, Yimei Zhu, George Maracas. February 25–26, 2014 [24].**

    This workshop encompassed all aspects of electron scattering and development with three types of microscopes was identified as one of the future goals, i.e. (i) Atomic-Resolution Multi-Dimensional Transmission Electron Microscope, (ii) Ultrafast Electron Diffraction and Microscope, and (iii) "Lab-in-Gap" dynamic microscope. The last part of the workshop is relevant to in-situ and operando TEM measurements. The challenges identified for "lab-in-gap" microscopes included imaging light elements such as Li or H, ion transport for energy material applications [25], imaging surface adsorbed molecules for catalytic reaction [26], improving temporal resolution as bond breaking and formation frequency [27] are on the order of sub-100 fs, observation of light-induced superconductivity [28]. One of the recommendations, common with the NIST workshop, was to open the objective pole-piece gap to 10 mm such that sufficient space is available to accommodate versatile TEM holders with multimodal functionality to include optical, thermal, electrical, and electrochemical probing of materials under operando conditions.

    Recommendation also included to develop correlative platforms encompassing TEM/STEM and X-rays (see Chapter 8 for its relevance to in-situ) and multimodal microscopies toward realizing the goals outlined in Materials Genome Initiative.

    Development of cryogenic platforms for imaging beam-sensitive and soft materials, application of machine learning and neural networks for data processing was also discussed. These ideas continue to be central theme of ongoing research and development technologies, both at academic and industrial scale.

    Relevant progress to date

    Novel ideas for the "lab-on-chip" continue to develop, especially for incorporating multiple stimuli on the same microfabricated chip (see Figure 10.4).

One of the experimental achievements is the reported observation of oxygen vacancy diffusion in ceria and Sm-doped ceria [30] and oxygen vacancy formation and annihilation on the surfaces of CeO_2 nanoparticles [31]. While the former utilized the lattice expansions as visualized by 4D-STEM nanobeam electron diffraction as a function of temperature [30], latter applied HREM images to reveal cation fluxional behavior at pico-scale resolution [31].

Major advancements have also been made in the field of improving the cryogenic techniques and its applications as discussed in Chapter 4. Similarly, significant progress has been made in developing neural networks and machine learning, which is continuing to develop at a fast pace (see Chapter 9 and Section 10.4.3).

4. **"Enabling Transformative Advances in Energy and Quantum Materials through Development of Novel Approaches to Electron Microscopy", 2020, Sponsored by NSF, Peter Crozier and Lena Kourkoutis (Organizers) September 4–6, 2020 (Virtual) [32].**

Although in-situ and operando TEM measurements were not the targeted theme of this workshop, many of the challenges identified in the area of sustainable energy and quantum science technology benefit from such measurements. For example, imaging and control of in-situ gas–solid interaction can lead to revealing reaction pathways on the particle surfaces. Similarly, liquid cell TEM can be used to probe electrochemical processes for energy storage and conversion systems. Also, quantum effects as function of temperature or magnetic field can help us in developing new devices. Infrastructure discussed in this report, especially employing ML and neural network, is also relevant to develop robust and fast processing methods for in-situ data management and analysis as discussed in Chapter 9 and Section 10.4.3.

Similarly, sample holders with more than one stimulus have become commercially available such that multimodal measurements on the same sample can be made simultaneously. Also, methods to directly measure, the temperature at the nanoscale [18, 20–22, 33, 34] and drift correction algorithms [9] are now available. However, as we move forward – find solutions to existing problems – address current challenges, new questions arise that need to be addressed. Moreover, there are several limitations identified in these workshop reports that still need to be addressed and as the continued technological progress provides new possibilities, it also raises new challenges.

These workshop are important for collective brainstorming and the reports provide guidelines for future technology developments. Fortunately, not only the research groups but both the TEM and TEM holder manufactures have responded positively to address the challenges identified during the past workshops. Examples include, but are not limited to, improving functionality of TEM/STEM units by incorporating high brightness monochromated electron source with high energy resolution, probe, and image aberration correctors to obtain picometer image resolution, modifying TEM column to achieve higher gas pressures around the sample (e.g. Titan 80-300 ESTEM and Hitachi).

> A happy man is too satisfied with the present to dwell too much on the future.
>
> – Albert Einstein

10.3.2 Open Challenges and Technical Roadmaps

Each of the above-described workshops, and some of the smaller ones organized by Department of Energy (DOE) national labs (not included in the list above), has identified challenges/limitation of in-situ TEM experiments and generated ideas for remedying them. Some of them are concrete and some abstract (dreams) but have been carefully considered by researchers and vendors with same amount of vigor, and as explained above, progress has been made. However, as expected, we are always finding new problems to solve that generate new challenges to overcome. We know that there are unknown facts that need to be revealed, and we also need to transform "unknown- unknowns" to "known-unknowns." Unfortunately, all of the problems and challenges are not universal, but some of them are specific to a specific research field. In Sections 10.3.2.1–10.3.2.4, we describe a few examples relevant to specific research field that may be related to other projects not included here.

10.3.2.1 Specific for Battery Research

The simple function of a battery is to convert chemical energy to electrical energy. During a battery's operation, ions are shuttled between anode and cathode during cycling, via solid or liquid electrolyte, where electron generated travel through an external circuit to flow electric current. Chemical reactions at the electrode/electrolyte interface do occur and form deposits of organic and/or inorganic species that limit the transport of ions from the anode and cathode during continued charge and discharge cycling, thereby affecting the lifetime of a battery. Moreover, lithium dendrites may form, which could short-circuit the battery and lead to catastrophic failure. In-situ TEM/STEM has successfully monitored the effect of deposit formation on the battery performance using an electrochemical liquid cell TEM holder with biasing for quantitative electroanalytical measurement capability [35]. Dendritic growth of deposits on the electrodes has also been confirmed [36–38]. However, atomic-resolution imaging and spectroscopy and electron beam–induced changes to the electrolyte composition remain to be problematic.

On the other hand, Li ion batteries function via intercalation and deintercalation mechanism, and the number of successful cycles determines their life. In-situ imaging of intercalation mechanism in SnO_2 and Si nanowires, using ionic liquid as electrolyte, has revealed the volume expansion during intercalation as a source of battery failure [39, 40]. However, it has not been possible to image the Li transport and thereby intercalation and deintercalation due to low contrast of Li atoms. Also, they are knocked out of the lattice by high energy electrons. Cryo TEM has been used to mitigate the electron beam damage to obtain images of solid–electrolyte interface and lithium dendrites of fresh and used batteries but not in-situ [41–43].

In Chapter 4, we have shown that cryo samples can be extracted from a reaction cell at set time intervals to examine a reaction mechanism. Of course, it does not provide the dynamic information of the entire process but can be used to extract intermediated steps. Ideally, following developments are needed for in-situ characterization of battery operations:

1. Process to freeze thin samples during operation in their inherent state.
2. Make thin samples without damaging and keeping them in frozen state.

3. Stable low-temperature and biasing TEM holder, preferably at liquid He temperature to obtain atomic-resolution images and spectroscopy data.
4. Advances in data processing, especially utilizing artificial intelligence, are needed (also see Chapter 9).
5. Developing cryo-biasing holder using thermoelectric materials.
6. Develop 4K TEM for keeping the samples frozen during observation.

A cryo-biasing holder is commercially available from Hummingbird Scientific, but limited applications have been reported so far (https://hummingbirdscientific.com/products/cryo-biasing).

10.3.2.2 Specific for Liquid-Cell TEM

Electron beam is a necessary evil for liquid cell experiments that has been used to initiate the nanoparticle growth and manufacturing of polymeric materials [44, 45]. Cryo TEM has been used to mitigate the electron beam effects at the cost of capturing dynamic observations. Future needs include to be able to access the reaction condition not currently reachable. For example, many electrocatalytic systems – such as the hydrogen evolution reaction during water splitting – only occurs in strong caustic or acidic electrolytes and often at very high current densities [46], conditions that are incompatible with currently available in-situ liquid cell holders.

10.3.2.3 Specific for Catalysis

In-situ observations of gas–solid reactions have successfully revealed a large number of atomic-scale mechanisms under the reactive environments and have made immense contribution toward the growth of the field (Chapter 7). It is desirable to measure the kinetics of a catalytic reaction at nanoscale and compare it with bulk processes. However, current technologies available, ETEM or gas cell, cannot be used under reactor (operando) conditions, mainly due to pressure gap, i.e. reactors operate in higher than one atmospheric pressure. Moreover, there are certain other reactor condition, i.e. plug flow reactor, that are not possible to achieve. Therefore, measuring chemical kinetics of catalytic reactions is not practical.

Recently, design of a novel MEMS based gas microreactor for in-situ observations and measurements, to obtain quantitative structure-property relationship (QSPR), have been reported [47]. The microreactor consists of a main and an auxiliary microfabricated chips that can be assembled together, after loading the sample, using O-ring seals (Figure 10.1). This microreactor can be placed in a commercial TEM gas-cell holder and the gas flow is obtained using holder's gas inlet and outlet lines and controlled through a gas delivery system. An observation and measurement cantilevers are integrated on the main chip to measure the mass change during the reaction and sample is loaded on both cantilevers. A resonant-cantilever gravimetric sensor is also integrated inside the reactor to detect the reaction-induced mass change by measuring the frequency change. A piezo-resistor is integrated on the measurement cantilever to read the mass-change induced frequency-shift with a detection limit of 1 pico-gram [48]. This QSPR reactor was successfully used to reveal the morphological and mass changes associated with sulfurization stages of Cu nanowire in

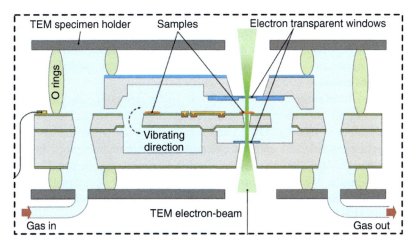

Figure 10.1 Working principle of the in-situ quantitative structure-property relationship microreactor for nanomorphology & physicochemical parameters interrelated characterization. Source: Li et al. [47]/with permission of Elsevier.

H_2S environment and measure the enthalpy $(\Delta H°)$ and the activation energy (E_a) for weak physio-adsorption and irreversible chemisorption of SO_2 adsorption on ZnO nanowires, respectively [47]. They have further modified this microreactor to integrate thermogravimetric analysis (TGA) with in-situ TEM measurements [49]

Also, currently imaging, diffraction pattern (DP), and spectroscopies are acquired in TEM mode as it has better temporal resolution than STEM mode. 4D-STEM and ptychography are attractive alternatives due to the flexibility of post-processing and make it possible to trade real space for reciprocal space containing more information. But the current detectors are not fast enough to acquire 4D-STEM images at ms time resolution. Therefore, we need fast detectors that can be used to synchronize the stimulus cycle with readouts, thus track the dynamic process.

Adsorption and desorption of gas molecules on catalyst surface are a fundamental step of catalytic reaction. Moreover, identifying active sites is important in synthesizing catalyst by design. In principle, in-situ TEM/STEM platform is most suitable to reveal both the gas adsorption and active sites, but the contrast in the images is not strong enough to obtain this information unambiguously and routinely. There are few reports showing gas adsorption sites, using imaging or EELS [50–52], but the methods used are not straightforward and are only achievable under specific condition (see Section 7.5.6).

10.3.2.4 Specific for Quantum Materials

Quantum materials, with reduced dimensionality, such as quantum confinement, quantum coherence, the topology of wavefunctions, quantum fluctuations, electron–electron interactions, and spin–orbit coupling, have found application in quantum computing, communication, sensing, and metrology [53]. In-situ TEM has been used to interrogate the process and conditions to generate a 10 nm Ge quantum disk at Al–Ge contact by heating two (Al and Ge) parallelly aligned nanowires [54]. However, imaging of the quantum dots or quantum behavior (charge ordering) could only be realized using cryo TEM real-space imaging [55–57]. In future, with appropriate technical and data processing developments, in-situ TEM can be used to understand the quantum confinement, orbital ordering, location of light emission sources, etc.

Therefore, we need an instrumental setup with spatial, temporal, and energy resolution of 0.4 nm, 1×10^{-12} seconds, <0.5 meV, respectively. Moreover, TEM holders to heat the sample to 1500 K (1227 °C) and cool down to 4 K (−269 °C) with precise temperature control and stability are required to obtain unambiguous data for quantum materials. 4D-STEM ptychography with post processing and fast detectors is promising solution. We also need to incorporate optical and vibrational spectroscopies for multimodal experiments. Designs for quantum electron microscope and multi-pass TEM are also being developed [58, 59].

> As for a future life, every man must judge for himself between conflicting vague vague probabilities.
> – Charles Darwin

10.4 Developing Relevant Strategies

We need to continue to find new ways to decode the known "unknowns" described above, by developing new technologies, modifying existing technologies, and/or developing intelligent data processing algorithms. First and foremost, we should ask ourselves the questions, commonly asked to the workshop participants, that should help us in developing future visions. For example,

1. What are the characterization, measurement, and analytical challenges and how can we address them?
2. What is keeping us from achieving our ultimate goals?
3. What are the data acquisition and data management issues that need to be resolved?
4. What steps should we take to increase the impact of in-situ TEM in materials research particularly in the advancement of nano- and quantum technology?
5. We have holders combining up to three stimuli in different environments commercially available, do we need to combine more of them on the same platform? If yes, which may be the most popular ones?
6. What is our dream microscope that we want to develop or get developed in next 5 to 10 years?

Based on the answers to these questions, we need to generate new visions and take a novel approach to combine key components such as extensive and multifaceted control of the environment at the sample. We need to move toward building more synergetic platform that include observation of materials at multi-scale (nano to macro) under multiple stimuli in controlled environments [60]. We can divide the areas future developments into three main categories: (i) modifying TEM column to achieve state of the art "lab-in-gap," (ii) developing new technologies to incorporate multiple stimuli, environment, and measurement capabilities on the same sample concurrently, i.e. "lab-on-chip," (iii) developing AI and ML based approaches for data processing, on the fly pattern recognition, TEM alignments, and "live" experimental parameter control (automation and autonomous experiments).

It is important to note that most of the currently available commercially technologies were started at lab scale by research groups, from basic TEM [61], including ETEM [62–65], to modified TEM holders [66–69]. Therefore, it is not abnormal for individual researcher or a group to develop new techniques or modify existing ones. In Sections 10.4.1 to 10.4.3.2 describe examples of some of the reported new modifications and proposed developmental ideas currently being pursued including my personal input for the three categories mentioned above.

> Nothing promotes the advancement of science so much as a new instrument.
> – H. Davey 1806

10.4.1 Modifying Base TEM/STEM Unit

Our capability to perform experiments in the TEM is very much controlled by the base instruments available in the market. Modifying a TEM in different configuration is often ignored as (i) it requires a lot of time and effort, during which a TEM may not be available for regular operations, and (ii) possibility to damage such an expensive instrument. However, as mentioned earlier, such modifications, which started in the research labs, have been made in the past, i.e. ETEM [62–65], UEM [2], DTEM [70], PIES [71], incorporation of Raman-CL [15], and ion radiation [72–76]. While ETEM is now commercially available, others are still in developmental stages, with each lab making modifications to suit their needs. Incorporation of other stimuli such as ion and photon source in the sample chamber has also been successfully accomplished [15, 77–81].

Another example of modifying TEM column by a research group is to synchronize electron beam with the phase shift induced by AC current while making electrical property measurements [82]. Soma et al. describe a modification made to obtain phase locked images, to remove the blurring due to the frequency and amplitude of the applied AC voltage, mimic laser-induced stroboscopic electron generation. However, it is comparatively simple modification that employs external controls to the pre-sample beam blanking system to pulse the electrons through the sample for imaging (Figure 10.2). The shutter speed τ_{min} is determined from the hysteresis of the magnetic AC scan coils and is estimated to be 12.5 μs. It is noted that the value of a sinusoidal function varies the most at $\phi = \pi \pm m\pi$, while it becomes stationary at $\pi/2 \pm m\pi$ (m is an integer). Therefore, the blur in TEM images is likely to be minimized by locking at $\phi = \pi/2 \pm m\pi$. The pulse signal is synchronized with an applied AC voltage, used for making electrical measurements. A number of phase-locked images, typically of the order of 10 000, are recorded using a charge-coupled detector (CCD) camera and integrated to obtain final phase-locked image. This modified TEM was also used for holography and to obtain EELS data [82]. Another example

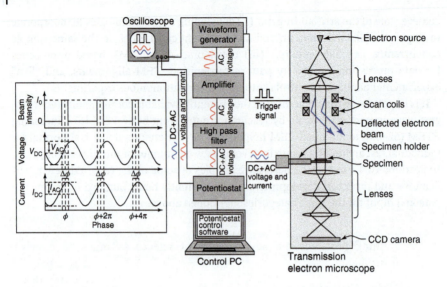

Figure 10.2 Block diagram of an apparatus for phase-locked strobe transmission electron microscopy (TEM). A specimen is observed by synchronization with an electric measurement under an applied AC voltage, V_{AC} (with or without DC bias voltage, V_{DC}) using a potentiostat connected to a control PC. To generate a synchronized pulsed electron beam, a shutter pulse signal is generated from an applied AC voltage using a waveform generator, filter, and amplifier. AC current I_{AC} and DC current I_{DC} as well as V_{AC} and V_{DC} are measured by the potentiostat and displayed on a multichannel oscilloscope. The specimen is irradiated by a phase-locked strobe electron beam that is characterized by beam intensity (electron current density), targeted phase ϕ, and phase width $\Delta\phi$. Source: Soma et al. [82]/Elsevier.

is the proposed design of a multi-pass TEM of beam sensitive materials where the sample is placed between objective lens and field lens encompassed by two mirrors. The feasibility of this approach, which could under-idealized conditions be considered interaction-free, is demonstrated using simulations [59].

A group from National Research Council–Canadian National Research Council (NRC-CNRC) from Canada is currently designing an open access and flexible platform to build and maintain affordable educational SEM, TEM. STEM instrument, called NanoMi [83]. Their plan is to buy hardware components from various sources, build the column, and employ ML-based software controls for operation. This innovative project is under development and open for collaborators, i.e. others can join the group with their ideas and build a microscope to their own specifications [83].

Still, some of these modified TEM/STEM platforms have restricted applications for in-situ observations, especially with respect to spatial and temporal resolutions. For example, while observing dynamic changes at sub-nanometer resolution are routinely made using aberration corrected microscopes with monochromated electron source, temporal resolution is limited to ms using a direct electron detector. On the other hand, femto second resolution is possible using different modes of a UEM, DTEM, or stroboscopic microscope, but spatial resolution in limited in nanometer range.

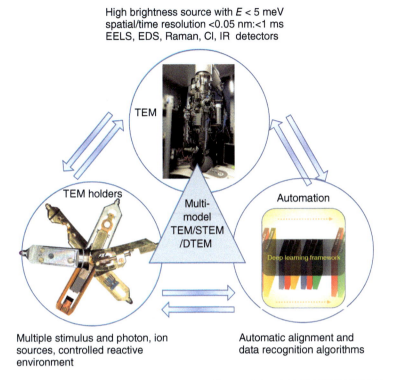

Figure 10.3 Elements to be incorporated in a multimodal TEM/STEM/DTEM instrument. Source: Adapted from Ziatdinov et al. [121] [2017].

Ideally, we want to have a TEM/STEM with high-brightness source, >10 mm pole-piece gap, <0.05 nm spatial resolution, <1 ms temporal resolution, multiple ports, <5 meV energy resolution at 80 keV. Other key performance designs include bidirectional optical access, fast-switching electron-optical elements, high-resolution angle-resolved energy loss spectroscopy and integrated control system for auto alignment, fast data transfer for data analysis (Figure 10.3).

Hattar and Jungjohann have proposed to develop a new setup that combines multiple modes such that TEM, STEM, and DTEM operations can be switched easily [29]. Figure 10.4 shows their proposed design that is based on the fact that the three modes differ in the way electrons/photoelectrons are generated and focused on the sample, but the optics of image formation and signal collection is the same for all three modes. The proposed prototype will provide rapid switching between imaging, diffraction, and spectroscopy techniques by interconnecting the different illumination systems by electrostatic lens directly above the objective pole piece. It is clear from Figure 10.4 that the illumination systems of these microscopes are too different from each other to be combined in a single column, so they will need to be integrated by coupling them to the objective lens pole piece through rapid switchable electrostatic lenses. Furthermore, the specimen assembly needs to be flexible enough to enable custom designed experiments to be performed in the pole-piece gap. Last but not the least, the detectors should be designed to collect all electrons, photons, ions,

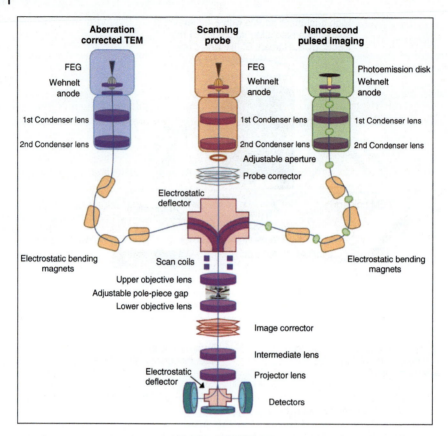

Figure 10.4 Proposed schematic of the integrated transmission electron microscope. Comprised of a multifaceted illumination system, an open-designed sample region, and a detection assembly design to synchronize imaging modes to detectors/cameras/spectrometer. Source: Hattar et al. [29]/Springer Nature/CC BY 4.0.

or other signals to obtain ideal SNR for imaging and spectroscopy. It can be combined with direct detection EELS to push the limits of sensitivity and energy resolution for a Schotky-FEG to 30 meV [62]. Incorporation of a GC-MS system attached to the gas outlet is highly desirable to monitor gas compositions at ppm level during catalytic reactions [84, 85]. The proposed setup has multiple practical issues associated, and it will be major undertaking to reduce it to practice, but it is worth thinking about.

Now we discuss other possible modifications related to the TEM column to realize the dream of "lab-in-gap." As we can imagine, the most crucial part utilized for in-situ observations is the sample chamber and the gap between the objective pole pieces. This gap is directly related the C_s of the objective lens, and thereby the spatial resolution, i.e. smaller the gap, smaller C_s value and higher the spatial resolution. Which means that sample holders with thin tips can only be used with limited tilt angle for aligning the sample along zone axis or for collecting EDS signal in a TEM with small (≈ 2 mm) pole-piece gap. Therefore, multicomponent liquid or gas cell holder may be too thick to fit between a pole-piece gap of 2 mm or less.

In order to keep reasonable tilt capability and sample holders with thick tip, the objective pole-piece gap needs to be at least 5 mm or larger. Also, currently multiple compromises have to be made to introduce other functionalities, such as photon or ion sources in the small gap [71, 81]. As explained in Chapter 1, generally the pole-piece gap for high-resolution TEM/STEM is smaller than for an ATEM. But it is expansive to have two or three microscopes in an individual lab, and ideally, we would like to open the pole-piece gap without compromising high-resolution capability and it is possible.

First, it has been shown, theoretically and practically, that increased C_s due to the larger objective pole-piece gap can be compensated by incorporating aberration correctors [86]. Yamamoto has installed a high-resolution CL system by opening the gap a JEOL microscope to 9 mm and can still acheive 1 nA current in a 1 nm probe for STEM applications [87]. We have obtained 0.06 nm resolution at 80 keV using a TEM with 5.5 mm gap equipped with a monochromated source and aberration corrector.

An alternative solution, proposed by Dr. Lewy Jones group that is under development, is very attractive [88]. They propose building a user-adjustable pole piece (UAP), where the modular design will allow users to adjust the gap according to their experimental requirement without any need to open the column to disassemble/remove any other lenses, detectors, goniometer (Figure 10.5). This unit must be stable, easy to realign the objective lens optics, and is currently designed for JEOL 200/300 keV microscope. The group has checked the feasibility and the functionality of this UAP by performing finite element simulations with collaborators from Sandia National Laboratory [89]. Reducing these design ideas to practice will revolutionize our capabilities of performing experiments and making in-situ measurements in TEM.

Yet another simple change to the sample chamber is to increase the number and the locations of ports such that incorporating external stimuli through a port will be straightforward. For example, incorporation of in-situ Raman or CL setup in an ETEM required us to remove the objective aperture rod as it was the only port available to introduce photon source and collect Raman, CL, and PL signals [15].

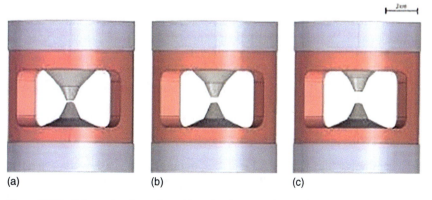

Figure 10.5 Schematic of an adjustable pole piece gap, (a) small, (b) medium, and (c) larger designed for a JEOL microscope such that to accommodate different holders including high tilt angle and in-situ holders with thick tip. Source: Adapted from McBean et al. [88, 89].

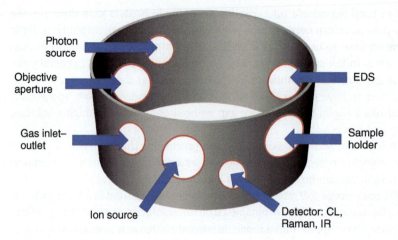

Figure 10.6 Modified specimen chamber with multiple ports to incorporate other capabilities.

Similarly, the location of an available port to introduce light or ion sources should be at an angle such as to have direct line of sight to the sample without hitting the pole pieces [81]. Increasing the number of ports on the sample chamber, combined with larger pole-piece gap, will allow us to incorporate a number of independent stimuli and/or detectors (Figure 10.6). For example, collecting CL signal using a small parabola that can fit under the sample takes long time as most of the signal is not collected. Enclosing the sample between two parabolas, one above and the other below the sample, can allow us the collect most of the signal emitted from the sample, thereby improving the time resolution. Similarly, a tuning fork-based IR detector can be introduced through one of the ports. Inclusion of ion and/or photon source and detectors, gas sensors [90], through available port will also lower the burden of jamming multiple stimuli within the limited space available on sample holder's tips. It is important to make sure that the stimuli or detectors introduced through these multi-modal ports can be aligned along line of site to the sample without causing geometric distortions. Feasibility of such modification has been demonstrated by Hotz et al., who have added three new ports to the sample chamber for a NION UHV STEM that can be used to incorporate gas or ion injection source and secondary electron detector [91, 92].

Automatic and/or remote alignment control, similar to electron beam translation control for STEM [93] and remote control of sample tilt [94], should also be considered.

10.4.2 TEM Holders with Multiple Stimuli

Now let us discuss the ideas for "lab-on-chip" outlined in the DOE-workshop report referred above in Section 10.3.1. If we can modify the TEM column to increase the available space in the sample chamber, as described above, increased tip thickness of a multimodal sample holders can be used without sacrificing the tilt capabilities of the holder. As we know, TEM holders have been successfully modified to apply external stimuli on the sample, e.g. heating (Chapter 3), cooling (Chapter 4), mechanical (Section 1.7.5.1) biasing (Section 1.7.5.3), magnetic (Section 1.7.5.4), control of the environment around the sample such as liquid (Chapter 6) or gas (Chapter 7). Moreover, following is a list of sample holders with more than one stimulus that are commercially available:

- Heating and biasing (electrical)
- Heating and biasing in liquids
- Mechanical and heating
- Heating in gas or liquid
- Photon and electrical in liquids
- Thermal and Mechanical
- Cryo with biasing

The website link for the commercial vendors for these holders is provided in Refs. [95–98].

However, TEM holder with all combinations of stimuli (desirable) in a controlled environment is still not commercially available. Such TEM holder design will be directed toward realizing our dream to have "lab-in-gap" with "lab-on-chip," i.e. perform experiments within the objective pole-piece gap of the TEM using a microfabricated chip. Overall, we would like to combine thermal, electrical, mechanical, magnetic, stimuli under electron, photon, and ion radiation in controlled liquid or gaseous environments and the detectors to make dynamic time-resolved measurements. A short summary of desired combination of stimuli – environment and their enabling technologies is given in Table 10.1.

Researchers are making efforts to design and/or build holder to combine multiple stimuli, environments, and/or property measurement capabilities on a microfabricated chip. For example, Hattar and Jungjohann have outlined a scenario to combine, thermal, mechanical, electrical, and ion radiation in controlled liquid and/or gaseous environment on one platform (Figure 10.7a) to provide a flexibility to choose any combination of experimental conditions [29]. Their group have also proposed to design and develop a microfabricated device that can be used for mechanical testing of materials under variable environments [99]. Figure 10.7b shows the schematic of such microfabricated device, enclosed by SiN_x windows, to allow sample to be immersed in gaseous or liquid environments and the chip includes a heater, an actuator, sensor, and biasing electrode. It is sized to fit in the tip area of any TEM holder. The small dimensions (≈ 1 mm $\times$ 4 mm $\times$ 6 mm) of the device allow it to be used for correlative in-situ characterization using TEM, STEM, X-ray transmission electron microscope (XTM), and X-ray diffraction (XRD) [99].

Another design including multiple connections on a chip such that the sample in a liquid environment can be heated, electrically biased, and/or subjected any other stimulus is shown in Figure 10.7c [100]. The microfabricated Si chip (4 mm $\times$ 15 mm) includes four reservoirs attached to liquid flow channels (Figure 10.7d) and a set of electrical contacts is used to flow the liquid into the sample area, and fits in 2 mm thick tip of a home-built holder. Another main feature of this design is the monolithic liquid cell (Figure 10.7e) that mitigates the window

Table 10.1 Possible combination of stimuli in controlled environment and technologies enabled.

Stimuli	Environment	Structure-properties measured	Technology enabled	Current status/References
Mechanical–thermal	Enclosed gas cell	Deformation as function of heat, mechanical force, and environment	Aerospace, turbines	N/A
Mechanical–electrical–thermal	gas	Structural modification leading to embrittlement of alloys and ceramics	Solid state batteries and solid oxide fuel cells	N/A
Thermal–mechanical–electrical	Liquid	Corrosion as a function these stimuli	Li ion battery operation and their stability	N/A
Thermal – photon – electrical	liquid	Photo-electro catalysis as a function of temperature	Hydrogen production, CO_2 reduction	N/A
Thermal – ion	Gas/liquid	Deterioration of nuclear materials with temperature in controlled environments	Nuclear technology	N/A
Thermal – magnetic	gas	Magnetic properties as a function of temperature in ambient	Memory devices	N/A
Thermal–weight change	gas	Measure reaction kinetics	Catalysis/operando	Lab model [48]

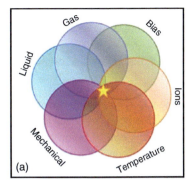
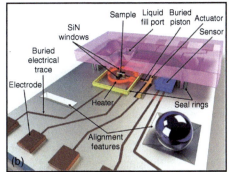
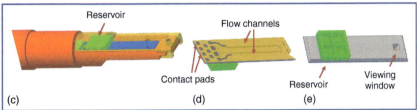

Figure 10.7 (a) Schematic showing the possible overlapping between the combined environments obtainable with the integrated TEM for operando experiments. Source: Hattar et al. [29]/Springer Nature/CC BY 4.0. (b) Labeled schematic of mechanical–electrical–thermal MEMS platform base. Source: Bhowmick et al. [99]/from Claire Chisholm. (c, d) Design of integrated monolithic liquid/gas cell holder, (c) front end of the holder with liquid reservoir (green) and microchip, (d) microchip with 11 contact pads and flow channels, (e) dissected view of the chip showing the location of the reservoir and viewing window. Source: Adapted from Liddle et al. [100].

bulging and can tolerate high pressure [101]. Although the functioning of the monolithic liquid cell has been tested, the entire setup is still a work in progress.

We also need Cryo holders with stable operation at temperatures as low as 4 K to make atomic-resolution in-situ measurements at low temperatures. Moreover, a fast temperature stabilization will allow us to perform time and temperature-resolved measurement below RT. The design proposed by Zandbergen (HennyZ) is a step in right direction [102], but (i) we need it to be improved for continuous He flow and (ii) make it commercially available. Cryo TEM coupled with other in-situ capabilities including biasing, heating, photoexcitation will allow the tuning of the quantum states.

Other research groups have also designed and built or proposed designs of new combinatorial approaches to include several stimuli on the same microfabricated chip connected to independent controllers outside the TEM column. For example, mechanical testing observations can be made in vacuum [103, 104], or low gas pressures achievable in an ETEM at variable temperatures [105]. Some of these multimodal approaches, not yet commercially available, are listed below:

o A double-probe, multifunction holder to introduce photons, induce static charge and probe current (see Figure 8.8) [106].

- TEM holder generating in-plane magnetic field [107].
- A design to fabricate a microchip to follow effect of thermal and mechanical stimuli (Figure 10.7b) [99].
- A liquid cell design with 11 connection pads to control liquid flow and other stimuli is also at the design stage (Figure 10.7c–e) [100].

One of the interesting suggestions from the workshop reports is to design and develop cartridge-based approach for sample loading. It has been employed for making a microfabricated gas cell such that it can be used in TEM, SEM, or on a synchrotron beam line [108]. A slightly different design has been reported for 3 mm grids such that the sample can be tilted using a wireless inclinometer, which has a precision of 0.1°, to monitor the sample holder tilt angle independent of the microscope goniometer readout [94]. We should consider combining these two designs to incorporate wireless controls through the TEM sample chamber ports, instead of holder, to mitigate the wired connections to the controllers located outside the column. We know that these wires are prone to introduce vibrations in the sample that could be difficult to mitigate.

Last but not the least are the detectors that control the time resolution unless we are using pump and probe mode. Introduction of direct electron detectors has made a difference, but further improvements are needed.

10.4.3 Automation and Autonomous Operation

In-situ TEM experiments, irrelevant of the mode or technique used, are notorious for generating large datasets, of the order of $GB\,s^{-1}$ and create a number of issues, i.e. read and write rates, data transfer and storage, and analysis. In Chapter 9, we have described some of the currently available paths for loss-less data reduction, computer-controlled data recognition algorithms to help us discard the unusable and redundant data, automatic image processing and measurement schemes, incorporation of ML to improve the analysis time, etc. We will not repeat them in this chapter but would look at how we can take advantage or improve existing methods for automation and finally for autonomous operation of in-situ TEM experiments.

First, let us clarify the terms "automation" and "autonomous operation" with respect to in-situ TEM experiments. Automation includes aligning the TEM and controlling the reaction parameters to (i) save time and (ii) maintain consistency between experiments. On the other hand, a fully autonomous operation means that desired experimental parameters and expected results are fed to the computer that controls the TEM alignment and experimental conditions, analyzes the results, tweaks the parameters until the desired result is obtained. We can opt for a partial autonomous operation where we can look at the results and decide how to proceed as described later in this section.

10.4.3.1 Automation

The need for automation for in-situ TEM experiments is primarily to save time during data collection and analysis. We have discussed the data management and automation for data processing in Chapter 9. In this section, we will concentrate on the need for automation for TEM/STEM alignments, current efforts, and path toward performing autonomous in-situ experiments [109, 110].

TEM/STEM alignment is a tedious process and must be done before starting an in-situ experiment. Moreover, we need electron beam to follow the alignment protocol and for certain samples/experiments, this time can have detrimental effects. The data collection during in-situ is also time-consuming and requires continuous human input, mainly to correct the subtle changes such as defocus, objective astigmatism, electron beam alignment, etc. Thus, the data collection requires constant attention from the operator for hours that is very tiring and could lead to making mistakes. An automation algorithm can not only save time and effort but lead to producing more coherent and reproducible results.

Biologists have developed automation algorithms and are employing them for some time [111]. But unlike biological samples, it is not as straight for materials as our samples differ in morphology, structure composition, and physical properties. Historically, partial automation for TEM alignments has been available for GIF and aberration correctors, where human intervention is needed, but is faster and consistency can be achieved. Recently, microscope companies and researchers have started to develop ML-based algorithms to control and align electro-optical components of TEM. For example, NION has developed an open-source software for integrating hardware control, data acquisition, visualization, processing, and analysis using Python [112]. Similarly, Thermo Fisher Scientific has released Smart EPU software for intelligent cryo-TEM data collection and processing [113]. They have also developed an auto-control software for their TEM/STEM alignments.

At the same time, research groups are also developing methods for independently control the optics of the TEM and develop auto-alignment algorithms. For example, Sang et al. have implemented a LabView to control the position, scan rate, and scan direction of the electron probe for STEM imaging [93]. They have employed this technique to generate position-controlled direct-write in liquids [114] and used spiral scans to acquire distortion-free STEM images [115].

Recently, LeBeau's group at Massachusetts Institute of Technology (MIT) has implemented reinforcement learning (RL) to automate the STEM operations and data collection [116]. They report the design and implementation of a virtual RL environment to develop and test a network to autonomously align the electron beam position without prior knowledge. The TEM alignments require to control the continuously changing parameters; therefore, Soft Actor-Critic (SAC) algorithm from the Stable Baselines3 Python package was implemented [117]. The flow and the microscope environment are shown in Figure 10.8. The RL agent is used to select an action, receive observed information and a reward value from the interaction with the microscope environments, and update the internal neural network weights for best action needed. The sensitivity, reward design, environment subsampling, goal tolerance, number of neurons, and step size of RL algorithm were evaluated to augment STEM workflow. The final environment template and RL training script can be downloaded from (https://github.com/LeBeauGroup/DiffAlignRL) [116].

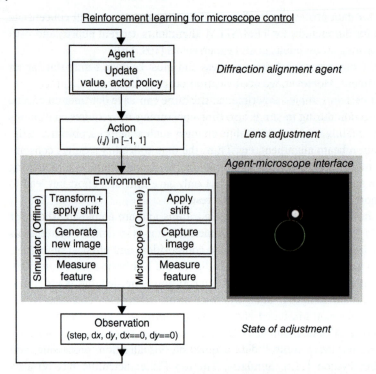

Figure 10.8 Overview of the reinforcement learning process for diffraction alignment. The agent repeatedly samples and sends actions in the form of a diffraction shift, which in turn produces a simulated (offline) or measured (online) environment response that can be observed. The resulting observation is used to update the value and actor policies, which dictate future decision making. A custom Python program was developed to access the phosphor screen FluCam and send commands to the microscope. Image acquisitions from the camera are overlaid with the screen center (green) as well as the current image center of mass (red). Source: Xu et al. [116]/Cambridge University Press.

10.4.3.2 Autonomous Experiments

The interest in autonomous in-situ experiments arises from the fact that the operator has to acquire, control, and record many parameters simultaneously. Also, slight change in the TEM alignment, such as beam tilt and/or image focus, can affect the accuracy of structural interpretation. Recent developments for automatic TEM alignment and logging of experimental parameters have the potential to reduce the number of chores and improve alignment consistency. The time it takes for us to recognize the change and react to make amends is much slower than for a trained neuron. Moreover, we are always looking for specific features or changes that introduce a large degree of bias in data collection. Those of us who have recorded videos can vouch for times when we have missed an interesting phenomenon that was happening just outside the recorded region, or we missed the focus change during the observation period! Also, our discriminative capability is limited and can be strongly affected by the chosen contrast setting or color scale.

As mentioned above, consistency and reproducibility are needed to obtain reliable results. Universal reproducibility of in-situ TEM results is dependent on the precise control on the large number of variables include design of the experiment and TEM alignment protocols. Although we can find qualitative consensus in literature for some of the in-situ experiments, such as charge and discharge mechanism of Li ion batteries, Si nanowire growth, ceria reduction or carbon nanotube growth, but there is scatter in quantitative parameters such as temperature, pressure, and/or reaction rates. Therefore, there has been a push to utilize machine learning schemes, similar to driving an autonomous car or robot, for autonomous in-situ experiments or autonomous experiments (AE) in short [118].

A group of researchers at Oak Ridge National Lab, led by Sergei Kalinin, have made some progress toward achieving autonomous analysis for EELS, STEM, and STM data [119–121]. They have proposed a method for automated structural analysis of STEM images using a priori knowledge, gained either from directly scanned experimental images or from a library of images, such as ImageNet [122]. The workflow, directed toward structure discovery, as illustrated in Figure 10.9 for three different scenarios, is divided in three parts: (i) image segmentation, (ii) linear or nonlinear dimensionality reduction (potentially allowing physics-based invariances), and (iii) output data collection. Although there are different ways to address each step, they report using two methods for image segmentation and decomposition, respectively, and Bayesian optimization for output data collection [123]. The methods used for image segmentation are pre-trained deep convolutional neural networks (DCNNs) or using fast Fourier transform (FFT) of sub-images. For decomposition they use non-negative matrix factorization (NMF)/single value detection (SVD) or unsupervised rotationally invariant variational autoencoders (rVAEs). They report that the workflow is highly dependent on the data set with some common trends. Overall, a sliding window and image deconvolution is ideally

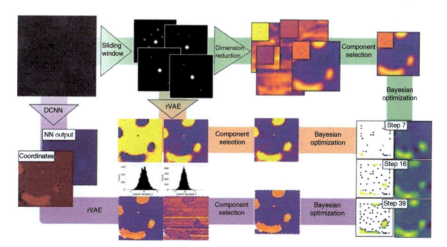

Figure 10.9 Workflow for the training and discovery phase for automated experiments with three different methods: human-driven (green), partially human-driven (orange), and computer-driven (purple). Source: Creange et al. [123]/IOP Publishing/CC BY 4.0.

suited for images with well-defined features, while workflow employing DCNN and rVAE is best for atomic-scale features. Further details about the samples used and the outcome can be found in Ref. [123].

There are several other reports of utilizing ML-based algorithms to conduct automated or AE using STEM [10] but not as many for TEM, probably due to the fact that contrast is more pronounced in STEM than in TEM images. However, it may be possible to implement these ideas and/or algorithms to conduct AE using TEM. While designing universal control for experimental factors is complicated, there are efforts made in this direction and hope that running autonomous in-situ TEM experiments will be realized soon that will also provide a control on correlative experiments.

10.5 Take Home Messages

At the end, we can conclude that despite of the significant progress made to conduct successful in-situ TEM/STEM experiments, there is more to be done. Currently, our success is limited by the technical and analytical capabilities available. In-situ experiments belong to the "high risk, high reward" category of research. It is clear that we have been successful in making a difference in understanding of the catalytic reaction mechanisms, nucleation and growth of nanomaterials, establishing electrical, mechanical, and magnetic structure–property relationships, and deciphered the origin of novel physical properties of 2D materials, to name a few. However, we need to continue to find solutions to current roadblocks.

In this chapter, we have reviewed possible technical developments for modifying TEM column and the sample chamber, building multimodal microfabricated sample holders, finding pathways for automation and AE using ML and AI. It is important for us to continue to dream big and think outside the box, as it pays off in the long run.

References

1 Montgomery, E., Leonhardt, D., and Roehling, J. (2021). Ultrafast transmission electron microscopy: techniques and applications. *Microscopy Today* 29 (5): 46–54.
2 Zewail, A.H. (2006). 4D ultrafast electron diffraction, crystallography, and microscopy. *Annual Review of Physical Chemistry* 57 (1): 65–103.
3 LaGrange, T., Armstrong, M.R., Boyden, K. et al. (2006). Single-shot dynamic transmission electron microscopy. *Applied Physics Letters* 89 (4): 044105.
4 Lau, J.W., Schliep, K.B., Katz, M.B. et al. (2020). Laser-free GHz stroboscopic transmission electron microscope: components, system integration, and practical considerations for pump–probe measurements. *Review of Scientific Instruments* 91 (2): 021301.
5 Eswara, S., Pshenova, A., Yedra, L. et al. (2019). Correlative microscopy combining transmission electron microscopy and secondary ion mass spectrometry: a general review on the state-of-the-art, recent developments, and prospects. *Applied Physics Reviews* 6 (2): 021312.

6 Sharma, R., Crozier, P.A., and Treacy, M.M.J. (2006). Dynamic in-situ electron microscopy as a tool to meet the challenges of the nanoworld. http://www.asu.edu/clas/csss/workshops/NSF_WS.

7 Lobastov, V.A., Srinivasan, R., and Zewail, A.H. (2005). Four-dimensional ultrafast electron microscopy. *Proceedings of the National Academy of Sciences of the United States of America* 102 (20): 7069.

8 Taheri, M.L., Stach, E.A., Arslan, I. et al. (2016). Current status and future directions for in situ transmission electron microscopy. *Ultramicroscopy* 170: 86–95.

9 Ophus, C., Ciston, J., and Nelson, C.T. (2016). Correcting nonlinear drift distortion of scanning probe and scanning transmission electron microscopies from image pairs with orthogonal scan directions. *Ultramicroscopy* 162: 1–9.

10 Ophus, C. (2019). Four-dimensional scanning transmission electron microscopy (4D-STEM): from scanning nanodiffraction to ptychography and beyond. *Microscopy and Microanalysis* 25 (3): 563–582.

11 Gammer, C., Ozdol, V.B., Liebscher, C.H., and Minor, A.M. (2015). Diffraction contrast imaging using virtual apertures. *Ultramicroscopy* 155: 1–10.

12 Ryll, H., Simson, M., Hartmann, R. et al. (2016). A pnCCD-based, fast direct single electron imaging camera for TEM and STEM. *Journal of Instrumentation* 11 (04): P04006–P04006.

13 McMullan, G., Faruqi, A.R., Clare, D., and Henderson, R. (2014). Comparison of optimal performance at 300 keV of three direct electron detectors for use in low dose electron microscopy. *Ultramicroscopy* 147: 156–163.

14 Liao, H.G., Zherebetskyy, D., Xin, H. et al. (2014). Facet development during platinum nanocube growth. *Science* 345 (6199): 916–919.

15 Picher, M., Mazzucco, S., Blankenship, S., and Sharma, R. (2015). Vibrational and optical spectroscopies integrated with environmental transmission electron microscopy. *Ultramicroscopy* 150: 10–15.

16 Kim, T.-H., Bae, J.-H., Lee, J.-W. et al. (2015). Temperature calibration of a specimen-heating holder for transmission electron microscopy. *Applied Microscopy* 45 (2): 95–100.

17 Li, M., Xie, D.-G., Zhang, X.-X. et al. (2021). Quantifying real-time sample temperature under the gas environment in the transmission electron microscope using a novel MEMS heater. *Microscopy and Microanalysis* 27 (4): 758–766.

18 Mecklenburg, M., Hubbard, W.A., White, E. et al. (2015). Nanoscale temperature mapping in operating microelectronic devices. *Science* 347 (6222): 629–632.

19 Miller, B., Perez Garza, H.H., and Mecklenburg, M. (2018). Local in situ temperature measurements from aluminum nanoparticles. *Microscopy and Microanalysis* 24: 1924–1925.

20 Niekiel, F., Kraschewski, S.M., Müller, J. et al. (2017). Local temperature measurement in TEM by parallel beam electron diffraction. *Ultramicroscopy* 176: 161–169.

21 Yan, X., Liu, C., Gadre, C.A. et al. (2019). Unexpected strong thermally induced phonon energy shift for mapping local temperature. *Nano Letters* 19 (10): 7494–7502.

22 Idrobo, J.C., Lupini, A.R., Feng, T. et al. (2018). Temperature measurement by a nanoscale electron probe using energy gain and loss spectroscopy. *Physical Review Letters* 120 (9): 095901.

23 Zuo, J.M., Vartanyants, I., Gao, M. et al. (2003). Atomic resolution imaging of a carbon nanotube from diffraction intensities. *Science* 300 (5624): 1419–1421.

24 Zheng, H., Yimei, Z., and George, M. (2014). Future of electron scattering and diffraction. https://www.osti.gov/biblio/1287380.

25 Xu, K., von Cresce, A., and Lee, U. (2010). Differentiating contributions to "Ion Transfer" barrier from interphasial resistance and Li^+ desolvation at electrolyte/graphite interface. *Langmuir* 26 (13): 11538–11543.

26 Liao, H.-G., Cui, L., Whitelam, S., and Zheng, H. (2012). Real-time imaging of Pt_3Fe nanorod growth in solution. *Science* 336 (6084): 1011–1014.

27 de Oteyza, D.G., Gorman, P., Chen, Y.-C. et al. (2013). Direct imaging of covalent bond structure in single-molecule chemical reactions. *Science* 340 (6139): 1434–1437.

28 Fausti, D., Tobey, R.I., Dean, N. et al. (2011). Light-induced superconductivity in a stripe-ordered cuprate. *Science* 331 (6014): 189–191.

29 Hattar, K. and Jungjohann, K.L. (2021). Possibility of an integrated transmission electron microscope: enabling complex in-situ experiments. *Journal of Materials Science* 56 (9): 5309–5320.

30 Ding, Y., Choi, Y., Chen, Y. et al. (2020). Quantitative nanoscale tracking of oxygen vacancy diffusion inside single ceria grains by in situ transmission electron microscopy. *Materials Today* 38: 24–34.

31 Lawrence, E.L., Levin, B.D.A., Boland, T. et al. (2021). Atomic scale characterization of fluxional cation behavior on nanoparticle surfaces: probing oxygen vacancy creation/annihilation at surface sites. *ACS Nano* 15 (2): 2624–2634.

32 Crozier, P.A. and Lena, K. (2020). Enabling transformative advances in energy and quantum materials through development of novel approaches to electron microscopy. https://www.temfrontiers.com.

33 Winterstein, J.P., Lin, P.A., and Sharma, R. (2015). Temperature calibration for in situ environmental transmission electron microscopy experiments. *Microscopy and Microanalysis* 21 (6): 1622–1628.

34 Vijayan, S. and Aindow, M. (2019). Temperature calibration of TEM specimen heating holders by isothermal sublimation of silver nanocubes. *Ultramicroscopy* 196: 142–153.

35 Unocic, R.R., Sacci, R.L., Brown, G.M. et al. (2014). Quantitative electrochemical measurements using in situ ec-S/TEM devices. *Microscopy and Microanalysis* 20 (2): 452–461.

36 Sacci, R.L., Black, J.M., Balke, N. et al. (2015). Nanoscale imaging of fundamental Li battery chemistry: solid-electrolyte interphase formation and preferential growth of lithium metal nanoclusters. *Nano Letters* 15 (3): 2011–2018.

37 Mehdi, B.L., Qian, J., Nasybulin, E. et al. (2015). Observation and quantification of nanoscale processes in lithium batteries by operando electrochemical (S)TEM. *Nano Letters* 15 (3): 2168–2173.

38 Leenheer, A.J., Jungjohann, K.L., Zavadil, K.R. et al. (2015). Lithium electrodeposition dynamics in aprotic electrolyte observed in situ via transmission electron microscopy. *ACS Nano* 9 (4): 4379–4389.

39 Huang, J.Y., Zhong, L., Wang, C.M. et al. (2010). In situ observation of the electrochemical lithiation of a single SnO_2 nanowire electrode. *Science* 330 (6010): 1515.

40 Wang, C.M., Xu, W., Liu, J. et al. (2010). In situ transmission electron microscopy and spectroscopy studies of interfaces in Li ion batteries: challenges and opportunities. *Journal of Materials Research* 25 (8): 1541–1547.

41 Wang, X., Zhang, M., Alvarado, J. et al. (2017). New insights on the structure of electrochemically deposited lithium metal and its solid electrolyte interphases via cryogenic TEM. *Nano Letters* 17 (12): 7606–7612.

42 Zachman, M.J., Tu, Z., Choudhury, S. et al. (2018). Cryo-STEM mapping of solid–liquid interfaces and dendrites in lithium-metal batteries. *Nature* 560 (7718): 345–349.

43 Li, Y., Li, Y., Pei, A. et al. (2017). Atomic structure of sensitive battery materials and interfaces revealed by cryo-electron microscopy. *Science* 358 (6362): 506–510.

44 Gibson, W. and Patterson, J. (2021). Liquid phase electron microscopy provides opportunities in polymer synthesis and manufacturing. *Macromolecules* 54 (11): 4986–4996.

45 Liao, H.-G. and Zheng, H. (2016). Liquid cell transmission electron microscopy. *Annual Review of Physical Chemistry* 67 (1): 719–747.

46 Osterloh, F.E. (2013). Inorganic nanostructures for photoelectrochemical and photocatalytic water splitting. *Chemical Society Reviews* 42 (6): 2294–2320.

47 Li, W., Li, M., Wang, X. et al. (2020). An in-situ TEM microreactor for real-time nanomorphology & physicochemical parameters interrelated characterization. *Nano Today* 35: 100932. (9pages).

48 Li, W., Yu, H., Yao, F., et al. A MEMS Reactor for Observing Morphology-Evolution in Tem and Simultaneously Detecting Involved Molecule-Number Change During Nano-Constructing Reaction. *in 2019 IEEE 32nd International Conference on Micro Electro Mechanical Systems (MEMS)*, 2019.

49 Yao, F., Xu, P., Li, M. et al. (2022). Microreactor-based TG–TEM synchronous analysis. *Analytical Chemistry* 94 (25): 9009–9017.

50 Yoshida, H., Kuwauchi, Y., Jinschek, J.R. et al. (2012). Visualizing gas molecules interacting with supported nanoparticulate catalysts at reaction conditions. *Science* 335 (6066): 317–319.

51 Yuan, W., Zhu, B., Li, X.-Y. et al. (2020). Visualizing H_2O molecules reacting at TiO_2 active sites with transmission electron microscopy. *Science* 367 (6476): 428–430.

52 Yang, W.-C.D., Wang, C., Fredin, L.A. et al. (2019). Site-selective CO disproportionation mediated by localized surface plasmon resonance excited by electron beam. *Nature Materials* 18 (6): 614–619.

53 Cava, R., de Leon, N., and Xie, W. (2021). Introduction: quantum materials. *Chemical Reviews* 121 (5): 2777–2779.

54 Luong, M.A., Robin, E., Pauc, N. et al. (2020). In-situ transmission electron microscopy imaging of aluminum diffusion in germanium nanowires for the fabrication of sub-10 nm Ge quantum disks. *ACS Applied Nano Materials* 3 (2): 1891–1899.

55 Hart, J.L. and Cha, J.J. (2021). Seeing quantum materials with cryogenic transmission electron microscopy. *Nano Letters* 21 (13): 5449–5452.

56 El Baggari, I., Savitzky, B.H., Admasu, A.S. et al. (2018). Nature and evolution of incommensurate charge order in manganites visualized with cryogenic scanning transmission electron microscopy. *Proceedings of the National Academy of Sciences of the United States of America* 115 (7): 1445–1450.

57 Bourrellier, R., Meuret, S., Tararan, A. et al. (2016). Bright UV single photon emission at point defects in h-BN. *Nano Letters* 16 (7): 4317–4321.

58 Kruit, P., Hobbs, R.G., Kim, C.S. et al. (2016). Designs for a quantum electron microscope. *Ultramicroscopy* 164: 31–45.

59 Juffmann, T., Koppell, S.A., Klopfer, B.B. et al. (2017). Multi-pass transmission electron microscopy. *Scientific Reports* 7 (1): 1699.

60 Robertson, I.M., Schuh, C.A., Vetrano, J.S. et al. (2011). Towards an integrated materials characterization toolbox. *Journal of Materials Research* 26 (11): 1341–1383.

61 Knoll, M. and Ruska, E. (1932). Das Elektronenmikroskop. *Zeitschrift für Physik* 78: 318–339.

62 Parkinson, G.M. (1989). High resolution, in-situ controlled atmosphere transmission electron microscopy (CTEM) of heteroheneous catalysts. *Catalysis Letters* 2: 303–307.

63 Lee, T.C., Dewald, D.K., Eades, J.A. et al. (1991). An environmental cell transmission electron microscope. *Review of Scientific Instruments* 62 (6): 1438–1444.

64 Boyes, E.D. and Gai, P.L. (1997). Environmental high resolution electron microscopy and applications to chemical science. *Ultramicroscopy* 67 (1): 219–232.

65 Sharma, R. and Weiss, K. (1998). Development of a TEM to study in situ structural and chemical changes at atomic level during gas solid interaction at elevated temperatures. *Microscopy Research and Technique* 42 (4): 270–280.

66 Allard, L.F., Bigelow, W.C., Jose-Yacaman, M. et al. (2009). A new MEMS-based system for ultra-high-resolution imaging at elevated temperatures. *Microscopy Research and Technique* 72 (3): 208–215.

67 Creemer, J.F., Helveg, S., Hoveling, G.H. et al. (2008). Atomic-scale electron microscopy at ambient pressure. *Ultramicroscopy* 108 (9): 993–998.

68 Ross, F.M, and Searson, P.C. (1995). *Dynamic Observation of Electrochemical Etching in Silicon*. IOP Publishing Ltd.

69 Radisic, A., Vereecken, P.M., Searson, P.C., and Ross, F.M. (2006). The morphology and nucleation kinetics of copper islands during electrodeposition. *Surface Science* 600 (9): 1817–1826.

70 LaGrange, T., Campbell, G.H., Reed, B.W. et al. (2008). Nanosecond time-resolved investigations using the in situ of dynamic transmission electron microscope (DTEM). *Ultramicroscopy* 108 (11): 1441–1449.

71 Wirtz, T., Philipp, P., Audinot, J.-N. et al. (2015). High-resolution high-sensitivity elemental imaging by secondary ion mass spectrometry: from traditional 2D and 3D imaging to correlative microscopy. *Nanotechnology* 26 (43): 434001.

72 Gentils, A. and Cabet, C. (2019). Investigating radiation damage in nuclear energy materials using JANNuS multiple ion beams. *Nuclear Instruments and Methods in Physics Research Section B: Beam Interactions with Materials and Atoms* 447: 107–112.

73 Yi, X., Jenkins, M.L., Hattar, K. et al. (2015). Characterisation of radiation damage in W and W-based alloys from 2 MeV self-ion near-bulk implantations. *Acta Materialia* 92: 163–177.

74 Taylor, C.A., Bufford, D.C., Muntifering, B.R. et al. (2017). In situ TEM multi-beam ion irradiation as a technique for elucidating synergistic radiation effects. *Materials (Basel)* 10 (10): 1148.

75 Kirk, M., Baldo, P., Liu, A.Y. et al. (2009). In situ transmission electron microscopy and ion irradiation of ferritic materials. *Microscopy Research and Technique* 72 (3): 182–186.

76 Parrish, R.J., Bufford, D.C., Frazer, D.M. et al. (2021). Exploring coupled extreme environments via in-situ transmission electron microscopy. *Microscopy Today* 29 (1): 28–34.

77 Birtcher, R.C., Kirk, M.A., Furuya, K. et al. (2005). In situ transmission electron microscopy investigation of radiation effects. *Journal of Materials Research* 20 (7): 1654–1683.

78 Hinks, J.A. (2009). A review of transmission electron microscopes with in situ ion irradiation. *Nuclear Instruments and Methods in Physics Research Section B: Beam Interactions with Materials and Atoms* 267 (23): 3652–3662.

79 Ohno, Y. and Takeda, S. (1995). A new apparatus for in situ photoluminescence spectroscopy in a transmission electron microscope. *Review of Scientific Instruments* 66 (10): 4866–4869.

80 Allen, F.I., Kim, E., Andresen, N.C. et al. (2017). In situ TEM Raman spectroscopy and laser-based materials modification. *Ultramicroscopy* 178: 33–37.

81 Miller, B.K. and Crozier, P.A. (2013). System for in situ UV-visible illumination of environmental transmission electron microscopy samples. *Microscopy and Microanalysis* 19 (2): 461–469.

82 Soma, K., Konings, S., Aso, R. et al. (2017). Detecting dynamic responses of materials and devices under an alternating electric potential by phase-locked transmission electron microscopy. *Ultramicroscopy* 181: 27–41.

83 NRC-CNRC (2022). This is where the code for the NANOmi open source electron microscope lives. https://github.com/NRC-NANOmi/nanomi.

84 Vendelbo, S.B., Elkjær, C.F., Falsig, H. et al. (2014). Visualization of oscillatory behaviour of Pt nanoparticles catalysing CO oxidation. *Nature Materials* 13 (9): 884–890.

85 Wang, C., Yang, W.-C.D., Bruma, A. et al. (2021). Endothermic reaction at room temperature enabled by deep-ultraviolet plasmons. *Nature Materials* 20 (3): 346–352.

86 Kabius, B., Hartel, P., Haider, M. et al. (2009). First application of Cc-corrected imaging for high-resolution and energy-filtered TEM. *Journal of Electron Microscopy* 58 (3): 147–155.

87 Yamamoto, N. (2016). Development of high-resolution cathodoluminescence system for STEM and application to plasmonic nanostructures. *Microscopy* 65 (4): 282–295.

88 McBean, P., Murphy, P., Sagawa, R., and Jones, L. (2022). The user adjustable pole-piece: expanding TEM functionality without compromise. *Microscopy and Microanalysis* 28 (S1): 2636–2638.

89 McBean, P., Milne, Z., Kanthawar, A. et al. (2022). Multiphysics simulation for TEM objective lens evaluation & design. *Microscopy and Microanalysis* 28 (S1): 2494–2495.

90 Steinhauer, S., Vernieres, J., Krainer, J. et al. (2017). In situ chemoresistive sensing in the environmental TEM: probing functional devices and their nanoscale morphology. *Nanoscale* 9 (22): 7380–7384.

91 Hotz, M.T., Corbin, G.J., Dellby, N. et al. (2016). Ultra-high vacuum aberration-corrected STEM for in-situ studies. *Microscopy and Microanalysis* 22 (S3): 34–35.

92 Leuthner, G.T., Hummel, S., Mangler, C. et al. (2019). Scanning transmission electron microscopy under controlled low-pressure atmospheres. *Ultramicroscopy* 203: 76–81.

93 Sang, X., Lupini, A.R., Unocic, R.R. et al. (2016). Dynamic scan control in STEM: spiral scans. *Advanced Structural and Chemical Imaging* 2 (1): 6.

94 Diehle, P., Kovács, A., Duden, T. et al. (2021). A cartridge-based turning specimen holder with wireless tilt angle measurement for magnetic induction mapping in the transmission electron microscope. *Ultramicroscopy* 220: 113098–113098.

95 DENSsolutions. https://denssolutions.com.

96 Hummingbird Scientific. https://hummingbirdscientific.com.

97 Mei-Build. https://www.melbuild.com.

98 Protochips Inc. https://www.protochips.com.

99 Bhowmick, S., Espinosa, H., Jungjohann, K. et al. (2019). Advanced microelectromechanical systems-based nanomechanical testing: beyond stress and strain measurements. *MRS Bulletin* 44 (6): 487–493. https://doi.org/10.1557/mrs.2019.123.

100 Liddle, J.A., Stavis, S.M., and Holland, G.E. (2019). Vacuum compatible fluid sampler. US 10,265,699 B2, June 1, 2017, April 23, 2019, USA. pp. 1–78.

101 Tanase, M., Winterstein, J., Sharma, R. et al. (2015). High-resolution imaging and spectroscopy at high pressure: a novel liquid cell for the transmission electron microscope. *Microscopy and Microanalysis* 21 (06): 1629–1638.

102 Goodge, B.H., Goodge, B.H., Bianco, E. et al. (2020). Atomic-resolution cryo-STEM across continuously variable temperatures. *Microscopy and Microanalysis* 26 (3): 439–446.

103 Yu, Q., Legros, M., and Minor, A.M. (2015). In situ TEM nanomechanics. *MRS Bulletin* 40 (1): 62–70.

104 Minor, A.M., Syed Asif, S.A., Shan, Z. et al. (2006). A new view of the onset of plasticity during the nanoindentation of aluminium. *Nature Materials* 5 (9): 697–702.

105 Gouldstone, A., Chollacoop, N., Dao, M. et al. (2007). Indentation across size scales and disciplines: recent developments in experimentation and modeling. *Acta Materialia* 55 (12): 4015–4039.

106 Shindo, D., Takahashi, K., Murakami, Y. et al. (2009). Development of a multifunctional TEM specimen holder equipped with a piezodriving probe and a laser irradiation port. *Journal of Electron Microscopy* 58 (4): 245–249.

107 Arita, M., Tokuda, R., Hamada, K., and Takahashi, Y. (2014). Development of TEM holder generating in-plane magnetic field used for in-situ TEM observation. *Materials Transactions* 55 (3): 403–409.

108 Zhao, S., Li, Y., Stavitski, E. et al. (2015). Operando characterization of catalysts through use of a portable microreactor. *ChemCatChem* 7 (22): 3683–3691.

109 Choudhary, K., Garrity, K.F., Reid, A.C.E. et al. (2020). The joint automated repository for various integrated simulations (JARVIS) for data-driven materials design. *npj Computational Materials* 6 (1): 173.

110 Hong, S., Liow, C.H., Yuk, J.M. et al. (2021). Reducing time to discovery: materials and molecular modeling, imaging, informatics, and integration. *ACS Nano* 15 (3): 3971–3995.

111 Yin, W., Brittain, D., Borseth, J. et al. (2020). A petascale automated imaging pipeline for mapping neuronal circuits with high-throughput transmission electron microscopy. *Nature Communications* 11 (1): 4949.

112 Nion Swift (2022). https://github.com/nion-software/nionswift.

113 Thermo Fisher Scientific (2022). EPU 2 EM Software. http://www.thermofisher.com/us/en/home/electron-microscopy/products/softwareem-3d-vis/epu-software.html.

114 Unocic, R.R., Lupini, A.R., Borisevich, A.Y. et al. (2016). Direct-write liquid phase transformations with a scanning transmission electron microscope. *Nanoscale* 8 (34): 15581–15588.

115 Sang, X., Lupini, A.R., Ding, J. et al. (2017). Precision controlled atomic resolution scanning transmission electron microscopy using spiral scan pathways. *Scientific Reports* 7 (1): 43585.

116 Xu, M., Kumar, A., and LeBeau, J.M. (2022). Towards augmented microscopy with reinforcement learning-enhanced workflows. *Microscopy and Microanalysis* 28 (6): 1952–1960.

117 Raffin, A.H.A., Ernestus, M., Gleave, A. et al. (2019). Stable baselines3. https://github.com/DLR-RM/stable-baselines3.

118 Stach, E., DeCost, B., Kusne, A.G. et al. (2021). Autonomous experimentation systems for materials development: a community perspective. *Matter* 4 (9): 2702–2726.

119 Roccapriore, K.M., Dyck, O., Oxley, M.P. et al. (2022). Automated experiment in 4D-STEM: exploring emergent physics and structural behaviors. *ACS Nano* 16 (5): 7605–7614.

120 Roccapriore, K.M., Kalinin, S.V., and Ziatdinov, M. (2021). Physics discovery in nanoplasmonic systems via autonomous experiments in scanning transmission electron microscopy. [arXiv:2108.03290 p.], Available from: https://ui.adsabs.harvard.edu/abs/2021arXiv210803290R.

121 Ziatdinov, M., Dyck, O., Maksov, A. et al. (2017). Deep learning of atomically resolved scanning transmission electron microscopy images: chemical identification and tracking local transformations. *ACS Nano* 11 (12): 12742–12752.

122 Krizhevsky, A., Sutskever, I., and Hinton, G.E. (2017). ImageNet classification with deep convolutional neural networks. *Communications of the ACM* 60 (6): 84–90.

123 Creange, N., Dyck, O., Vasudevan, R.K. et al. (2022). Towards automating structural discovery in scanning transmission electron microscopy. *Machine Learning: Science and Technology* 3 (1): 015024.

Index

a

aberration correctors 9, 13, 318, 322, 337
active pixel detectors (APS) 321
AC voltage 327–328
adsorption 196, 230, 241–243, 325
alignment protocol 337, 339
alloys 25, 50, 100, 132, 135, 165, 192, 239–241, 302
alternatives to free space approach 260–263
analytical TEM
　chemical analysis
　　EDS 15–16
　　EELS 17–18
　　EFTEM 19–20
　spectrum imaging (SI) 20
annealing metal hydroxides precipitates 99
anticontamination device 185, 202, 233–234
artificial intelligence age 308
Artificial neural network (ANN) 295
ATOMAI 299
atomic resolution imaging 47, 48, 67, 323
Atomic-Resolution Multi-Dimensional Transmission Electron Microscope 321
atomic resolution STEM images 96
auto-alignment algorithms 337
automation 307, 336–338, 340

autonomous experiments (AE) 308, 338–340
auxiliary microfabricated chips 324

b

battery electrode materials 318
battery operations 3, 323
battery research 198–201, 323
　closed liquid cell 200–201
　open cell 199–200
Bayesian optimization 339
biasing 3, 27–28, 63, 84, 154, 161, 181–182, 184, 200, 318
butterfly' configuration 164

c

carbon nanotube (CNT) growth 236–237
catalytic reactions 67, 215, 263, 318, 324, 330
charge density waves (CDW) 129
chromatic aberrations 149
citrination platform 299
closed cells
　graphene cells 175–178
　microfabricated window cell 178
closed liquid cell 200–201
closed vs. open cell configurations 244
clustering 294, 308
condenser lens 7, 46
controlled atmosphere 216
convolutional neural network (CNN) 295, 299, 301, 339

In-Situ Transmission Electron Microscopy Experiments: Design and Practice, First Edition. Renu Sharma.
© 2023 WILEY-VCH GmbH. Published 2023 by WILEY-VCH GmbH.

core-shell structures 100–102
correlative approaches
  electron and X-ray microscopies and spectroscopies 273–274
  independent corelative measurements 278–279
  portable reactor 274–278
  TEM and SEM 270–272
cryo holders 117, 335
cryo samples 54, 119, 122, 323
cryo techniques 115, 124
cryo TEM 323, 324, 335
Crystal Ball Plus algorithm 302
cyclic voltammetry (CV) 166, 197

d

darkfield images 321
data acquisition 10, 46, 185, 320, 326, 337
data management 291–292
data processing 285, 292–300, 321, 337
data transfer rate 307
deep convolutional neural networks (DCNNs) 339
deep kernel learning (DKL) 295, 299
DefectSegNet 299, 305
DefectTrack 299, 306
1,2-dichlorobenzene (oDCB) 124
diffraction patterns 1, 11, 103, 131, 302–303, 321, 325
digital micrograph (DM) 287
direct electron detectors (DED) 10, 125, 317, 321, 328
direct heating holder 79
dislocation motion 93–94
dynamic transmission electron microscope (DTEM) 317–319, 327

e

electrode-electrolyte interfaces 318, 323
electron beam 86–88, 324
  effects 187, 229
  translation control 332
electron-beam-induced deposition (EBID) 238–239
electron diffraction 11–12
electron-electron interactions 17, 325
electron energy-loss spectroscopy (EELS) 17–18, 217
electron optics 4, 18, 317
electron scattering and development 321
electron source or electron gun 5–6
electrons-sample interactions 4–5
electron transparent windows 84, 146, 174, 216
electropolishing 50, 51
electrostatic lens 329
embedded electrodes 166
energy-dispersive X-ray spectroscopy (EDS) 15, 318
energy materials 126–128
environmental microscopes (open cell)
  differentially pumped TEM 224–227
  ETEM combined with gas injection sample holder 223–224
environmental scanning transmission electron microscope (ESTEM) 318
environmental TEM (ETEM) 318, 327
eucentric height 7, 148
extended X-ray absorption fine structures (EXAFS) 255, 277

f

Findable, Accessible, Interoperable, and Reusable (FAIR) 292
flow cells 157, 159–160, 178, 202
focused ion beam (FIB) 52–53, 221
4D data cube 321
4D-STEM 325
4D-STEM nanobeam electron diffraction 322
fully convolutional networks (FCN) 295
furnace heating holders 79–82, 85

g

galvanic replacement reactions 193–194
gas adsorption 234, 241–243, 325

gas-cell holders 318, 330
    benefits and limitations of 222–223
    flow rate and pressure equilibrium 221–222
    reactivity of gases and sample with windows 222
gas-cell reactors 274
gaseous environment 15, 57–59, 67, 77, 215, 218, 230, 234, 318, 333
gas pressure and resolution 231–232
Gatan imaging filter (GIF) 19, 217
Gaussian Filter 287
GC-MS system 223, 277, 330
geometric distortions 332
geometric phase analysis (GPA) technique 287
graphene 175–178
graphene sandwich superstructure (GSS) 177

h

hexagonal boron nitride (hBN) 150, 196
high-angle annular dark-field (HAADF) 14, 175
high energy resolution EELS 89
high resolution EELS 318
holey or lacy 48
holography
    in-line holography 22
    off-axis holography 22
hybrid microscope 258–260
hybrid pixelated array detectors (PAD) 321
hydrogen evolution reaction 324
HyperSpy 298, 299

i

image and DP data base 308
ImageNet 299–300, 339
image processing
    history of 285–289
    progress in 290–291
    techniques 317
image segmentation 339

image simulations
    history of 285–286
    progress for 289–290
indentation 145, 244, 318
independent component analysis (ICA) 293
indirect heating holders
    furnace 79–82
    MEMS based heating holders 82–84
inelastic mean-free path 18
information recovery 308
in-line Holography 22
in situ cryo-TEM
    benefits and limitations 137–138
    biology 122
    cooling 121–122
    correlative in-situ experiments at low temperature
        magnetic field 136–137
        mechanical testing 135–136
    historical perspective 116
    materials science 122
    mitigating radiation damage
        energy materials 126–128
        MOF and zeolites structure 125–126
        polymers structure 124–158
        quantum and 2-D materials 129–131
        reactions in liquids 128–129
    phase transformations below RT 132
    physical science 122
    reduced radiation damage 122
    specimen design and preparation 119–121
    specimen holder design and function 116–119
in-situ gas-solid interactions
    anticontamination device 233–234
    applications
        carbon nanotube (CNT) growth 236–237
        electron-beam-induced deposition (EBID) 238–239
        gas adsorption 241–243

in-situ gas-solid interactions (*contd.*)
  gas environment on catalyst
    nanoparticles   234–235
    nanowire growth   237–238
    REDOX reactions   239–241
  benefits and limitations   243–244
  electron beam effects   229–231
  environmental microscopes (open cell)
    223–227
  example   229
  gas manifold design and construction
    227–228
  gas pressure and resolution   231–232
  historical perspective   215–218
  sample temperature and cell pressure
    232–233
  window holders
    Browning's group   220
    gas-cell holders   221–223
    incorporation of other stimuli   221
    SiN_x windows   219
    specimen design and preparation
      221
    windowed E-cells or nano-reactors
      219
in-situ heating
  direct heating holder   79
  dislocation motion   93–94
  experimental considerations
    electron beam   86–88
    general   84–86
    sample temperature at nanoscale
      88–90
    specimen design and selection
      90–91
    thermal drift   91–92
  history   78
  indirect heating holders
    furnace   79–82
    MEMS based heating holders
      82–84
  limitations and possibilities   105–106
  nucleation, precipitation, and
    crystallization   94–96
  sintering   98–100

  thermal stability of materials
    alloys   100
    core-shell structures   100–102
    materials synthesis   104–105
    phase transformation   102–104
    2-D materials   102
in-situ solid-liquid interactions
  applications
    battery research   198–201
    corrosion/oxidation   192–193
    galvanic replacement reactions
      193–194
    growth of core-shell nanoparticles
      194–195
    nucleation and growth of
      nanoparticles   190–192
    quantitative electrochemical
      measurements   197–198
    soft nanomaterials analyzed
      195–197
  closed cells
    graphene cells   175–178
    limitations of   178
    microfabricated window cell   178
  data acquisition   185
  electrochemical cell, biasing   181–182
  electron beam effects   187–188
  flow reactors, microfluidic design
    178–181
  heating in liquids   182–184
  historical perspective   173
  interaction of sample with windows
    189–190
  limitations   201–202
  sample loading   185–187
  specimen design and preparation
    184–185
  windows bulging   188–189
in-situ TEM   318
  advantages   3
  applications of quantitative data
    catalysis   67
    physical and materials science
      66–67
  development or modification of   47

electrical, mechanical, radiation and
   magnetization 44
electron beam effects 55–56
experiments using other stimuli 63
functioning and applications of 24
grid and support material
   reactivity of gasses/liquids with the
      TEM holder parts 59–60
   reactivity of liquids with the
      windows 59
   reactivity of TEM Grids in gaseous
      environment 58–59
   reactivity of TEM grids upon heating
      57–58
growth of GaN nanowires using ETEM
   63–64
liquid cell experiments 62–63
microscope selection
   image acquisition system and
      detectors 46–47
   operating voltage 45–46
   TEM/STEM and pole-piece gap 46
purity of gasses 60–62
specimen design and preparation
   cryo sample preparation 54–55
   direct dispersion 48–49
   electropolishing 50
   focused ion beam (FIB) 52–53
   mechanical and ion milling 50–52
   sintering pallets 49–50
   tripod polishing 54
   ultramicrotomy 50
stimulus and technique selection
   44–45
ionic liquids 174, 199, 323
isopropanol (IPA) 177
isothermal sublimation of Ag nanocubes
   89

j
Johnson-Mehl-Avrami-Kolmogorov
   (JMAK) equation 67, 96
Joint Automated Repository for Various
   Integrated Simulations (JARVIS)
   300

l
Lab-in-Gap' dynamic microscope 321
lab on a chip 145, 219
laser alignment 269
lens aberrations 7–9, 13
lenses 7
LiberTEM 299
light induced superconductivity 321
Li ion batteries 126, 199, 201, 323,
   339
linear/non-linear dimensionality
   reduction 339
liquid cell experiments 62–63
liquid cell holders 128, 318
liquid cell TEM (LC-TEM) 173
liquid-phase TEM (LP-TEM) 173
lithium dendrites 323
load-deflection behavior 151
Lorentz microscopy 20–22, 29
low vapor chemical vapor deposition
   (LPCVD) 155

m
machine learning 322
   limitation 309
   motivation 296–298
   supervised ML 294–296
   unsupervised ML 293–294
magnetization 28–29, 318
materials community 215
Materials Genome Initiative 321
materials synthesis 104–105
MATLAB 289, 290, 295
matminer 300
mean filter 287
mechanical and ion milling 50–52
mechanical testing observations 335
metal-organic-frameworks (MOFs)
   125–126
microchip fabrication process 154
microchips 85
microchips or chips 82
microelectromechanical systems (MEMS)
   175, 218
   based cryo-holder 118

microelectromechanical systems (MEMS) (contd.)
  based gas microreactor 324
  based heating holders 82–84
microfabricated chip 159, 202, 274, 276, 277, 333
microfabricated device 333
microfabricated Si chip 333
microfabricated window-cell (microchips)
  advantage 154
  flow cells 159–160
  incorporation of other stimuli 161–162
  microchip fabrication process 154
  microsphere ball lenses 154
  monolithic microchips 162–163
  SiNx windows 154
  static cells 157–159
micromachining 167
microreactor 274, 276, 324, 325
micro-scale in-situ characterization techniques 255
microscope goniometer readout 336
mimics laser induced stroboscopic electron generation 327
modern TEM
  aberration correctors 9
  electron source or electron gun 5–6
  lens aberrations 7–9
  lenses 7
monochromated electron source 317, 318
monolithic liquid cell 333, 335
monolithic microchips 162–163
multimodal approaches 187, 256, 269, 335
multimodal measurements 322
multimodal microscopy
  alternatives to free space approach 260–263
  hybrid microscope 258–260
  laser alignment 269
  parallel ion electron spectrometry (PIES) 257–258

photon (PL or CL) 263
  through sample chamber port 263–264
  through sample holder 264–269

n
nano-aquarium 166
nanofactory 166, 268
nanowire growth 237–238
neural networks 294, 295, 301, 322
Nion Swift 299
noise reduction 300–301
non-negative matrix factorization (NMF) 293

o
objective lens 7, 9, 11, 13, 21, 29, 46, 329
off-axis holography 22
one-class support vector machine (OCSVM) 304
open cell 199–200
open circuit potential (OCP) 165
operating voltage 30, 45–46, 106, 231
original closed cells 175
oxidative dissolution mechanism of Pd nanoparticles 191

p
parallel ion electron spectrometry (PIES) 257–258, 319, 327
periodic lattice distortion (PLD) 129
phase locked images 327
phase transformation 102–104
picometer image resolution 322
pixelated detectors 318
plug flow reactor 324
plunge freezing 119
pole-piece gap 331, 332
poly(3-hexylthiophene) 124
polymeric materials 324
polymethyl methacrylate (PMMA) 155
potentiodynamic polarization 165
principal component analysis (PCA) 293
projector or magnifying lens 7
protective layer 155

Protochips   165, 292
ptychography   325
pulse signal   327
py4DSTEM   298
Python   289, 290, 337
PyTorch   299
Pyxem   298

q

quantitative electrochemical measurements   197–198
quantitative parameters   339
quantum coherence   325
quantum computing   325
quantum confinement   325, 326
quantum dots   326
quantum fluctuations   325
quantum materials   325

r

REDOX reactions   239–241
reinforcement learning (RL)   337
research groups   327

s

sample loading   185–187
scanning transmission electron microscope (STEM)   14
scikit-learn   299
signal to noise ratio (SNR)   149
Si nanowires   60, 64, 237, 268, 323
single-layer graphene (SLG)   175
sintering   98
  pallets   49–50
SiNx window   168
Smart EPU software   337
SnO_2 323
sodium yttrium fluoride (NaYF4)   279
Soft Actor-Critic (SAC) algorithm   337
solid-electrolyte interface (SEI)   127
spatial resolution   330
spectrum imaging (SI)   20, 194, 293
speed direct electron cameras   318
spin-orbit coupling   325
Spotiton robot   119

static cells   157–159
stroboscopic TEM   317, 319
structure determination
  applications   305–307
  atomic column heights   305
  diffraction pattern analysis   302–303
  image analysis   303–305
supervised ML   294–296

t

TEM holders
  design philosophy   146–148
  historical perspective   146
  modified window holders   166
    combination   166
    microchips for commercial holder   164–166
    non-window cell holder to incorporate other stimuli   166–167
  windows
    image resolution thickness and material properties   149–150
    inert or corrosion resistant   153–154
    microfabricated window-cell (microchips)   154–157
    pressure difference tolerance   151–153
    strength and flexibility   150–151
TensorFlow   299
thermal drift   91–92
thermal stability of materials
  alloys   100
  core-shell structures   100–102
  materials synthesis   104–105
  phase transformation   102–104
  2-D materials   102
thermocouple   59, 78, 85, 88, 89, 118
Thermo Fisher Scientific   337
time resolution   10, 23, 30, 317, 332
Top Hat filter   287
transition metal dichalcogenides (TMD)   131

transmission electron microscope (TEM)
  analytical   14–20
  automation   336–338
  biasing   27–28
  characterization   2
  data acquisition systems   9–10
  electron diffraction   11–12
  electrons-sample interactions   4–5
  historical perspective   4
  Holography   22
  imaging modes   12–13
  ion radiation/implantation   25–27
  Lorentz microscopy   20–22
  magnetization   28–29
  mechanical testing   25
  modern   5–9
  potential limitations and cautions   29–30
  sample chamber ports   336
  UEM and DTEM   23–24

tri-n-octylphosphine (TOP)   191
tripod polishing   54
tuning fork-based IR detector   332
2-D materials   102

u

ultrafast electron diffraction and microscopy   321
ultrafast electron microscopes (UEM)   23, 317–319, 327
ultramicrotomy   50, 54
unsupervised ML   293–294, 298
user adjustable pole-piece (UAP)   331

w

water splitting   24, 195, 263, 324
wet cells   173
white lines   19, 93
window cells   148, 154, 157, 164, 166
windows bulging   188–189, 202